Revision

NEW GCSE SCIENCE

Science and Additional Science

for AQA A Higher

Authors: **Nicky Thomas**
Rob Wensley
Gemma Young

Revision guide +
Exam practice workbook

William Collins' dream of knowledge for all began with the publication of his first book in 1819. A self-educated mill worker, he not only enriched millions of lives, but also founded a flourishing publishing house. Today, staying true to this spirit, Collins books are packed with inspiration, innovation and practical expertise. They place you at the centre of a world of possibility and give you exactly what you need to explore it.

Collins. Freedom to teach

Published by Collins
An imprint of HarperCollinsPublishers
77–85 Fulham Palace Road
Hammersmith
London
W6 8JB

Browse the complete Collins catalogue at
www.collins.co.uk

10 9 8 7 6

ISBN 978-0-00-741601-1

British Library Cataloguing in Publication Data
A Catalogue record for this publication is available from the British Library

Project managed by Hart McLeod

Edited, proofread, indexed and designed by HartMcLeod

Printed and bound in China by South China Printing Co.Ltd

Acknowledgements
The Authors and Publishers are grateful to the following for permission to reproduce photographs.
p8 ©Hank Morgan/Science Photo Library; p15 ©Roman Kobzarev/istock.com; p35 ©Craig Smith/istock.com; p47 ©Shutterstock; p53 ©Alan Williams/Alamy; p58 ©Michael Eichelberger, Visuals Unlimited/Science Photo Library; p103 fig 4 ©Shutterstock; p103 fig 5 ©Shutterstock; p104 ©Shutterstock; p117 ©Michael Eichelberger, Visuals Unlimited/Science Photo Library; p157 left © Gabor Izso/istock.com; p157 right © Klemiantsou Kanstantsin/istock.com; p162 © Karl Dolenc/istock.com

MIX
Paper from
responsible sources
FSC™ C007454

About this book

This book covers the content you will need to revise for GCSE Science and Additional Science AQA A Higher. It is designed to help you get the best grade in your GCSE Science Higher Exam.

The content exactly matches the topics you will be studying for your examinations. The book is divided into two major parts: **Revision guide** and **Workbook**.

Begin by revising a topic in the Revision guide section, then test yourself by answering the exam-style questions for that topic in the Workbook section.

Workbook answers are provided in a detachable section at the end of the book.

Revision guide

The Revision guide (pages 6–110) summarises the content of the exam specification and acts as a memory jogger. The material is divided into grades. There is a question (**Improve your grade**) on each page that will help you to check your progress. Typical answers to these questions and examiner's comments, are provided at the end of the Revision guide section (pages 111–123) for you to compare with your responses. This will help you to improve your answers in the future.

At the end of each module, you will find a **Summary** page. This highlights some important facts from each module.

Workbook

The Workbook (pages 145–251) allows you to work at your own pace on some typical exam-style questions. You will find that the actual GCSE questions are more likely to test knowledge and understanding across topics. However, the aim of the Revision guide and Workbook is to guide you through each topic so that you can identify your areas of strength and weakness.

The Workbook also contains example questions that require longer answers (**Extended response questions**). You will find one question that is similar to these in each section of your written exam papers. The quality of your written communication will be assessed when you answer these questions in the exam, so practise writing longer answers, using sentences. The **Answers** to all the questions in the Workbook are detachable for flexible practice and can be found on pages 257–280.

At the end of the Workbook there is a series of **Revision checklists** that you can use to tick off the topics when you are confident about them and understand certain key ideas.

Additional features

Throughout the Revision Guide there are **Exam tips** to give additional exam advice, **Remember boxes** pick out key facts and a series of **How Science Works** features, all to aid your revision.

The **Glossary** allows quick reference to the definitions of scientific terms covered in the Revision guide.

Contents

B2 Biology

C2 Chemistry

P2 Physics

Diet and energy

Diet and metabolic rate

D–C

- People should eat a balanced diet which contains some of all these different kinds of nutrients:
  - carbohydrates, fats and protein for energy
  - small amounts of vitamins and mineral ions for keeping healthy.

- If your diet is not balanced, you may become malnourished (become too fat or too thin, or suffer from deficiency diseases).

- To lose body mass, people may go on a slimming diet where they eat less. Exercising more also helps. These both lead to more energy being used up than is taken in, and the body is forced to use up some of its stored fat for energy.

- Metabolic rate is the rate at which chemical reactions take place in your cells.

- The greater the proportion of muscle to fat in the body, the higher the metabolic rate is likely to be. It also increases during exercise.

- Metabolic rate can be affected by your genes, which you inherit from your parents.

Remember!
A balanced diet will contain all nutrients in the correct amounts.

How Science Works

- Some slimming programmes and products may make claims that are unrealistic. Also it is important to remember that people will put back on the mass they lost if they don't change their eating habits for good.

Fatty foods

B–A*

- One gram of fat releases almost twice as much energy as one gram of carbohydrate, or one gram of protein.

- Proteins are not usually a major source of energy for the body because they are used for the more important functions of growth and repair.

Diet, exercise and health

Diet and cholesterol

D–C

- A high level of cholesterol in the blood increases the risk of developing plaques in the walls of the arteries. Figure 1 shows how this can happen.

- Sometimes, a clot blocks one of the arteries that take oxygenated blood to the heart muscle. This causes a heart attack – the muscle cannot work, so the heart cannot beat properly.

- Eating saturated fats (those found in animal products) raises blood cholesterol levels. Unsaturated fats, found in plants, seem to lower blood cholesterol level.

- Some people's bodies are better than others at keeping low levels of cholesterol in their blood. They have inherited this from their parents.

A healthy artery has a stretchy wall and a space in the middle for blood to pass through.

Sometimes, a substance called plaque builds up in the wall. This is more likely to happen if you have a lot of cholesterol in your blood.

The plaque slows down the blood and a clot may form. A part of the plaque may break away.

Figure 1: How a plaque develops in an artery

Good and bad cholesterol

B–A*

- Cholesterol is carried in your blood in two ways, as:
  - low-density lipoprotein (LDL) cholesterol, which is 'bad' and can cause heart disease
  - high-density lipoprotein (HDL) cholesterol, which is 'good' as it can protect against heart disease by helping to remove cholesterol from the walls of blood vessels.

Improve your grade

Harry's diet is very high in saturated fats. Suggest two ways that this could affect his health. **AO2 (3 marks)**

Pathogens and infections

Disease

- Microorganisms that cause disease are called pathogens.

- Bacteria can reproduce rapidly inside the body. They may produce toxins (poisons) that make us feel ill.

- Viruses reproduce inside a body cell then destroy it when they burst out. The viruses then invade other cells.

- An epidemic occurs when a wide spread of people have a disease. A pandemic is when the disease affects a whole country or goes worldwide.

- In the 1840s, a doctor called Semmelweis used evidence from the death rates of women to work out that they were dying because doctors were transferring something to them from dead bodies. He made all the doctors wash their hands in chlorine water and, within a very short time, the death rate plummeted. We now know the infection that killed the women was caused by bacteria.

D–C

A new hypothesis

- Scientists are still learning about pathogens.

How Science Works

- Doctors once thought that stomach ulcers were caused by stress or over-secretion of acid in the stomach. In 1982, two Australian researchers, Marshall and Warren, found bacteria in the stomachs of people with ulcers. Their hypothesis was that these bacteria caused the ulcers. The evidence to prove it came when Warren swallowed some of the bacteria and developed a stomach ulcer.

B–A*

Fighting infection

Phagocytosis and lymphocytes

- Figure 2 shows how a type of white blood cell, called a phagocyte, can surround and ingest bacteria. This activity is called phagocytosis.

- Lymphocytes produce chemicals called antibodies. The antibodies group round and stick to the pathogen. This may kill it directly, or stick it to other pathogens in clumps so that the phagocytes can destroy them more easily.

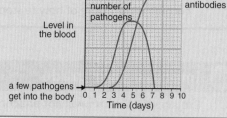

1 A phagocyte moves towards a bacterium 2 The phagocyte pushes a sleeve of cytoplasm outwards to surround the bacterium 3 The bacterium is now enclosed in a vacuole inside the cell. It is then killed and digested by enzymes

Figure 2: Phagocytosis

D–C

- Some lymphocytes make antitoxins that can stick to the toxins given off by bacteria, and destroy them.

- Both antibodies and antitoxins are very specific – each kind only works against a particular pathogen or toxin.

More about antibodies

- This is an antibody molecule. The bits on the end of the Y arms can come in millions of different shapes. Each lymphocyte can make just one kind. The end bits fit onto molecules on the pathogen. Each shape only fits onto one kind of pathogen.

Figure 3: A simplified antibody molecule

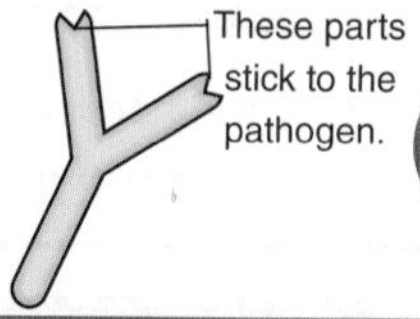

These parts stick to the pathogen.

B–A*

Improve your grade

Use the graph on the right to explain the change in the number of pathogens. **AO2 (4 marks)**

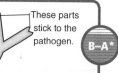

number of pathogens

number of antibodies

Level in the blood

a few pathogens get into the body

Time (days)

Drugs against disease

Antibiotics

- Antibiotics are drugs that kill bacteria inside your body, without killing your own cells. Examples are penicillin and streptomycin.
- Antibiotics do not all work equally well against all the kinds of bacteria.
- Figure 1 shows how we find the best antibiotic to kill a bacterium. Bacteria are spread onto a jelly. Paper discs soaked in different antibiotics are placed on the jelly and the antibiotics diffuse out. If the antibiotic kills the bacteria, they do not grow around the disc.

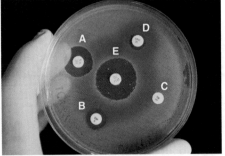

Remember!
Antibiotics cannot destroy viruses. Some viral infections can be treated by taking antiviral drugs.

EXAM TIP

You may be shown an image of a test like this and be asked what it shows. The clear jelly shows no bacterial growth: so the bigger this area, the more effective the antibiotic. In this case, antibiotic E is the most effective.

Figure 1: Testing antibiotics

Prescribing antibiotics

- Scientists now know that people must not use antibiotics unnecessarily. Overuse makes it more likely that bacteria will become resistant to them.

Antibiotic resistance

Reducing the risk

- Bacteria do not become resistant to antibiotics on purpose. It happens by natural selection. Figure 2 shows how.

- As new antibiotic-resistant strains of bacteria emerge, we have to find new antibiotics to kill them.

This is a population of bacteria in someone's body. By chance, one of them has mutated and is slightly different from the others.

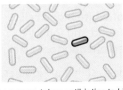

The person takes antibiotics to kill the bacteria. It works – except on the single odd bacterium. This one is resistant to the antibiotic.

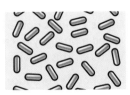

The bacterium now has no competitors and grows rapidly. It divides and makes lots of identical copies of itself. There is now a population of bacteria that the antibiotic cannot kill. This process is an example of **natural selection**.

Figure 2: How antibiotic resistance arises

- To reduce the chance of new strains forming, we need to reduce the use of antibiotics.

- Whenever antibiotics are used, it gives an advantage to any mutant bacterium that happens to be resistant to them.

- If they are not used, then a mutant bacterium does not have any advantage: it is no more likely to reproduce than any other bacterium.

MRSA deaths

- MRSA (methicillin-resistant *Staphylococcus aureus*) is sometimes called a superbug because it is resistant to most antibiotics.
- The number of deaths from MRSA rose between 1993 and 2006, but it is now gradually decreasing as people become more aware of how to stop it spreading.

Improve your grade

Explain why a doctor will not prescribe you antibiotics for a bacterial throat infection. **AO2 (3 marks)**

Vaccination

MMR

- In the UK, children have the MMR vaccination which makes them immune to measles, mumps and rubella.

- A small amount of the dead or inactive viruses that cause the diseases are injected into the blood.

- The white blood cells attack them, just as they would attack living pathogens. They remember how to make the antibody, so the child is now immune to the diseases without having to suffer them first.

How Science Works

- In 1998, a group of scientists published an article suggesting that the MMR vaccination might cause autism. Many parents decided not to let their child have the MMR vaccination, even though there was no evidence in the article. Many studies have been carried out since. No one has found any link between the MMR vaccination and autism.

D–C

New diseases

- New infectious diseases appear when mutations occur in bacteria or viruses.

- For example in 2009, a new kind of flu virus, called swine flu, spread quickly to all parts of the world. As existing flu vaccines would not work against it, new ones had to be developed.

Remember!
Some diseases cannot be vaccinated against because there are too many strains of the pathogen. An example of this is the common cold, which is caused by over 250 different types of virus.

B–A*

Growing bacteria

Avoiding contamination

- The microorganisms growing on the nutrient medium are called a culture.
- You should use a sterile technique to stop unwanted microorganisms from entering the nutrient medium.
- All equipment, including the medium, should be sterilised before use.
- Metal equipment, such as a wire inoculating loop, can be held in a flame.
- You must not touch the nutrient jelly (agar) with your fingers or breathe over it.
- The dish containing the agar should be sealed with tape. This prevents microorganisms, from the air, contaminating the culture.
- You should keep the cultures lower than 25 °C. If you keep them warmer then this might encourage the growth of microorganisms that live in the body, which are more likely to be pathogens.

D–C

Pathology and industrial laboratories

- Bacteria that are causing an illness in a patient may be grown in a hospital pathology lab.
- The Petri dishes, on which the bacteria are growing, will be put into an incubator to keep them warm and to encourage rapid growth.
- In industrial labs, harmless bacteria are grown commercially, for example, to make enzymes.

B–A*

⊙ Improve your grade

You wish to grow some harmless *E. coli* bacteria on some agar jelly. You use an inoculating loop to transfer the bacteria to the jelly. Explain why it is important to hold the inoculating loop in a flame first. **AO1 (2 marks)**

Co-ordination, nerves and hormones

Nerves and hormones

D–C

- Nerves contain special cells called nerve cells, which transmit impulses to and from the brain and spinal cord (the central nervous system).
- Glands secrete chemicals called hormones into the blood. The bloodstream carries the hormones around the body.
- Most hormones affect just a few different organs. These are called their target organs.
- An example of a hormone is adrenaline, which affects the heart, breathing muscles, eyes and digestive system.

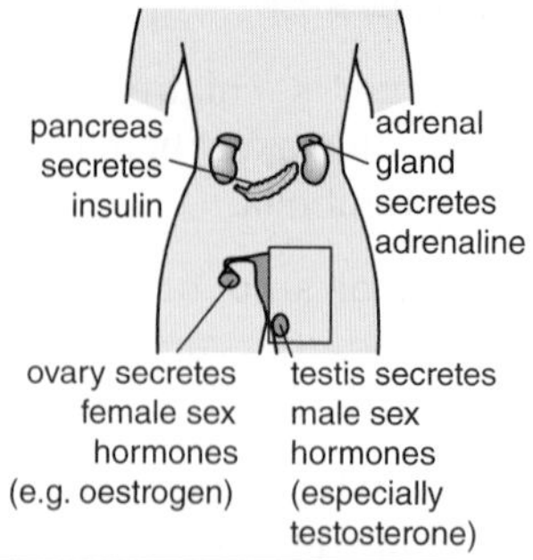

pancreas secretes insulin

adrenal gland secretes adrenaline

ovary secretes female sex hormones (e.g. oestrogen)

testis secretes male sex hormones (especially testosterone)

Figure 1: Four glands that secrete hormones

Response duration

B–A*

- A nervous response, for example touching a football, is fast and short-lived because impulses travelling along nerves only take a short time.
- Where a longer-term response is needed, hormones are often a more appropriate method of communication.

Receptors

Neurones

D–C

- Information is carried in the nervous system as electrical impulses. The cells that transmit these impulses are called nerve cells or neurones.
- The neurones that transmit impulses from receptors to the central nervous system are called sensory neurones.
- The neurones that transmit impulses from the central nervous system to effectors are called motor neurones.

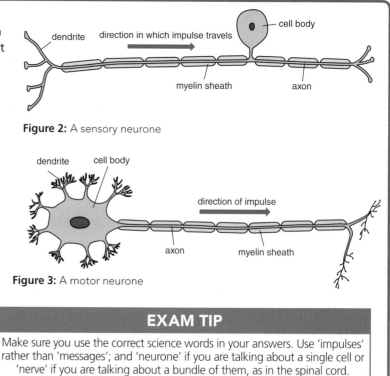

dendrite direction in which impulse travels cell body

myelin sheath axon

Figure 2: A sensory neurone

dendrite cell body

direction of impulse

axon myelin sheath

Figure 3: A motor neurone

Remember!

Antibiotics cannot destroy. Receptors detect stimuli and send an impulse along a neurone. Effectors (muscles and glands) respond to the stimuli.

EXAM TIP

Make sure you use the correct science words in your answers. Use 'impulses' rather than 'messages'; and 'neurone' if you are talking about a single cell or 'nerve' if you are talking about a bundle of them, as in the spinal cord.

Rod cells

B–A*

- Rod cells at the back of the eye (retina) are sensitive to light.
- If only one photon of light falls onto a rod cell, this is enough to make it generate an electrical impulse which is sent along the optic nerve to the brain.
- The brain uses the pattern of impulses, arriving from different parts of the retina, to construct a 'picture' of the world you are looking at.

Improve your grade

An injury that results in breaking of the spine may result in the person being paralysed. Explain why. **AO2 (3 marks)**

Reflex actions

Impulse pathways

- A reflex arc is the pathway taken by a nerve impulse as it passes from a receptor, through the central nervous system, and finally to an effector.

- Figure 4 shows the path the impulses take.

- It takes a nerve impulse only a fraction of a second to go along this route. That is why reflex actions are so quick.

- The gaps between neurones are called synapses.

- Electrical impulses cannot jump across synapses. When an impulse gets to the end of a neurone, it causes a chemical to be secreted. The chemical diffuses across the gap and starts an electrical impulse along the next neurone.

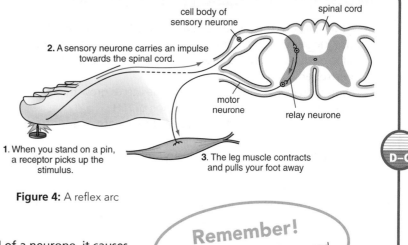

Figure 4: A reflex arc

cell body of sensory neurone

spinal cord

2. A sensory neurone carries an impulse towards the spinal cord.

motor neurone

relay neurone

1. When you stand on a pin, a receptor picks up the stimulus.

3. The leg muscle contracts and pulls your foot away

D–C

> **Remember!**
> Synapses slow down the speed of the impulse.

Conscious control

- Synapses enable us to respond to a stimulus in more than one way. For example, the relay neurone, in the spinal cord, will have synapses to other neurones that can carry nerve impulses down from the brain.

- This allows us to take conscious control of our response to a stimulus.

B–A*

Controlling the body

Temperature control

- You gain water from food and drink. You lose water in your breath, sweat and urine.

- Your blood has ions dissolved in it, such as those found in salt.

- The kidneys help to keep the balance of water and ions by varying the amount of water and salt excreted from your body in urine.

- Human body temperature needs to be kept at around 37 °C, as this is the temperature at which our enzymes work best.

- The body loses heat by radiation from the skin, and from the evaporation of sweat.

- The body also has mechanisms to keep the concentration of sugar in the blood constant.

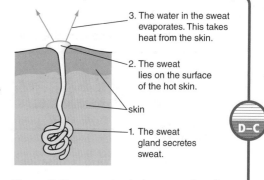

3. The water in the sweat evaporates. This takes heat from the skin.

2. The sweat lies on the surface of the hot skin.

skin

1. The sweat gland secretes sweat.

D–C

Figure 5: How sweating helps you to lose heat

Survival in the desert

- An SAS survival manual gives this advice to conserve water: avoid exertion, keep cool and stay in the shade; don't lie on the hot ground; don't eat, because digestion uses up fluids; talk as little as possible; and breathe through your nose rather than your mouth.

B–A*

Improve your grade

James is dancing in a nightclub. He starts to sweat. Explain how sweating helps to cool him down. **AO2 (2 marks)**

Reproductive hormones

Hormones and the menstrual cycle

- At the start of the menstrual cycle, the pituitary gland secretes FSH (follicle-stimulating hormone). This causes an egg to mature in one of the woman's ovaries and also stimulates the ovary to secrete oestrogen.
- Oestrogen makes the inner lining of the uterus grow thicker. High levels of oestrogen stop the production of FSH.
- Luteinising hormone (LH) stimulates the release of eggs from the ovary.
- As the level of FSH drops, the ovary stops secreting oestrogen. This cuts off the inhibition of FSH secretion, so the cycle starts all over again.
- The contraceptive pill may contain oestrogen. It stops FSH being produced, so that eggs do not mature.

How Science Works

- When the pill was first used it produced many side effects, for example people put on weight. Nowadays, the pills contain much less oestrogen than they used to (some have no oestrogen at all but contain another hormone called progesterone), so there are fewer side effects.

In the pituitary gland	In the ovary
FSH is secreted.	
	FSH causes the ovary to secrete oestrogen.
Oestrogen reduces the amount of FSH secreted.	
	The low amount of FSH stops oestrogen being secreted.
FSH is secreted again.	

Figure 1: FSH and oestrogen secretion

Remember!

FSH and LH are produced in the pituitary gland which is in the brain. They reach their target organs (the uterus and the ovary) through the bloodstream.

Co-ordinating the menstrual cycle

- The change in concentration of oestrogen causes changes in the thickness of the uterus lining. The rise in oestrogen concentration causes the uterus lining to thicken. When oestrogen concentration falls below a certain point, the uterus lining breaks down.
- The release of an egg from the ovary usually happens at about day 14 of the cycle. It is called ovulation.

Controlling fertility

IVF

- IVF stands for 'in vitro fertilisation'.
- The woman is given hormones, such as FSH, to make her ovaries produce several eggs.
- The eggs are removed and are mixed with her partner's sperm for fertilisation to occur.
- One of the embryos is chosen and placed in the woman's uterus. With luck, it will sink into the uterus lining and develop as a foetus.

Multiple births

- Twins or triplets are more likely to have problems developing in the uterus than a single foetus. They are also more likely to be underweight at birth.

EXAM TIP

Multiple births are more common with women who have used fertility treatments. You should be able to explain why this is.

Improve your grade

FSH is found in fertility drugs. Explain how taking FSH will increase a woman's fertility. **AO2 (2 marks)**

Plant responses and hormones

Phototropism and gravitropism

- A growth response to light is a tropism called phototropism.

- Auxin is a plant hormone which makes cells in shoots get longer. When light shines onto a shoot, the auxin builds up on the shady side. This makes the cells on that side get longer. So, the shoot bends towards the light.

- A growth response to gravity is called gravitropism or geotropism.

- Auxin tends to accumulate on the lower side of a root. In roots, auxin reduces the rate of growth. So, the lower side of the shoot grows more slowly than the upper surface. This causes the root to bend downwards.

- Gardeners dip the base of a cutting into a powder or gel called rooting hormone, which makes the cutting grow roots.

- Plant hormones are also used as weedkillers. The hormones make the weeds grow very fast and then die. The hormones only affect weeds because they have a different metabolism.

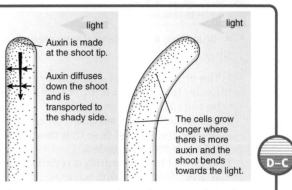

Figure 2: How auxin makes a shoot grow towards the light

D–C

Where are the receptors in a plant?

- Scientists in the past carried out experiments to work out how plants detect stimuli.

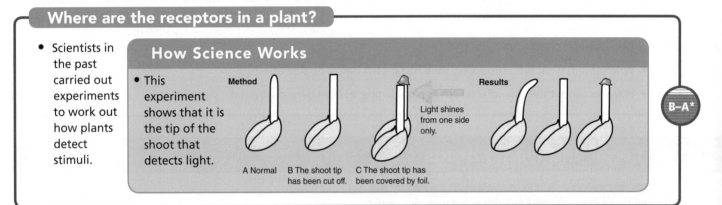

How Science Works

- This experiment shows that it is the tip of the shoot that detects light.

Method

Results

Light shines from one side only.

A Normal B The shoot tip has been cut off. C The shoot tip has been covered by foil.

B–A*

Drugs

Dangers of drugs

- Some people take drugs because they make them feel different. This is called recreational drug use.

- Many recreational drugs are legal but still can be harmful, such as alcohol.

- Some recreational drugs are illegal because of the harm they can cause. Cannabis is an example, it may cause mental illness.

- A drug addiction can have dangerous long-term effects. Over time, the lungs, brain and liver can be seriously damaged.

- If someone is addicted to a drug, they may suffer from very unpleasant withdrawal symptoms if they stop taking it.

D–C

Deaths from drug use

- Each year, thousands of people in Britain die from misusing drugs.

- Some deaths are from poisoning. Some happen because drugs can affect the brain, making people behave in a dangerous way.

B–A*

Improve your grade

Explain how the hormone auxin brings about a response to light called phototropism. **AO2 (4 marks)**

Developing new drugs

Drug trials

- A drug trial, on a potential new medicine, contains three stages.

  1 The drug is tested in a laboratory on human cells or tissues to find out if it is toxic (poisonous). It may be also be tested on live animals.
  2 Human volunteers are given different doses, to find out what is the highest dose that can be taken safely. Any side-effects are recorded.
  3 In clinical trials, the drug is tested on its target illness. It is given to people who have the illness, to see if it makes them better. Some patients are given placebos which do not contain the drug. Neither the patient nor their doctor knows whether they have a placebo or the real drug (a double-blind trial). This helps determine whether the drug really works.

D–C

How Science Works

- In the 1960s, many pregnant women were prescribed the drug thalidomide to treat 'morning sickness' in pregnancy.
  The drug had been thoroughly trialled as a sleeping pill, but no one had thought to test it on pregnant women. Women who took thalidomide in early pregnancy often gave birth to babies with short arms or no arms.
  Thalidomide was banned worldwide, but it is now being used to treat serious diseases such as leprosy.

Evaluating statins

B–A*

- Statins are drugs that help people to reduce their blood cholesterol level and therefore greatly reduce their risk of getting heart disease.

- When statins were first introduced, trials had shown almost no side-effects.

- However, since then, side-effects such as painful muscles have been discovered.

Legal and illegal drugs

Dangers from recreational drugs

D–C

- Alcohol and nicotine cause far more illness and death each year than all the illegal drugs put together.

- People do not worry so much about them because they have been around for so long, and because so many people use them.

Drugs in sport

B–A*

- Professional sports people are banned from taking certain performance-enhancing drugs because they could give them an unfair advantage. For example:
  - steroids can stimulate the body to grow larger, stronger muscles
  - beta blockers can help someone to stay calm and steady
  - stimulants can increase the heart rate.

- These drugs can often cause long-lasting damage or even death.

Improve your grade

Some doctors believe that everyone over the age of 50 should be prescribed statins on the NHS for free. Evaluate this decision. **AO2 (4 marks)**

Competition

Competing

- Organisms may have to compete for resources if they are in short supply.

- For example, plants compete for light – the ones that grow tallest win the competition.

- The individuals best at competing are the most likely to survive. Those not good at getting resources are the most likely to die.

- If there are not enough females to go around, then males will compete for a mate.

- Animals may also compete for a territory – a space in which they can find food and a place to breed.

Remember!
Competition does not usually mean that organisms actually fight over resources. They just have to find ways of being better at getting them than others are.

Figure 1: Stags fight over the right to mate with the females in the herd

D–C

Avoiding competition

- Many organisms have become able to live in places where few others can survive. Although these places make survival tough, there is no need to share resources.

- This can increase chances of surviving and having large numbers of offspring.

B–A*

Adaptations for survival

Living in difficult climates

- Plants that live in dry places usually have:
    - long, wide-spreading roots – the roots grow deep into the soil, to reach water
    - small or no leaves – the smaller the leaf surface area, the less the amount of water evaporating away
    - tissues that can store water.

Remember!
Plants and animals must have features that allow them to survive in their habitat. These features are adaptations.

- Animals that live in dry places must be able to manage without much water. For example, camels' stomachs can hold over 20 litres of water, they can drink very quickly, store water as fat in their humps, and they produce very little urine.

- Desert animals often have large ears. A large surface area helps the animal lose body heat and stay cool.

- Animals that live in very cold places, such as the Arctic, often have thick fur and thick layers of fat. This insulation helps the animal reduce heat loss. They are coloured white, for camouflage against snow.

- Many plants and animals have thorns, poisons and warning colours to deter predators.

D–C

Extreme environments

- Organisms that can live in very difficult environments are called extremophiles. They are usually microorganisms.

- For most organisms, conditions such as high temperatures and high pressure would be lethal.

- Extremophiles must have very stable protein molecules that are not affected by these conditions.

EXAM TIP

Make sure you can describe how the camel and one animal from the Arctic, for example a polar bear, are adapted.

B–A*

Improve your grade

Explain how cacti are adapted to living in the dry desert. **AO2 (3 marks)**

Environmental change

Causes of change

D–C

- Environmental changes are caused by living and non-living factors. For example:
  - non-living factors include global warming, which has caused rainfall in central Australia to decrease
  - living factors include the introduction of the grey squirrel into Britain, which caused a decrease in the population of the native red squirrel.

The disappearing bees

B–A*

- Honeybees help pollinate flowers that will develop into food crops.

- In recent years, there has been a decline in the numbers of honeybees.

- We are not sure of the cause but various suggestions have been put forward to explain it.

Pollution indicators

Measuring changes in the environment

D–C

- In the UK, the composition of the air and of the water in rivers and streams; and the air temperature and rainfall, are constantly being measured. This makes sure that any changes can be tracked.

- Oxygen meters measure the concentration of dissolved oxygen in the water. Unpolluted water contains a lot of dissolved oxygen.

- Thermometers measure temperature. Rain gauges measure rainfall.

- Scientists can use the distribution of living organisms to find out about pollution. For example:
  - if there is a lot of sulfur dioxide in the air, many species of lichens will not be able to grow
  - if there is not very much oxygen in a river, there will be no oxygen-loving mayfly larvae in the water, instead there will be just rat-tailed maggots and bloodworms.

Remember!
The higher the number of different lichen species that can grow in an area, the lower the levels of sulfur dioxide in the air.

Sewage pollution and invertebrates

B–A*

- Polluted water often contains very little dissolved oxygen. Some species of invertebrate are able to live in this water, but others are not. Figure 1 shows some of these species.

EXAM TIP

You do not have to memorise this diagram but you may be asked to apply the information it contains.

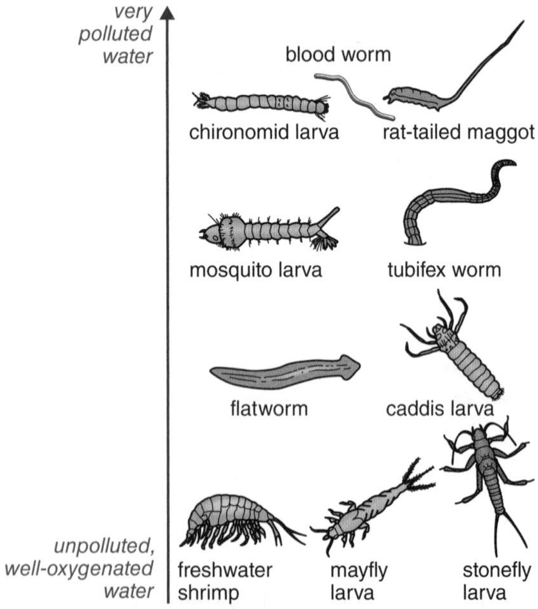

Figure 1: Invertebrates that can be used as pollution indicators

Improve your grade

Sewage contains a lot of microorganisms. Explain why mayfly larva cannot live in water polluted with sewage.
AO2 (2 marks)

Food chains and energy flow

Energy wastage

- Green plants capture only a small amount of energy from the light that falls onto them. This is because some light:
  - misses the leaves altogether
  - hits the leaf and reflects back from the leaf surface
  - hits the leaf, but goes all the way through without hitting any chlorophyll
  - hits the chlorophyll, but is not absorbed because it is of the wrong wavelength (colour).

- As a result, very little of the light energy that falls on a plant can be used for photosynthesis and get transferred into chemical energy in carbohydrates and other substances.

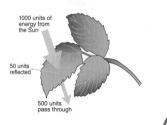

1000 units of energy from the Sun

50 units reflected

500 units pass through

Figure 2: This is what happens to energy from the Sun that falls onto a leaf

D–C

Remember!

A food chain shows that energy is passed from producer (plants) to consumers. Consumers gain their energy by eating the organism before it in the food chain.

Efficiency

- Whenever energy is transferred, some of it is wasted.

- To calculate the efficiency of energy transfer, use the formula:

  $$\text{efficiency} = \frac{\text{useful energy transferred} \times 100\%}{\text{original amount of energy}}$$

B–A*

EXAM TIP

If you are asked to calculate efficiency, you will be supplied with this formula. However, you should practice using it in preparation for the exam.

Biomass

Energy losses

- Figure 3 shows a pyramid of biomass drawn to scale. Its shape is explained by the fact that whenever energy is transferred, some is wasted (not used for useful work). So at each step, there is less energy available for the organisms to use. Less energy means less biomass.

- The food chain loses energy because:
  - some materials and energy are lost in the waste materials produced by each organism, such as carbon dioxide, urine and faeces
  - respiration in each organism's cells releases energy from nutrients to be used for movement and other purposes, so much of this energy is eventually lost as heat to the surroundings
  - not all of the organism's tissues are eaten, for example the antelope does not eat the roots of the grass as they are under the ground.

Figure 3: A pyramid of biomass drawn to scale

mass of lion

mass of antelope

mass of grass

D–C

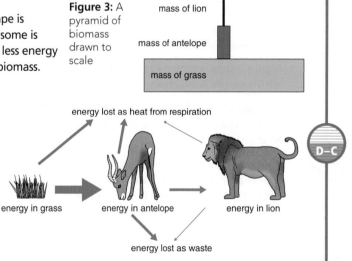

energy lost as heat from respiration

energy in grass → energy in antelope → energy in lion

energy lost as waste

Figure 4: How energy is wasted as it passes along a food chain

More about energy loss

- Mammals and birds use glucose to provide energy to keep their body temperature high.

- This means that energy loss from birds and mammals is high.

- Other animals, such as snakes, frogs and fish, just stay the same temperature as their environment.

B–A*

Improve your grade

Explain why biomass decreases as you move along a food chain. **AO2 (3 marks)**

Decay

Speeding or slowing decay

D–C

- Most of the bacteria and fungi that carry out decay need:
  - oxygen for aerobic respiration
  - a warm temperature for their enzymes to work at an optimum rate
  - moisture for reproduction.
- Increasing the temperature of microorganisms slows or stops decay.

Remember!

Compost heaps are kept warm, moist and aerated to speed up the decay of plant waste into compost. Compost is high in nutrients and is used to promote plant growth.

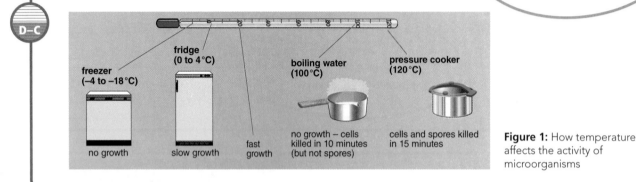

freezer (−4 to −18°C) no growth

fridge (0 to 4°C) slow growth

fast growth

boiling water (100°C) no growth – cells killed in 10 minutes (but not spores)

pressure cooker (120°C) cells and spores killed in 15 minutes

Figure 1: How temperature affects the activity of microorganisms

Preventing decay

B–A*

- If food is not to decay, it can be treated so as to slow down or stop the activity of microorganisms.
- Examples of this include canning, pickling and drying food.

EXAM TIP

You should be able to apply your knowledge to explain how these methods of food preservation help to slow down decay.

Recycling

Recycling and food chains

D–C

- Figure 2 shows how microorganisms fit into a simple food chain.
- You can see that these decay microorganisms feed on every organism in the chain.
- They will break down most of the waste material that the plants and animals produce, and then their bodies will be broken down by others when they die.

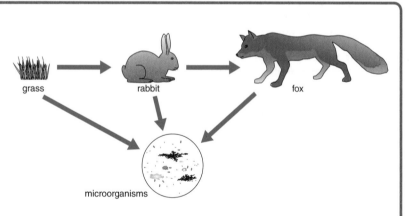

grass   rabbit   fox

microorganisms

Figure 2: A food chain, including the decay organisms

Dead whales

B–A*

- Whole communities of organisms use whale carcasses as food.
- Crabs, worms and fish eat the whale's body.
- Microorganisms gradually decay the whale's tissues. The whole process can take decades.

⦿ Improve your grade

Write a simple plan for an investigation to prove that the decay of bread by mould requires moisture.
AO2 (2 marks)

The carbon cycle

Processes in the carbon cycle

- Animals, plants and decomposers all interact with each other in the carbon cycle (Figure 3).

- Photosynthesis converts carbon dioxide into carbohydrates and other food molecules such as proteins:
carbon dioxide + water → glucose + oxygen

- When animals eat plants (or another animal) the food goes into their cells and is broken down by respiration: glucose + oxygen → carbon dioxide + water

- Carbon dioxide is returned to the air when the animal breathes out.

- Plants and microorganisms also respire.

- Some dead organisms do not decay. They become buried and compressed, deep underground and change into fossil fuels.

- Carbon dioxide is returned to the air when wood or fossil fuels are burnt (combustion).

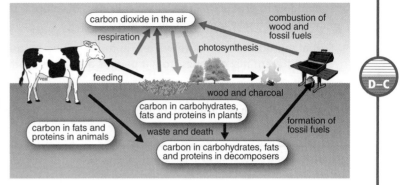

Figure 3: How carbon is recycled in the carbon cycle

D–C

Energy in the carbon cycle

- Energy is transferred in the carbon cycle.

- During photosynthesis, energy from sunlight is transferred to energy stores as chemicals in carbohydrates.

- Some of this energy is transferred to other organisms – such as animals or decomposers – when they feed on the plant.

- Some of the energy is wasted, heating the soil and air.

> **Remember!**
> Both energy and carbon atoms are never destroyed or created. They are constantly being recycled around the planet.

B–A*

Genes and chromosomes

Chromosome numbers

- Chromosomes are long sections of DNA.

- Most human cells have 46 chromosomes, that is 23 from the gamete of each parent (sperm and egg), which carry about 25 000 genes.

- Each gene contains coded information that controls one characteristic. For example, some genes control hair colour. Other genes control eye colour.

- Most of these genes come in two or more forms. For example, a gene that controls hair colour might have one form that produces brown hair and a different form that produces red hair.

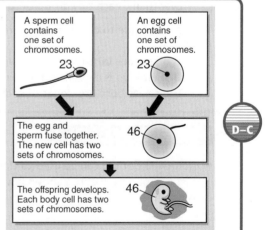

Figure 4: Chromosome numbers in gametes and other cells

D–C

Causes of variation

- Variation (differences) in organisms may be due to either: – the genes they have inherited (genetic causes) – the conditions in which they have developed (environmental causes) – or a combination of both.

B–A*

Improve your grade

Emma and Alice are identical twins. Emma has blonde hair and Alice has brown hair. Is this variation due to genetics or the environment? Explain your answer. **AO2 (2 marks)**

Reproduction

How it works

D–C

- In sexual reproduction, gametes and fertilisation are always involved.
- The new cell that is produced by fertilisation is a zygote. It divides repeatedly to produce a little ball of cells. This develops into an embryo and finally into an adult animal.
- Sexual reproduction produces variety in the offspring because each zygote has a different mix of genes from it parents and all its brothers and sisters.
- In asexual reproduction, an individual splits in two (as in bacteria) or a part divides off. This is the offspring.
- There is no variation. The new organisms all have exactly the same genes as their parent, and as each other. They are genetically identical (clones).

> **Remember!**
> Sexual reproduction does not always need two parents. Some plants have flowers that produce both male and female gametes, so they can fertilise themselves.

Different kinds of fertilisation

B–A*

- In birds and mammals, the male sperm are deposited and the egg is fertilised inside the female's body. This is called internal fertilisation.
- In other animals, such as fish, the male and female shed sperm and eggs into water. This is called external fertilisation. The fertilised eggs develop outside the female's body.

Cloning plants and animals

Cloning methods

D–C

- Taking cuttings is a way of making new plants from one original plant. Stems are cut from the parent plant; the ends dipped in hormone rooting powder and placed into soil. The cuttings will grow into new plants which are genetically identical to each other and the parent plant.
- Tissue culture can also be used to clone plants.
    - A small piece of tissue is taken from a root, stem or leaf of the parent plant. The tissue is then grown on a jelly containing all the nutrients it needs.
    - Everything has to be kept sterile, so this is usually done in a laboratory.
    - Eventually, each tiny group of cells grows into a complete adult plant.
- One technique to clone animals is called embryo transplants. This is sometimes done with farm animals, such as cows.
    - Egg cells are taken from a cow and fertilised with sperm from a bull.
    - One embryo is chosen and split into two (or more) and then each is placed into a host mother.
    - The calves born are clones of each other as they have the same genes.

Adult-cell cloning

B–A*

- Adult-cell cloning can be used to clone just one parent. Figure 1 shows how it is done.

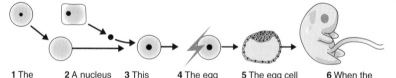

1 The nucleus is removed from an unfertilised egg cell.

2 A nucleus is taken from another cell in an adult animal's body.

3 This nucleus is put into the egg cell.

4 The egg cell is given a small electric shock. This persuades it to start dividing.

5 The egg cell grows into an embryo. Each of its cells contains exactly the same genes as the adult cell from which the nucleus was taken.

6 When the embryo is big enough, it is put into the uterus of a host mother, to continue its development.

Figure 1: Adult cell cloning

Improve your grade

Buttercup the cow produces the most milk in her herd. Her farmer is considering using her eggs for embryo transplants. Explain to him why cloning her might be an even better idea. **AO2 (3 marks)**

Genetic engineering

How it is done

- Bacteria have been genetically engineered to make human insulin. Figure 2 shows how this is done.

- Farmers spray bean fields with herbicides to kill weeds that compete with soya plants. The spray contains a chemical called glyphosate.

- Some soya bean varieties have been genetically engineered to give them a gene that makes them resistant to glyphosate.

- So when a farmer sprays the field with glyphosate, the weeds die but the bean plants do not.

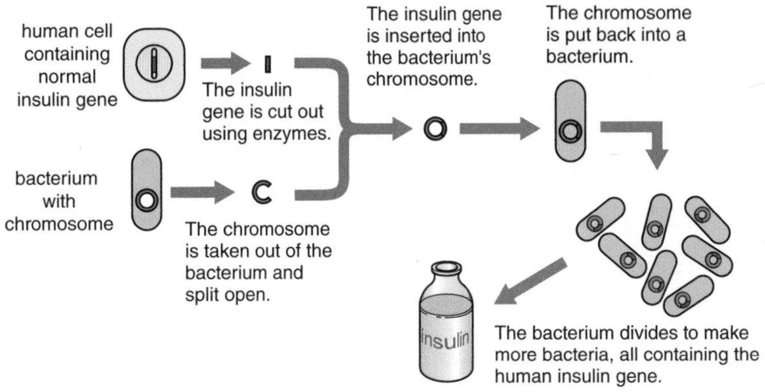

human cell containing normal insulin gene

The insulin gene is cut out using enzymes.

bacterium with chromosome

The chromosome is taken out of the bacterium and split open.

The insulin gene is inserted into the bacterium's chromosome.

The chromosome is put back into a bacterium.

The bacterium divides to make more bacteria, all containing the human insulin gene.

insulin

Figure 2: Bacteria have been genetically modified to produce human insulin

D–C

Genetic modification – good or bad?

- Some GM crop plants are resistant to attack by pests. This can greatly increase the yields and keep prices down. It also reduces the amount of pesticide that has to be sprayed.

- Some people have concerns about GM crops:
  - genes for a toxin to kill insects could be transferred to a wild plant, which could then disrupt natural food chains
  - there may be effects on humans of eating food from GM plants.

- GM crops have to be thoroughly tested before they are allowed to be grown on a large scale and there is no evidence that eating GM plants does any harm.

B–A*

Evolution

Accepting Darwin's ideas

- Jean-Baptiste Lamarck suggested that changes in organisms caused by their environment were passed on to their offspring. We now know that this is not correct.

- Charles Darwin suggested that species gradually changed from one form to another by natural selection. Darwin thought that, in each generation, only the best-adapted individuals survive and reproduce to pass on their characteristics to the next generation.

- Darwin's ideas challenged the established thinking of the day, so his ideas were not accepted at first. They undermined the idea that God made all animals and plants.

- In the late 19th century, there was not much scientific evidence to support the theories of evolution and natural selection. At that time, no one even knew that genes existed, let alone the way that they are inherited. This was not discovered until 50 years later. However, this is the theory that almost all scientists today believe as there is now a lot of scientific evidence to support it.

D–C

Simple to complex?

- The very earliest forms of life on Earth were almost certainly simple, single-celled organisms. Today, many organisms are much more complicated. Does this mean that evolution produces increasingly complex organisms?

- Some bacteria living today are almost the same as bacteria that lived billions of years ago. They were, and still are, supremely well adapted to their environment.

B–A*

Improve your grade

In 1859, Charles Darwin published a book containing his ideas about natural selection and evolution. Explain why many people at the time did not believe what it said. **AO2 (3 marks)**

Natural selection

How natural selection works

D–C

- This is how natural selection happens.
  - Living organisms produce many offspring.
  - The offspring vary from one another, because they have differences in their genes.
  - Some of them have genes that give them a better chance of survival. They are most likely to reproduce.
  - Their genes will be passed on to their offspring.
- Occasionally, unpredictable changes to chromosomes and genes, called mutations, happen.
- Occasionally, the new form of the gene increases an organism's chances of surviving and reproducing. It is therefore very likely to be passed on to the next generation. Over time, the new feature, produced by this gene, becomes more common in the species.
- The change in colour of the peppered moth from pale to dark is an example of evolution occurring because of a mutation.

> **EXAM TIP**
>
> You may be asked to explain how certain species evolved. Do this by applying the stages of natural selection: variation, competition, survival and reproduction.

The randomness of mutation

B–A*

- Some forms of bacteria have become resistant to antibiotics.
- This happened as a result of mutation in the bacteria producing a form of a gene that helped them survive, even when the antibiotic was present in their environment.
- This was just chance. The bacteria did not purposefully mutate to become resistant.

Evidence for evolution

Comparing living organisms

D–C

- You can get clues about evolution by looking carefully at organisms that are alive today.
- For example, your arm, a bat's wing and a bird's wing all have the same bones in the same places. Similarities like this suggest that humans, bats and birds are quite closely related and that, long ago, an animal lived from which humans, bats and birds have all evolved (a common ancestor).
- Evolutionary trees like this one (Figure 1) show the pathway along which different kinds of organisms may have evolved.
- Organisms that lived longest ago are at the bottom of the tree.
- Models like this help to show how different groups of organisms might be related, which helps scientists to classify them.

> **Remember!**
> There are five main classification groups: bacteria, protctists, fungi, plants and animals.

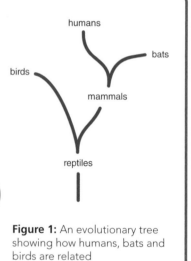

Figure 1: An evolutionary tree showing how humans, bats and birds are related

Classification

B–A*

- Scientists are still discovering new species which change their ideas about how organisms should be classified.
- Recently, they have found that there are two distinct groups of these microorganisms. They are as different from one another as animals are from bacteria.
- They have now been split into two big groups – the bacteria and the archaea.

⊙ Improve your grade

The cheetah is the fastest land animal on Earth. It uses its speed to catch its prey. Use natural selection to explain how cheetahs have got faster over time. **AO2 (5 marks)**

Keeping healthy

To lose mass you need to use more energy than you take in. You can do this by eating less and exercising more. Metabolic rate is the speed at which energy is used by the body to carry out chemical reactions.

White blood cells attack pathogens via phagocytosis, antibodies and antitoxins. Vaccination encourages white blood cells to make antibodies against a pathogen. If the person is later infected, the antibodies will be made much more quickly. We say they are **immune**.

High blood **cholesterol** can lead to heart disease. It encourages the build-up of **plaque** which can lead to the blockage of arteries. Blood cholesterol level can be influenced by diet and inherited factors.

Antibiotics are drugs which kill bacteria. Some bacteria are becoming resistant to anti-biotics. An example is MRSA.

You can grow cultures of bacteria and fungi in liquids or jellies called a **nutrient** medium. It is important that everything is kept sterile.

Nerves, hormones and drugs

Sensory neurones carry electrical **impulses** to the CNS from receptors. Motor neurones carry impulses from the CNS to effectors (muscles and glands) where they have an effect.

Hormones control the menstrual cycle:
- FSH causes an egg to mature and the ovary to secrete oestrogen
- oestrogen causes the uterus lining to thicken and stops the production of FSH
- LH causes ovulation.

Conditions surrounding cells must be kept constant. These include water, sugar and salt concentration, and temperature. Sweating is one mechanism that the body uses to reduce body temperature when it rises above 37 °C.

Plant hormones called auxins cause their shoots to grow towards the light (positive phototropism) and roots to grow towards gravity (positive gravitropism). Artificial hormones can be used as rooting powder and herbicide.

Recreational drugs can be legal or illegal. Many drugs are addictive and can be dangerous if abused. Medicinal drugs have to be trialled before they can be prescribed by doctors.

Interdependence and adaptation

Organisms have characteristics that enable them to live in their environment. This is called adaptation. The better the adaptation, the more likely it is that the organism will win the competition for resources and survive to reproduce.

Energy is wasted at each stage in a food chain, which results in less biomass at each trophic level. This is represented as a pyramid of biomass.

Decay by microorganisms is an important way of recycling nutrients. The carbon cycle shows how carbon is moved around the planet.

Pollution levels can be monitored by measuring factors such as temperature and pH; or by studying the distribution of living pollution indicators, such as lichens to show air pollution and invertebrates to show water pollution.

Genes and evolution

Organisms vary because of the genes they inherit from their parents. Asexual reproduction results in clones. The offspring from sexual reproduction are different from their parents.

Genetic engineering means taking a gene from one organism, and putting it into another. This is used to make human insulin from bacteria and to create GM crops.

Scientists believe that all life on earth evolved from single-celled organisms that lived 3.5 billion years ago.

Clones can be created artificially by taking cuttings and carrying out tissue culture, embryo transplants and adult cell cloning.

Darwin's ideas about evolution explained how living things evolved due to variation and survival of the best adapted. Many people did not believe his ideas until 50 years after they were published.

Atoms, elements and compounds

Composition and structure

D–C

- Mixtures can be separated into simpler substances with fewer parts.
- A compound can be broken down into simpler compounds or its elements.
- Elements are substances that cannot be broken down by chemical reactions into simpler substances.

Modelling structures

B–A*

- Symbols are not always the letters of the name of the element. Sodium has the symbol Na from Natrium, the Latin word for salt.
- Chemists use models to represent the 3D arrangement of atoms. Coloured spheres represent atoms.

Inside the atom

Sub-atomic particles

D–C

- Sub-atomic particles have both a charge and a mass. You can work out the mass of an atom by adding together the numbers of protons and neutrons.
- Since an atom has equal numbers of protons and electrons, the charges cancel out.
- Neutrons have no charge and are neutral.
- The sum of the number of protons and neutrons in an atom is its mass number.

How are the electrons arranged?

- The electronic configuration of an atom gives the number of its electrons and the arrangement of the shells.
- You can show electrons as dots or crosses.
- Each shell can hold a limited number of electrons.
- The first shell (lowest energy level) can hold up to 2 electrons.
- The second shell (next energy level) can hold up to 8 electrons.
- The third shell (third energy level) can hold up to 18 electrons, but fills up with only eight before the fourth shell is started.

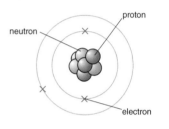

particle	relative mass	charge
proton	1	+
neutron	1	neutral
electron	almost 0	–

Figure 1: The structure of a lithium atom and sub-atomic particle chart

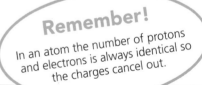

Remember!

In an atom the number of protons and electrons is always identical so the charges cancel out.

Why is electronic configuration so important?

B–A*

- Elements with the same number of outer electrons have similar chemical properties.

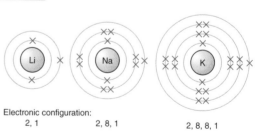

Electronic configuration:
2, 1   2, 8, 1   2, 8, 8, 1

Figure 2: Electronic configuration

Improve your grade

Draw the electronic structures of the following atoms with proton numbers 3, 9, 11, 16, and 20.
AO2 (5 Marks)

Element patterns

Reactive and unreactive

- The periodic table (see page 252) lists all the known elements by increasing proton number. It has horizontal rows called Periods, and vertical columns called Groups.

- The Group number tells you how many outer electrons the atom has.

- The number of outer electrons in the atom determines how an element reacts.

D–C

Why are the elements arranged in this pattern?

- The periodic table lists the elements in order of atomic number. Noble gases (Group 0) are unreactive because they have full outer electron shells.

- After each Noble gas, the next period (row) on the table starts.

B–A*

Combining atoms

Eight in a shell

- Atoms with fewer than eight outer electrons react in ways that give them a stable group of eight in their outer shell.

- In a water molecule (H_2O), an oxygen atom needs two more electrons to have eight in its outer shell. A hydrogen atom needs one more to become stable (two electrons in the first shell). Oxygen forms two covalent bonds, one with each of two hydrogen atoms. This way all the atoms achieve a noble gas configuration.

- To form an ionic bond in sodium chloride the chlorine atom gains an electron to get an outer shell of eight. It is now a negatively charged chloride ion, Cl^-. The electron comes from the sodium atom, leaving the second shell of the sodium atom as its outer shell. This has eight electrons, so it is stable. The opposite charges attract.

Figure 3: Covalent bonding in water molecules

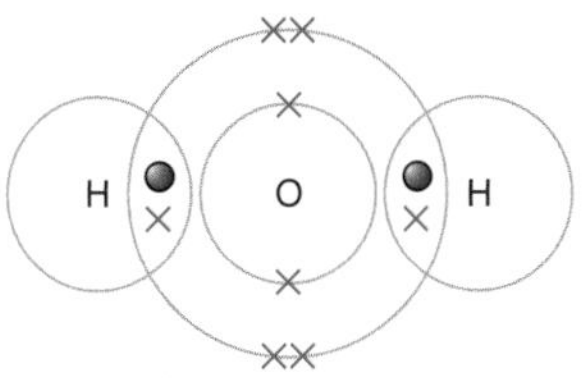

Figure 4: Ionic bonding in sodium chloride

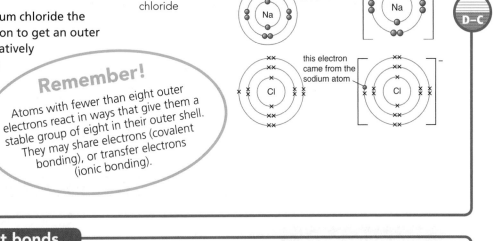

this electron is lost by the sodium atom

this electron came from the sodium atom

Remember!
Atoms with fewer than eight outer electrons react in ways that give them a stable group of eight in their outer shell. They may share electrons (covalent bonding), or transfer electrons (ionic bonding).

D–C

Representing covalent bonds

- Bonds are often represented as a short line between element symbols, for example F–F represents a fluorine molecule, a single bond joins the elements together.

- O=O represents an oxygen molecule where the two atoms are joined together by a covalent double bond.

EXAM TIP

Remember to use your periodic table to check how many outer electrons an atom has when answering questions on atomic structure and bonding. You do not need to learn this information.

B–A*

Improve your grade

Explain why oxygen has a molecule with a double covalent bond and fluorine has only a single covalent bond. You should refer to the number of outer electrons in both atoms in your answer. **AO2 (4 Marks)**

Chemical equations

Tracking atoms and molecules

- When methane burns, the atoms in methane (CH_4) and oxygen (O_2) react and turn into carbon dioxide (CO_2) and water (H_2O).

- Figure 1 shows the molecules, but doesn't account for all the reactant atoms as products. Figure 2 shows what happens to all the atoms, it is balanced.

- This is the balanced symbol equation;
  $CH_4 + 2 O_2 \rightarrow CO_2 + 2 H_2O$

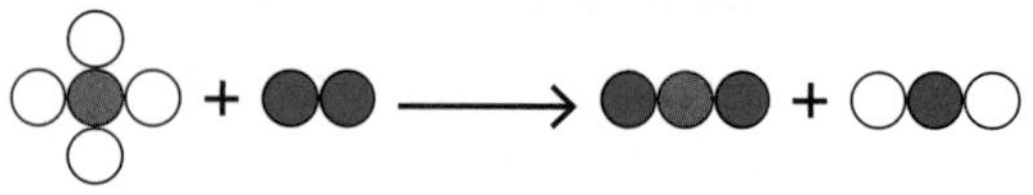

Figure 1: Methane burning with oxygen to form carbon dioxide and water

Figure 2: Methane burning with oxygen showing what happens to all the atoms

(D–C)

Balancing equations

(B–A*)

- To make an equation balance, you need to make sure you have the same number of each element on each side of the equation. To do this you add whole numbers in front of formulae as necessary. You must not change any of the formulae, because a different formula would represent a different substance.

- To balance an equation start with the metal atoms, and make sure you have the same number of reactants as the products. If needed put a whole number before the formula containing the atom.

- Now count the other atoms and balance them, leave hydrogen and oxygen to the end.

- Finish off by counting the hydrogens, then the oxygens, and correct them too. You should now have the same number of each element on each side of the equation.

Building with limestone

What are the effects of a limestone quarry?

(D–C)

- Limestone is obtained by quarrying. The table shows some advantages and disadvantages of quarrying limestone.

Table 1

Advantage	Disadvantage
Jobs for local people in an area with little industry	Damage to the landscape. Loss of wildlife habitats
More, better-paid jobs, so more money to boost local economy	Noise and vibration from blasting, machinery and vehicles
Better healthcare and leisure facilities, as more people move into the area	Dust pollution in the environment
Better transport links, needed for lorries or railways	Traffic congestion and vibration from heavy lorries

Limestone caves

(B–A*)

- Water and carbon dioxide make carbonic acid
  $H_2O(\ell) + CO_2(g) \rightarrow H_2CO_3(aq)$

- Rain containing carbonic acid dissolves limestone making a solution of **calcium** hydrogen carbonate
  $H_2CO_3(aq) + CaCO_3(s) \rightarrow Ca(HCO_3)_2(aq)$

- This produces caves and potholes. When the water evaporates it reverses the process producing stalactites and stalagmites (solid calcium carbonate).

Remember!
In formulae and equations capital letters are always the first letter of a symbol. The second letter is always small. Two capital letters do not make a symbol, it's a compound!

Improve your grade

Balance these two chemical equations
AO2 (2 Marks)

a $Mg(s) + O_2(g) \rightarrow MgO(s)$
b $C_3H_8(g) + O_2(g) \rightarrow CO_2(g) + H_2O(\ell)$

Heating limestone

Cement

- Calcium hydroxide solution (limewater) is used to test for carbon dioxide. If **carbon dioxide** is bubbled through limewater, it turns cloudy or 'milky' as tiny solid particles of white calcium carbonate form.

- Cement is made by heating limestone and clay. Cement is used widely – on its own, and in mortar and concrete.

- Mortar is made by mixing cement, sand and water. Mortar binds bricks together in brick walls.

- Concrete is made by mixing cement, sand, gravel (or crushed rock) and water. Concrete is very strong. It is used for the foundations of buildings and for structures such as bridges.

D–C

The difference between mortar and concrete

- Cement, mortar and concrete do not dry out. The cement reacts with the water to form crystals that 'cement' the mixtures together.

- The mix of small sand particles and various sized stones in concrete makes it much stronger than mortar.

- Concrete can be made even stronger by reinforcing it with steel.

B–A*

Metals from ores

Making use of ores

- Only minerals with enough metal to make it worth extracting are used as ores.

- Some ores are metal oxides. These can be smelted directly.

- At the smelter, the ore is crushed and concentrated, to remove rock with little or no metal.

- Other ores are converted to the metal oxide before or during smelting.

- To convert the metal oxide to the metal, the oxygen must be removed. This is called reduction.

- The metal oxide is reduced by heating it in a furnace with carbon if the metal is below carbon in the reactivity series. Originally, the carbon was charcoal, but now it is coke (a nearly pure form of carbon, from coal).

- Limestone is often added, to remove impurities in the ore forming **slag**.

Table 2: The reactivity series

potassium	K
sodium	Na
calcium	Ca
magnesium	Mg
aluminium	Al
carbon	C
zinc	Zn
iron	Fe
lead	Pb
hydrogen	H
copper	Cu
silver	Ag
gold	Au

Increasing reactivity →

D–C

Extracting more reactive metals

- Many metals were not discovered until the discovery of electricity in the 1800s.

- Metals above carbon in the reactivity series are extracted using electrolysis.

- Electrolysis involves passing an electric current through a molten metal compound, splitting it into its metal and non-metal elements.

Electricity
Molten aluminium oxide → molten aluminium + oxygen gas

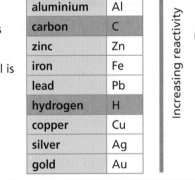

Remember!
When asked about the environmental or social impact of quarrying limestone make sure you give both advantages and disadvantages if you are to get full marks.

B–A*

EXAM TIP

Always check the reactivity series before answering questions about metal extraction. Metals below carbon always use carbon to reduce them, metals above are reduced by electrolysis.

◉ Improve your grade

Draw a table to show which of these metals is extracted by carbon and which by electrolysis. **AO1 (2 Marks)**
iron, magnesium, aluminium, copper, zinc, lead, potassium, and calcium

Extracting iron

Inside a blast furnace

- Iron ore (Fe_2O_3), coke and limestone are fed in at the top of the furnace.

- Molten iron and slag collect at the bottom.

- As the hot air is blown through, the coke burns in oxygen to produce carbon monoxide:

  carbon + oxygen → carbon dioxide
  $$C(s) + O_2(g) → CO_2(g)$$
  carbon dioxide + carbon → carbon monoxide
  $$CO_2(g) + C(s) → 2CO(g)$$

- The carbon monoxide then reduces the iron oxide to iron, and is oxidised to carbon dioxide:
  $$Fe_2O_3(s) + 3CO(g) → 2Fe(\ell) + 3CO_2(g)$$

- The limestone reacts with impurities in the ore to make slag, this floats on top of the iron.

Figure 1: Blast furnace

waste gases

iron ore, coke and limestone

hot air blast

slag tapped off

iron tapped off

Oxidation and reduction

- Oxidation reactions occur when oxygen is added to a substance.

- Reduction reactions occur when oxygen is removed from a substance.

- Oxidation is the reverse of reduction and both must always occur together.

Remember!
Iron is lower than carbon in the reactivity series. This is why carbon reduces (removes the oxygen) from iron oxide. The carbon becomes oxidised (gains oxygen) at the same time.

Metals are useful

Atoms and alloys

- Metal atoms are arranged in giant structures, in regular rows and layers.

- Metals can bend because these layers can slide over each other. The shape can change but the atoms remain bonded together.

- Metals conduct electricity because some of their outer electrons are free to move through the layers.

- To make metals harder they can be mixed to form alloys.

- An alloy is not a compound but is a mixture of elements, usually metals. Steel is an alloy of iron and carbon. Brass is an alloy of copper and zinc.

- The proportions of each element in an alloy can differ. This affects the alloy's properties and uses.

- The more carbon in steel the harder it is. Nine carat gold has more copper in it than 22 carat gold, and is harder wearing.

Figure 2: Layer of copper atoms

Smart alloys

- These alloys can remember their original shape, and if warmed return to that shape.

- To mend broken bones, strips of smart alloy are cooled, stretched and screwed onto the bone. As the strips warm up they shrink, pulling the bones back together so the break heals faster and in the correct position.

Improve your grade

Explain the difference between oxidation and reduction. **AO1 (2 Marks)**

Chemistry C1

Iron and steel

Designer steels

- Pure iron is too soft for most uses, so steels are used instead.

- Adding other metals to molten steel can give it special properties. The choice of metal depends on how the steel will be used, and therefore the properties required.

- Stainless steel is about 70% iron, 20% chromium and 10% nickel. Stainless steel is very resistant to corrosion and does not rust.

Table 1 How other metals change steel's properties

Metal added to steel	Improvement to steel properties
Chromium and/or nickel	More corrosion resistance
Manganese	More strength and hardness
Molybdenum and/or tungsten	More strength, hardness and toughness
Vanadium	More strength, less brittle

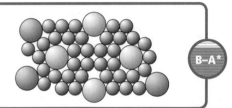

EXAM TIP

When asked about the use of an alloy always think about the properties the alloy needs to have for the suggested use.

D–C

Why is steel harder than iron?

- The atoms in pure metals have the structure shown in Figure 2. If other different sized atoms are added to make an alloy, as in Figure 3, the layers of atoms find it hard to slide past each other and so the metal is harder.

Figure 3: Different sized atoms give steels hardness and strength

B–A*

Copper

Physical and chemical changes during extraction

- Copper is purified using electrolysis.

- Electricity is passed through a copper sulfate solution. This is called electrolysis.

- Copper atoms in the impure positive electrodes lose electrons. They become copper ions (Cu^{2+}) and dissolve into the solution. The impurities fall to the bottom.

- At the pure copper negative electrodes, copper ions in the solution gain electrons and coat the electrode with copper atoms.

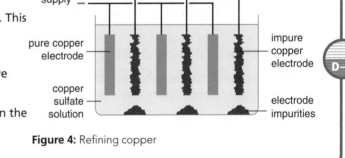

electricity supply +/−

pure copper electrode

copper sulfate solution

impure copper electrode

electrode impurities

Figure 4: Refining copper

D–C

Extracting copper from low grade ores

- High quality copper ores are very limited so new ways of cheaply producing copper ores are being developed.
  - Bioleaching – Bacteria convert insoluble copper compounds into soluble ones.
  - Phytomining – Plants absorb copper compounds through their roots. After harvesting, the plants are burnt leaving a copper-rich ash.

How Science Works

- When testing properties of new materials such as alloys, the test must be both repeatable (gives the same result every time you carry out the test), and reproducible (gives the same conclusion when done by someone else, or using a different method).

B–A*

Improve your grade

Explain why the waste heaps of an old copper mine would be a suitable site for phytomining or bioleaching.
AO2 (2 Marks)

Aluminium and titanium

Extraction

- Aluminium is very expensive as it can only be extracted using electricity.

- The ore bauxite (Al_2O_3) is dissolved in cryolite to allow it to melt at 900 °C.

- The molten aluminium oxide then is electrolysed:
  aluminium oxide → aluminium + oxygen
  $$2Al_2O_3 \rightarrow 4Al + 3O_2$$

- Each aluminium ion needs **three electrons** to become an atom. This is why so much electricity is needed.
  $$Al^{3+} + 3e^- \rightarrow Al$$

- Titanium cannot be extracted by carbon or electricity. First the ore, rutile (titanium dioxide) is converted to titanium chloride. Then titanium chloride is reacted with magnesium.

- Titanium chloride + magnesium → titanium + magnesium chloride.
  $$TiCl_4 + 2Mg \rightarrow Ti + 2MgCl_2$$
  This method is very costly which means that titanium is also very expensive.

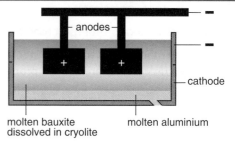

Figure 1: Extracting aluminium by electrolysis

Remember!
Using cryolite saves money as bauxite normally melts at 2000 °C. Reducing the temperature saves energy and money.

Corrosion resistance

- Both titanium and aluminium are very reactive metals, yet they resist corrosion.

- Both metals react easily with oxygen forming a tough oxide layer on the surface, preventing further reaction.

- Acids and alkalis do attack aluminium, because they react with the oxide layer. Titanium oxide does not react so titanium can be used inside our bodies for replacement joints and to hold broken bones together.

Metals and the environment

Effects on the environment

- Mining metal ores destroys the landscape, wildlife habitats, and displaces the local people, changing their way of life.

- Using carbon to reduce metal ores produces carbon dioxide. This adds to carbon dioxide levels in the atmosphere.

- Smelting ores containing sulfur produces sulfur dioxide which causes acid rain. If collected, the sulfur dioxide can be used to make useful sulfuric acid.

- Recycling aluminium and steel saves energy and the environment. Aluminium cans are cleaned and melted before reusing the metal. Iron and steel are added as scrap to steel-making furnaces.

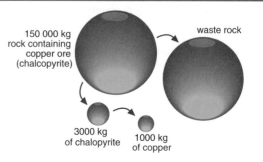

Figure 2: How much waste rock is mined, and later dumped, for each 1000 kg copper produced?

Recovering metals from brownfield sites

- Brownfield sites are areas of land that have been used before. They are often polluted by toxic metal compounds of cadmium, nickel and cobalt.

- Instead of removing the soil from these sites, developers now grow plants on them. These plants absorb the toxic metal ions making the land safe to use.

- When burnt the plant ash contains compounds of the toxic metals that can be used as ores. This is phytomining.

Remember!
Phytomining takes time, but it is cheap, and the metal compounds are easily obtained at the end.

Improve your grade

Explain why titanium and aluminium metals are both reactive and corrosion free. **AO1 (2 Marks)**

A burning problem

Global effects

- Global warming. Increasing carbon dioxide in the atmosphere traps more energy from the Sun making the Earth warm up.

- Global dimming. Dust particles in the air prevent sunlight reaching the ground. This reduces the energy available for photosynthesis, and may cool the Earth as well.

- Both may affect the weather patterns on the Earth.

- Acid rain is caused by nitrogen, sulfur and carbon oxides dissolving in rain water. The acid rain formed attacks limestone buildings much more quickly than normal rain. It can also cause serious damage to trees and to aquatic life in affected lakes.

D–C

Other problems with combustion

- During combustion, each carbon atom in a fuel needs two oxygen atoms to form carbon dioxide, CO_2.

- If there is not enough air, some carbon atoms get only one oxygen atom (carbon monoxide), or none at all (carbon or soot).

- Carbon monoxide is poisonous, it is easily absorbed by your red blood cells instead of oxygen, therefore depriving your body of oxygen.

B–A*

Reducing air pollution

Alternative solutions

- Vehicles that burn fossil fuels produce poisonous carbon monoxide, and they also produce nitrogen oxides which can cause acid rain.

- To reduce the amount of these compounds released by vehicles all new cars have catalytic converters fitted.

- Carbon monoxide is oxidised to carbon dioxide, and nitrogen oxides are reduced to nitrogen.

- Alternative fuels are also being used, such as biodiesel made from vegetable oils, which contain almost no sulfur. Ethanol produced by fermentation from sugar, and hydrogen, are sulfur free.

D–C

Problems with alternatives

- Rainforest is cut down to grow sugar and soya bean for biofuels. Elsewhere, land that could grow crops to feed the world's increasing population is, instead, producing fuels. This forces food prices up for every one.

- Hydrogen can only be obtained by using electricity to electrolyse water. Electric cars are emission free. However, producing electricity does produce harmful pollution, and whilst the vehicles do not produce pollution, making the electricity they need does.

B–A*

EXAM TIP

When asked to evaluate the benefits and risks make sure you consider both the benefits and also the drawbacks in your answer. Be specific, 'food prices will rise' is better than 'things will cost more', 'less sulfur oxides will be produced' is better than 'pollution will be less'.

Improve your grade

Suggest why using biofuels may create more problems than it solves. **AO2 (3 Marks)**

Crude oil

How does fractional distillation work?

- When crude oil is heated to about 400 °C, most of it boils and vaporises.

- Vapours consisting of hydrocarbons rise up the column, gradually cooling. When cooled below their boiling point, they condense back to liquid and are run off.

- Hydrocarbons with high boiling points condense first. The lower their boiling point, the higher up the column they rise before condensing.

- Hydrocarbons with different size molecules condense at different levels, separating the crude oil mixture into a series of fractions with similar numbers of carbon atoms and boiling points.

Name of fraction	Carbon atoms per molecule	Uses
petroleum gas	1 to 4	heating, cooking, LPG fuel
petrol	5 to 9	fuel (cars and lorries)
naphtha	6 to 10	to make other chemicals
kerosene	10 to 16	jet fuel, paraffin
diesel	14 to 20	fuel (cars, lorries and trains)
fuel oil	20 to 50	fuel for ships, factories and heating
bitumen	more than 50	tar for road making

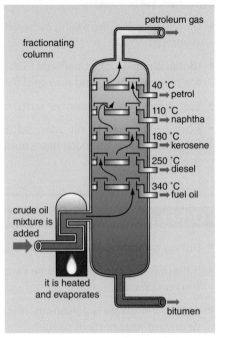

Figure 1: Fractional distillation of oil

Why do boiling points depend on size?

- There are strong covalent bonds between the atoms in hydrocarbon molecules.

- There are only weak attractions between the hydrocarbon molecules. With small molecules there are fewer attractions between the molecules.

- Bigger molecules have more bonds between them, and more attractions that need to be broken to make the molecules separate into a gas, so they need more energy (a higher temperature).

Alkanes

Patterns and properties

- The general formula for alkanes is C_nH_{2n+2} where n is the number of carbon atoms.

- All alkanes have similar chemical properties.

- As the number of carbon atoms increases;
  - molecules become larger and heavier
  - boiling point increases
  - flammability decreases (catch fire less easily)
  - viscosity increases (liquid becomes thicker).

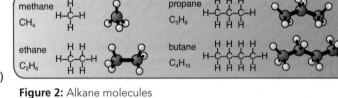

Figure 2: Alkane molecules

What makes alkanes similar to each other?

- Alkanes only have single covalent bonds joining each atom together. They are said to be saturated hydrocarbons.

- Alkanes have similar structures so they react in similar ways.

EXAM TIP

When drawing molecules of hydrocarbons, remember each carbon atom needs four bonds, and each hydrogen atom needs just one bond.

Improve your grade

Draw out the structure of the straight chained alkanes with 5 carbon atoms and 7 carbon atoms. **AO2 (2 Marks)**

Cracking

How cracking works

- Crude oils have different amounts of different hydrocarbons, but never enough petrol sized molecules for our needs.

- Heating the hydrocarbons makes each molecule waggle until a carbon–carbon bond breaks. They break in different places, giving a mixture of products. So $C_{10}H_{22}$ could crack to form;
$C_8H_{18} + C_2H_4$ or $C_7H_{16} + C_3H_6$ or $C_6H_{14} + C_4H_8$ and so on.

- One of the molecules made is an alkane, the other has two bonds (a double bond) shown as C=C between a pair of carbon atoms, and is called an alkene.

- Alkenes have the general formula C_nH_{2n}

- The double bond in alkenes makes them much more **reactive** than alkanes. This makes alkenes extremely useful.

- Ethene is a particularly important alkene product of cracking. It is the starting point for making polythene and many other plastics.

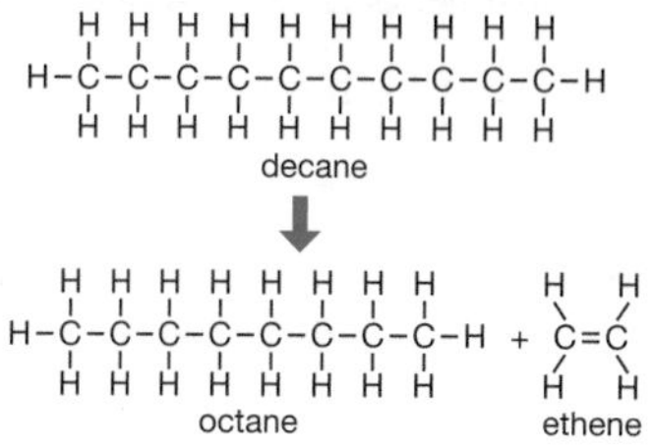

Figure 3: Cracking decane

Remember!
There is no alkene with only 1 carbon atom. They need at least two carbon atoms. Each carbon atom still needs to have four bonds, the double bond counts as two!

D–C

Types of cracking

- Fuel oil is mixed with steam in a furnace at about 850 °C. The hydrocarbons undergo thermal decomposition. Changing the amount of steam alters the products made.

- Fuel oil is vaporised and mixed with a catalyst at about 600 °C. Using a catalyst allows the cracking reaction to take place at a lower temperature than in steam cracking.

B–A*

Alkenes

Reactive alkenes

- A double bond is just two bonds.

- One of the two bonds can 'open up', allowing each carbon atom to form a bond with another atom. This means that alkenes are reactive compounds.

- Each carbon atom in an alkane already has bonds to four other atoms. So, unlike alkenes, alkanes cannot react by adding extra atoms. Alkanes are saturated – they cannot add any more atoms. Alkenes can, so alkenes are unsaturated.

- Fats are more complex than hydrocarbons, but we also refer to them as being saturated and unsaturated. Polyunsaturated fats have lots of double bonds.

D–C

Detecting double bonds

- Adding a few drops of orange-brown bromine water to a sample of a hydrocarbon shows whether the hydrocarbon is saturated.

- If the hydrocarbon contains double bonds, the orange brown colour will rapidly disappear, it is unsaturated. If the bromine water stays orange-brown, then the hydrocarbon is saturated.

Remember!
When your clothes are so wet they cannot accept any more water we say they are saturated. Hydrocarbons that cannot accept any more atoms, because all the bonds are single, are also called saturated.

B–A*

Improve your grade

Draw structural diagrams to show how $C_{12}H_{26}$ can be converted to C_3H_6, and another molecule. State which of the two products is unsaturated, and why. **AO1 and 2 (3 Marks)**

Making ethanol

Converting into ethanol

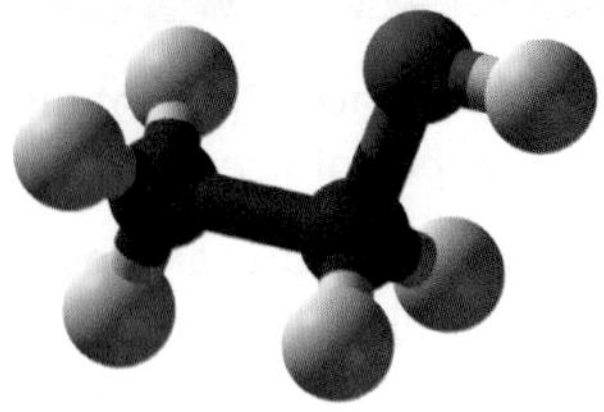

- Alcohol is the name of another chemical family.

- Alcohols are compounds with a hydroxyl group (–OH).

- Ethanol is a member of the alcohol family, ethanol has two carbons. Its formula is C_2H_5OH– written like that, rather than C_2H_6O, to show that it has the –OH group.
  - Ethanol is made by two methods:
  - from crude oil. Cracking large alkanes produces ethene. Blowing ethene and steam over a hot catalyst makes ethanol.
  - from sugar. Adding yeast to sugar dissolved in water causes fermentation and changes sugar into ethanol.

Figure 1: Molecular model of ethanol

- Both methods give a solution of ethanol in water. Ethanol is separated by fractional distillation.

- Fermentation happens when micro-organisms, called yeasts, feed on the sugar and convert it into ethanol.

- To manufacture ethanol for use as a fuel, a sugar solution made from sugar cane or maize (corn) is prepared. Yeast is added to cause fermentation. sugar → ethanol + carbon dioxide

- Fuels obtained from animals and plants are called biofuels, so ethanol made by fermentation is a biofuel.

- As more plants can be grown after harvesting, biofuels are a renewable source of energy. Unfortunately it uses land and crops that could be used to feed people.

Other uses of ethanol

- Some substances do not dissolve in water, but do dissolve in ethanol.

- This is useful when making pharmaceuticals (medicines), perfumes and aftershaves, inks and varnishes.

- It is the major ingredient in surgical spirit (an antiseptic that kills pathogens).

- It is used to make a wide range of other chemicals, including flavourings and perfumes.

Polymers from alkenes

A variety of polymers

- To make poly(ethene), ethene is heated under pressure. Ethene is known as the monomer. A catalyst sets off a chain reaction. It makes a C=C bond open up and join onto another ethene molecule. That double bond then opens up and joins onto the next, and so on, forming a polymer chain.

- The reaction involves only the C=C double bond. It does not matter what other atoms are attached to the carbons. By changing the other atoms attached to the carbons, it is possible to produce lots of other different polymers with different properties.

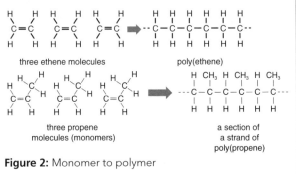

three ethene molecules   →   poly(ethene)

three propene molecules (monomers)   →   a section of a strand of poly(propene)

Figure 2: Monomer to polymer

Polymers and plastics

- Polymers are molecules consisting of very long chains made from carbon atoms, with various other atoms attached. Many natural materials are polymers, including silk, rubber, starch, proteins and DNA.

- Most polymers have low densities, are not brittle (and do not break easily), can withstand corrosion by chemicals and soften when heated (can easily be moulded into shape).

Improve your grade

PVC (polyvinylchloride) is a commonly used polymer. Its monomer has the structure:
Show, using three monomer molecules, the structure and bonding of a PVC polymer chain. **AO2 (2 Marks)**

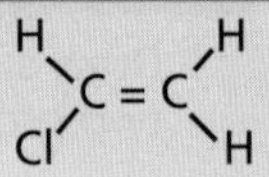

Designer polymers

Special polymers

- Plastic polymers are easily moulded into shape, low density (lightweight), waterproof and resistant to acids and alkalis.

- Properties such as strength, hardness and flexibility vary.

- Polymers can be designed to have the specific properties needed for a particular purpose.

D–C

What about the future?

Smart polymer materials can, for example, respond to changes such as temperature or voltage.

- Conducting polymers can conduct electricity.

- Light-emitting polymers give off light when electricity passes through.

- Shape memory polymers 'remember' the shape of the object.

- Biodegradable polymers rot away, unlike other polymers that cause waste problems.

B–A*

Polymers and waste

Sorting out waste

- Most disposable plastic items are labelled with a recycling symbol and code number to identify the polymer.

- Instead of dumping in landfills, household rubbish, including plastics, can be burned in incinerators. The heat produced may be used to generate electricity or heat local buildings. This is wasteful of polymers and can cause air pollution.

D–C

Biodegradable plastics

- Plastics in landfill do not rot, and the ground will be unusable for agriculture.

- A few plastics are water-soluble and/or biodegradable:
  - Biodegradable plastic carrier bags are made from polythene and corn starch. The cornstarch is decomposed by micro-organisms leaving the polythene in microscopic pieces.
  - Biopol® is a biodegradable plastic produced by micro-organisms. It is used in medicine to make stitches and hold bones together while they heal. It dissolves away so you don't need to have the stitches removed.

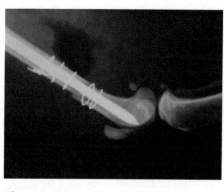

Figure 3: Polymer pins used to fix a broken bone

B–A*

> **Remember!**
> Smart polymers are used where a smart solution to a problem is needed. They are usually chosen because they are lighter or more robust than traditional materials.

⊙ Improve your grade

Describe the difference between recycling polymer waste and re-using it by incineration. Give two reasons why recycling is more environmentally friendly than incineration. **AO1 (3 Marks)**

Oils from Plants

Oils in food and fuel

D–C

- Plant oils store a lot of energy. Oils provide more energy than most other foods.
- Oils from rapeseed, soya beans and other crops are converted into biodiesel fuel.
- Cooking food in oils produces different flavours and textures; the food is cooked faster at a higher temperature than in water. The food absorbs some oil increasing its energy content.
- Oils also contain other nutrients we need.
  - Essential fatty acids (for example omega 3 and 6), for the heart, muscles and nervous system to function properly.
  - Vitamins. Seed and nut oils are particularly rich in vitamin E.
  - Minerals. Minerals are compounds of metals and non-metals such as potassium, calcium, iron and phosphorus.
  - Trace elements, but only in tiny amounts. For instance, a single Brazil nut provides the whole recommended daily allowance (RDA) of selenium.

Essential oils

B–A*

- Flowers contain essential oils, and are different from the natural oils in seeds and nuts. They have low boiling points, so evaporate easily, giving flowers their scents. They are used in perfumes.
- Essential oils are extracted using steam distillation. The less dense essential oil floats on top of the water.
- Vegetable oils have much more complex molecules than mineral oil, with many carbon, hydrogen and oxygen atoms.

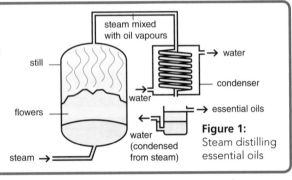

Figure 1: Steam distilling essential oils

Biofuels

Biofuel issues

Biofuels are made from animal or plant material, for example wood.

- Biofuels contain no sulfur and this advantage means that they do not cause sulfur dioxide pollution.

D–C

- Biofuels may be renewable, but they have problems of their own.
  Practical issues:
  - plant oils are very viscous so are hard to use directly in car engines
  - they need more air to burn properly than diesel and petrol
  - few garages sell biofuels as not many people currently use them.
  - Economic and ethical issues:
  - biofuels produce less energy per litre, so a greater volume is needed
  - growing crops needs energy for the machinery, fertilisers and transport, this may be more than is produced by the biofuel crop
  - land used to grow food crops may be used to grow biofuels instead, leading to food shortages and raised food prices.
  Environmental issues:
  - the demand for fuel is so great that huge amounts of land would be needed
  - changing land use can affect the plants and animals that live there, reducing biodiversity.

> **EXAM TIP**
>
> If asked to evaluate the impact of biofuels on the environment or another issue. Give both benefits and problems in your answer to get full marks.

Making biodiesel

B–A*

- Plant oils, animal fats or used cooking oil can be converted to biodiesel and then used in diesel engines.
- Reacting the oils or fats with methanol converts them to biodiesel, which is a mixture of chemicals called esters.

Improve your grade

Explain why biofuels are considered to be carbon neutral. **AO1 (2 Marks)**

Oils and fats

Unsaturated fats

- Testing an unsaturated fat is similar to testing for an alkene. They both contain double C=C bonds.

- The amount of bromine water decolorised by the fat indicates how many C=C bonds there are in the fat.

- Everyone needs some fats in their diet, for energy and essential nutrients, but too much fat is unhealthy.

- Saturated fats can increase the level of cholesterol in the blood. You should;
  - eat foods rich in polyunsaturates, such as sunflower oil, and monounsaturates, such as olive oil
  - avoid saturated fats, such as lard, limit the amount of fat of all types in your food,
  - eat fish, some fish oils contain omega-3 fatty acids. These have been shown to lower blood pressure.

D–C

Hardening vegetable oils

- Liquid vegetable oils can be hardened into solid spreads by hydrogenation.

- Hydrogenation adds hydrogen to the double bonds so the oil becomes saturated and solid.

- Hydrogenation uses a nickel catalyst and a temperature of about 150 °C.

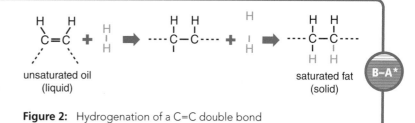

B–A*

Figure 2: Hydrogenation of a C=C double bond

Emulsions

Useful emulsions

- Oil and water are immiscible. However they can be made to mix by stirring or shaking, forming an emulsion.

- Emulsions are less viscous (less runny) than the oil and more viscous than the aqueous solution.

- Oil-in-water emulsions contain droplets of oil suspended in an aqueous solution.

D–C

Emulsifiers

- Many emulsions are unstable and rapidly separate back to oil and water.

- Emulsifiers are compounds whose molecules have opposite properties at each end.
  - One end is hydrophilic (water-loving) – it is attracted to water but not to oil.
  - The other end is hydrophobic (water-fearing) – it gets away from the water by sticking into an oil droplet.

- The oil droplets become surrounded by emulsifier molecules with their hydrophobic ends in the oil. The hydrophilic ends on the outside attract the droplets to the water. This also prevents the oil droplets joining together to form a separate oil layer again.

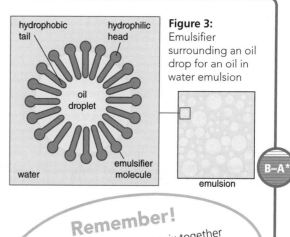

Figure 3: Emulsifier surrounding an oil drop for an oil in water emulsion

B–A*

How Science Works

- When deciding how much emulsifier to use to make a stable emulsion it is important to control all the other variables. You should always state how to control the variables, giving examples.

Remember!
Immiscible liquids do not mix together however hard you shake them. They simply separate into two layers, the less dense layer is the top layer, which is usually an oil, and the denser water layer is below.

⊙ Improve your grade

Draw a diagram to show how an emulsifier would surround a water droplet to make a water in oil emulsion.
AO2 (2 Marks)

Earth

What is Earth like inside?

D–C

- The crust is a thin outer layer of cold, solid rock. Its thickness is between 5 and 30 kilometres.

- The mantle is made of very hot molten rock that flows very slowly by convection currents.

- The core is at the centre; an inner core, in the centre, is solid iron and nickel and an outer core is a molten mixture of iron and nickel.

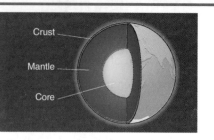

Figure 1: The layers of the Earth

How did the Earth's surface features form?

B–A*

- Earth's crust is broken up into separate parts called tectonic plates. These float on the mantle.

- Where plates push against each other, they force the land upwards, forming mountain ranges.

- In the same way, rocks such as limestone, which formed under the sea, were pushed up and became hills.

- Over millions of years, rivers gradually wear away rocks, forming valleys and spectacular canyons.

- Islands such as Iceland and Hawaii are formed when volcanoes on the sea bed throw out material which builds up.

Continents on the move

Continental drift

D–C

- Alfred Wegener in 1915 suggested that all the continents were once joined together in a super continent he called Pangea.

- When Wegener suggested his hypothesis, other scientists rejected it because no one could explain how huge continents could move. Later other scientists found evidence to support Wegener's theory.

200 million years ago   100 million years ago   50 million years ago

Figure 2: Pangea breaking up

- Earth's crust and the semi-solid upper part of the mantle make up the lithosphere.

- The tectonic plates float on the liquid rock of the mantle. Heat from radioactive processes within the Earth drives convection currents in the mantle. These currents carry the floating plates, and the continents very slowly, only a few centimetres a year, in a process called continental drift.

- When two tectonic plates the size of continents collide they push the land upwards, forming mountains. The world's largest mountain ranges are formed where continental plates collide.

What happens when continents move apart?

B–A*

- When tectonic plates move apart, magma (molten rock) escapes from the mantle, forming new crust.

- In the Atlantic Ocean as the plates move apart, magma rises through the gap creating an underwater mountain range called the Mid-Atlantic Ridge.

- Iceland is on the Mid-Atlantic ridge, where the ridge has become so high that it is no longer underwater. The two plates are still moving, so Iceland is getting wider each year.

> **EXAM TIP**
>
> Evidence for continents having been joined together includes: the shape of the continents; Africa and South America are like jigsaw pieces that can fit together; similar rock formations where the continents would have joined together; and similar fossils in rocks on both sides of the Atlantic.

Improve your grade

Explain how the Atlantic Ocean is getting wider each year. Suggest another part of the world where a similar process is taking place. **AO1 (3 Marks)**

Earthquakes and volcanoes

Volcanoes

- Volcanoes often occur at plate boundaries.

- Magma stays sealed under the crust, often for hundreds of years. Eventually the pressure builds up enough for magma to burst through a vent – a crack or weak spot in the crust. The blast creates a crater, lava and gas pour out, and the familiar cone shape of volcanoes forms.

- Scientists can monitor the movement of tectonic plates. They can tell when pressure is building and an earthquake is possible. However, they cannot obtain hard data about the forces involved, the friction between plates and structural weaknesses in plates. Eruptions are not the same as earthquakes, but the problems of predicting them are: the lack of sufficient valid data.

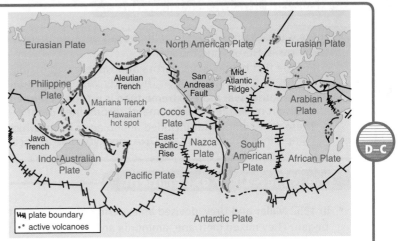

Figure 3: Main tectonic plate boundaries and active volcanoes

D–C

Subduction zones

- Where plates meet, the lighter plate, usually the oceanic one is pushed under the heavier continental plate. This is called subduction. As the lighter plate pushes into the mantle it forces magma to the surface through weaknesses in the crust (see Figure 3).

B–A*

The air we breathe

How did the atmosphere evolve?

- The Earth was formed about 4 600 million years ago. For the first 1 000 million years there was intense volcanic activity, releasing huge amounts of steam, carbon dioxide, and some ammonia (NH_3) and methane (CH_4).

- The steam condensed and eventually formed the seas and oceans.

- About 3 400 million years ago simple life that could photosynthesise had developed in the oceans and seas. These algae used the carbon dioxide and water and released oxygen. The oxygen reacted with the ammonia to make nitrogen gas.

- About 400 million years ago the atmosphere had enough oxygen to allow land plants and then animals to evolve.

- The atmosphere has stayed the same for about the last 200 million years with 78% nitrogen, 21% oxygen, and small amounts of other gases such as argon and carbon dioxide.

D–C

Separating gases from the air

- Air is dried and filtered to remove water vapour and dust.

- Air is cooled to about –200 °C. Since this is below the boiling points of nitrogen (–196 °C) and oxygen (–183 °C), the air liquefies.

- As the liquid air warms, the nitrogen boils off first, leaving liquid oxygen behind.

Figure 4: Fractional distillation of liquid air

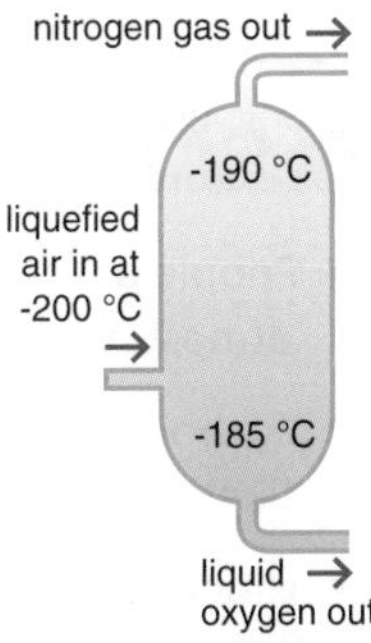

nitrogen gas out →

–190 °C

liquefied air in at -200 °C →

–185 °C

liquid → oxygen out

B–A*

How Science Works

- You should be able to explain why scientists cannot predict accurately when earthquakes and volcanic eruptions will take place.

Improve your grade

Sketch a timeline showing how the proportions of water vapour, carbon dioxide, oxygen, and nitrogen have changed since the Earth was formed. **AO2 (4 Marks)**

The atmosphere and life

How did life on Earth begin?

D–C

- There is uncertainty about how life began because there is no evidence. The first primitive life-forms did not form fossils. Here is one theory.
  - For the first billion years, Earth's atmosphere was mainly carbon dioxide, with some methane, ammonia, hydrogen and water vapour. The water vapour eventually condensed to form oceans.
  - The weather was more extreme than it is today. Frequent lightning provided energy to break chemical bonds and split molecules. The fragments recombined in different ways, forming new compounds.
  - These new compounds included amino acids (from which all proteins are built up), sugars and other carbon compounds needed to make DNA (deoxyribonucleic acid). These compounds are the basis of life.

The Miller–Urey experiment

B–A*

- In 1952 Miller and Urey, devised an experiment to test the theory of how life began. They mixed methane, ammonia and hydrogen in a sterile glass bulb. This simulated the early atmosphere.

- A flask of water represented the ocean. They circulated water vapour from the 'ocean' through the 'atmosphere', and made electric sparks to simulate lightning.

- After a week they analysed the mixture. They found many different organic compounds – carbon compounds that normally come from living organisms. These included glycine and several other amino acids, and sugars, including ribose. This was significant because D in DNA stands for deoxyribose, a closely related sugar.

- Miller and Urey did not create life, they showed that molecules essential for living things could be made by a natural process from Earth's early atmosphere.

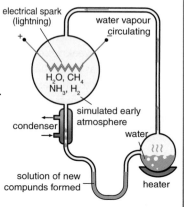

Figure 1: The Miller–Urey apparatus

Carbon dioxide levels

Carbon recycling

D–C

- Both animals and plants take in oxygen and give out carbon dioxide. (see Figure 2)

- Dead plants and animals decay. Oxygen from the air or water converts them back into carbon dioxide, with the help of bacteria, fungi, and other organisms.

- Most of the carbon dioxide in the early atmosphere became locked up as fossil fuels or in rocks such as limestone made from animal shells.

- Carbon dioxide is also absorbed by the oceans, removing it from the atmosphere.

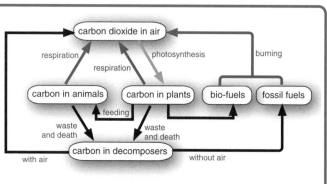

Figure 2: The carbon cycle

So what's the problem?

B–A*

- People are using more and more fossil fuels for energy. This is releasing far more carbon dioxide into the atmosphere than photosynthesis can remove.

- Carbon dioxide is a greenhouse gas. It traps energy from the Sun. Increasing amounts of carbon dioxide are likely to make the Earth hotter in future.

- Higher concentrations of carbon dioxide in the atmosphere mean that more dissolves in the seas and oceans. They become slightly more acidic (lower pH), upsetting ecosystems, killing coral for instance.

Improve your grade

Explain why the Miller–Urey experiment does not prove how life evolved. **AO3 (2 Marks)**

C1 Summary

Fundamental ideas

All substances are made from atoms, which have protons and neutrons in a central nucleus, with electrons in shells around the nucleus.

Electrons occupy specific energy levels or shells but always the lowest available level. Atoms with full outer electron shells (Group 0) are stable.

The atomic number = the number of protons in the nucleus. It is the same as the number of electrons in the atom. The mass number of an element = the number of protons + neutrons.

Metals within non-metal compounds have ions held together by ionic bonds. Non-metal compounds consist of molecules with covalent bonds.

Limestone and building materials

Limestone is calcium carbonate. When heated, it is thermally decomposed to calcium oxide, an alkali.

Carbonates react with acids to produce carbon dioxide gas, a salt, and water. Limestone is damaged by acid rain.

Limestone is quarried for use in cement, mortar, concrete and glass. The quarrying has economic, social and environmental implications.

Metals and their uses

Metals are extracted from rocks known as ores, which contain a large amount of the metal's compounds.

Metals that are more reactive than carbon are extracted by electrolysis. Metals that are less reactive than carbon are extracted by heating them with carbon.

Copper ores are of low quality. Phytomining and bioleaching are used to obtain copper compounds to use as ores.

Transition metals have many uses based on their properties. Many pure metals are improved by mixing with others to form alloys.

Crude oil and fuels

Crude oil is made up of hydrocarbon molecules. They can be separated into fractions by fractional distillation using boiling points.

Burning a hydrocarbon fuel produces carbon dioxide and water. It may also produce pollutants.

Most of crude oil's compounds are alkanes (C_nH_{2n+2}). The bonds between the atoms are covalent bonds.

Biofuels, such as biodiesel and ethanol, are produced from plants. There are economic, ethical and environmental issues with their production.

Other useful substances from crude oil

Hydrocarbons can be broken down (cracked) to produce smaller, more useful molecules.

Alkenes can be used to make polymers such as poly(ethene) and poly(propylene).

Alkenes are unsaturated molecules with double bonds represented as C=C.

Ethanol can be produced by reacting ethene with steam or by fermentation.

Many polymers are not broken down by microbes posing waste disposal problems.

Plant oils and their uses

Vegetable oils are important foods and fuels as they provide a lot of energy. They also provide us with nutrients.

Oils do not dissolve in water. They can be used to produce emulsions.

Vegetable oils that are unsaturated contain double carbon–carbon bonds.

Emulsifiers help oil and water molecules to mix. They have hydrophilic and hydrophobic properties.

Changes in the Earth and its atmosphere

The Earth consists of a core, mantle and crust, and is surrounded by the atmosphere.

The Earth's atmosphere has changed over time. It is now 78% nitrogen, 21% oxygen with traces of other gases.

The Earth's crust is cracked into a number of tectonic plates. Convection currents within the mantle cause the plates to move.

Air is a mixture of gases with different boiling points. It can be fractionally distilled to provide raw materials.

Energy

Storing and transferring energy

D–C

- Everything stores energy. The more energy things store, the more they can do. The amount of energy stored is measured in Joules (J).

- When anything happens, energy is transferred (moved) from one place to another. It cannot appear or disappear.

Evaporation and pressure

B–A*

- Energy is not a physical substance. In a chemical reaction, the mass of chemicals before and after is the same, but the amount of energy is different.

Remember!
We detect energy being transferred or stored in different places in different ways. However, energy cannot exist in different forms.

Figure 1: Examples of things that transfer energy

Infrared radiation

Emission absorption and uses

D–C

- All objects emit and absorb infrared radiation.

- Shiny light coloured surfaces absorb infrared radiation slower than black matt surfaces. They reflect infrared radiation well.

- Hotter surfaces emit radiation faster than cooler surfaces.

- We can design objects to reduce the rate of energy transfer.

Remember!
Objects are absorbing and emitting infrared radiation at the same time. If infrared radiation is emitted faster than it is absorbed, then the object cools down.

How Science Works

- Black surfaces inside an oven emit infrared radiation better than shiny surfaces. This helps heat to transfer to the food cooking inside the oven.

Figure 2: Comparing the emission and absorption of infrared radiation by different surfaces

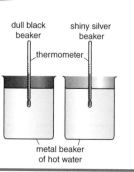

Infrared radiation and global warming

B–A*

- Many things affect the rate of global warming. One theory is that sea water is dark coloured and will absorb infrared radiation from the Sun faster than ice does.

Remember!
Objects warm up if they absorb more infrared radiation than they emit.

Improve your grade

Explain whether a kettle of hot water cools down quicker if its outer surface is coloured white or dark green.
AO2 (3 marks)

Kinetic theory

Bonds between particles

- Bonds between particles are strongest in solids and weakest in gases.
- Melting is when a solid changes to a liquid. When the solid is heated, particles vibrate more vigorously. Bonds between particles break and reform so particles can change places.
- Freezing is when a liquid cools and changes to a solid.
- Boiling is when a liquid changes to a gas when the liquid is heated. The particles break their bonds and can move around randomly.
- Condensing is when a gas cools and changes to a liquid.

D–C

What's a particle?

- In solids, particles vibrate and are held together by strong bonds.
- In liquids, particles move around each other but cannot escape due to weak bonds.
- In gases, particles move randomly as the bonds are very weak.

EXAM TIP

Remember to use the terms particles and bonds in answers to questions about changes of state.

B–A*

Remember!
The particles do not change as they change states, but they behave differently.

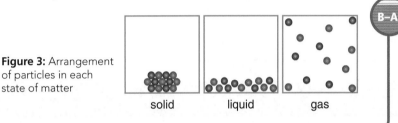

Figure 3: Arrangement of particles in each state of matter

solid   liquid   gas

Conduction and convection

Transferring the energy

- If one end of a solid is heated, particles vibrate more and shake their neighbours. Conduction transfers energy from one particle to another through the solid.
- If one part of a liquid or gas is warmed, the particles vibrate more taking up more space. The warm region of gas or liquid expands, becoming less dense and rising above cooler denser regions. This is convection.
- Convection currents spread heat through liquids or gases that are heated from the base, or cooled from the top.

Figure 4: The dye traces the convection current as the water warms

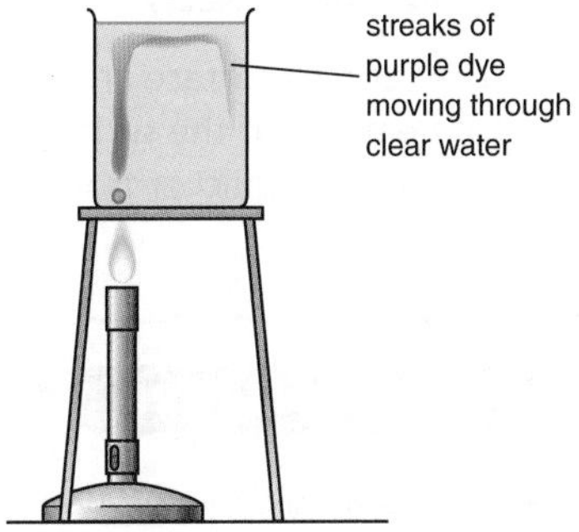

streaks of purple dye moving through clear water

D–C

Why are metals good conductors?

- Electrons in metals are free to move rapidly from hotter to cooler regions, transferring energy.

B–A*

Improve your grade

Explain why convection can take place in a liquid but not in a solid. **AO2 (3 marks)**

Evaporation and condensation

Evaporation and environment

- Evaporation is when a liquid changes to a gas at temperatures lower than the boiling point.

- Evaporation is quickest if it is warm (more particles have enough energy to break bonds linking them), if the liquid has a large surface area, or if it is windy (so vapour above the liquid's surface does not become saturated).

- Water evaporates from the Earth's surface, and cools forming clouds. Water vapour condenses into liquid rain. This is the water cycle.

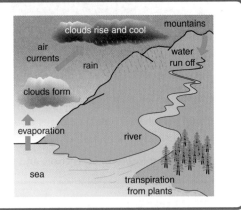

Figure 1: The water cycle

Evaporation and pressure

- At high altitudes, wind speeds are high and air pressure is low. It is easier for particles to evaporate.

> **Remember!**
> Evaporation occurs at the surface at low temperatures, but boiling occurs throughout the liquid at boiling point.

Rate of energy transfer

Conduction, infrared radiation and convection

- When hot objects cool, energy is transferred to the surface by conduction and from the surface by convection. Hotter objects lose energy more quickly than cooler objects.

- The ratio of surface area to volume affects the rate of heat loss. If the surface area is larger, energy is transferred quicker from the object.

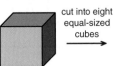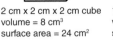

cut into eight equal-sized cubes

2 cm x 2 cm x 2 cm cube
volume = 8 cm³
surface area = 24 cm²

1 cm x 1 cm x 1 cm each cube
volume = 1 cm³
surface area = 6 cm²
total volume of the eight cubes = 8 cm³
total surface area of the eight cubes = 48 cm²

Figure 2: Cutting the block increases its surface area

Cooling curves

- The cooling curve shows the temperature drops more quickly at hotter temperatures.

Figure 3: A typical cooling curve for water. Room temperature is 20 degrees C.

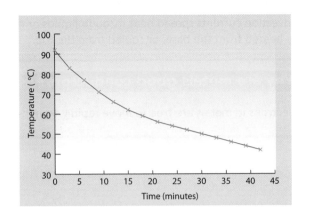

Improve your grade

Explain why a towel dries quicker on a windy summer day. **AO2 (3 marks)**

Insulating buildings

How much can we save

- Payback time is the time taken to save as much money as it costs to install an energy saving measure.

- Payback time in years = installation cost ÷ annual saving

Method of insulation	installation cost (£)	Annual saving (£)	Payback time (years)
loft insulation	240	60	4
cavity wall insulation	360	60	6
draught proofing doors and windows	45	15	3
double glazing	2400	30	80

Figure 4: Payback time for different energy saving measures

D–C

The wider picture

- The energy needed to manufacture and install insulation should be less than the energy saved by using it. Some councils give grants to install different types of insulation.

B–A*

Specific heat capacity

Calculating absorbed energy

- Energy is transferred to an object that heats up, or from an object that cools down.

- Energy transferred to or from a material (J) = mass (kg) × specific heat capacity (J/Kg °C) × temperature change (°C)

- For example, the specific heat capacity of water is 4200 J/Kg. The energy absorbed by 250 g of water heated from 20 °C to 100 °C is $0.25 \times 4200 \times 80 = 84\,000$ J

D–C

Oil filled radiators

- Oil filled radiators use oil with a specific heat capacity of 2000 J/Kg °C. They reach a higher temperature than water filled radiators for the same energy input.

Remember!
You must always use the temperature difference in specific heat capacity calculations.

B–A*

Improve your grade

The specific heat capacity of copper is 390 J/Kg °C. Explain whether copper heats up quicker than the same mass of water when they are put in a hot place. The specific heat capacity of water is 4200 J/Kg °C. **AO2 (3 marks)**

Energy transfer and waste

Wasted energy

- The Law of Conservation of Energy says energy cannot be created or destroyed when it is transferred. All energy is usefully transferred, dissipated or stored. The energy that spreads to the surroundings is called wasted energy.

- A Sankey diagram shows energy transfers in a device.

- The energy input is shown at the left side of the arrow.

- The arrow splits. Each section shows the output energy form.

- The width of each part of the arrow shows the proportion of energy it represents.

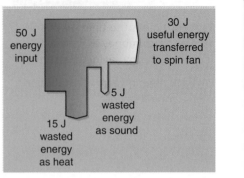

Figure 1: Sankey diagram for an electric fan

Comparing devices

- Sankey diagrams compare the proportion of energy transferred usefully by different devices.

Remember!
All energy must be accounted for during a transfer. It may be stored, usefully transferred or dissipated to the surroundings.

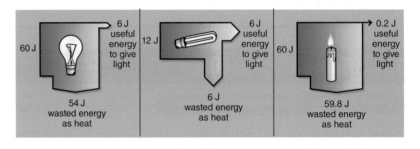

Figure 2: Sankey diagram comparing different light sources

Efficiency

Efficiency and wasted energy

- Efficient devices do not waste much energy. Efficiency = useful energy out ÷ total energy in. The answer is always a decimal less than 1.

- Convert decimals to percentages by multiplying by 100. Efficiency is always less than 100%.

- You can use information from a Sankey diagram to calculate efficiency.

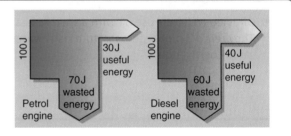

Figure 3: These diagrams show the petrol engine's efficiency is 30% and the diesel engine's efficiency is 40%

Perpetual motion machines

- Perpetual motion machines cannot be made as they would be 100% efficient, which is impossible.

Improve your grade

Explain what this Sankey diagram shows in as much detail as possible.
AO2 (3 marks)

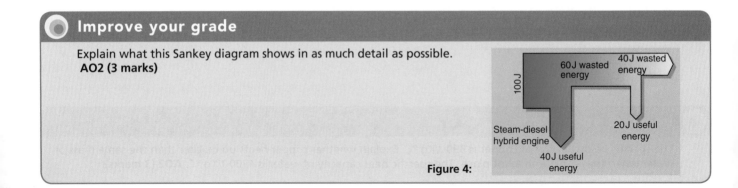

Figure 4:

Electrical appliances

Alternatives to electricity

- Electrical appliances transfer the energy supplied by electricity into something useful.

- Mains electricity is convenient, safe and pollution free at the point of use. However, generating electricity in power stations causes pollution.

- Batteries change chemical energy into electricity when the battery is part of a circuit.

- Alternatives to electricity include gas or biomass for cooking, lighting and transport.

- Some appliances work using solar power or energy stored in springs instead of batteries.

D–C

LEDs

- LEDs produce light but work in a different way to light bulbs. Electrons in materials used to make LEDs absorb energy from a supplied voltage. Then they release this energy as light. The light is a single colour because the same energy is absorbed and released by all electrons in an LED.

Figure 5: The colour of light from LEDs depends on the energy released by the electrons inside it

B–A*

Energy and appliances

How much energy is used?

- Power is the rate that something does work, or the rate of energy transfer. It is measured in Watts (W) or kilowatts (kW).

- Calculate the energy used by an appliance in joules using: power (W) × time (s)

D–C

Heat and current

- Larger currents carry more energy than smaller currents. More powerful equipment uses thicker cables than less powerful equipment.

- Connecting cables are designed to heat up as little as possible when they carry a current so that as little energy is transferred to the surroundings as possible.

- Other wires (for example, in heating or lighting filaments) are designed to become very hot when a current flows through them so they can transfer energy to the surroundings as light or heat.

EXAM TIP

Check you know the right units for time, power and energy. You will lose marks if you write **j** instead of **J** for example, or if you leave them out.

B–A*

Improve your grade

Two different bulbs are switched on for 10 minutes. Calculate the energy transferred by each one.
i)   a 60 W filament bulb
ii) a 10 W energy efficient bulb **(AO2 4 marks)**

The cost of electricity

How much does it cost?

D–C

- A kilowatt-hour is calculated using power (in kW) × time (in hours).
  - For example, a 0.1 kW light bulb switched on for 15 hours uses 1.5 kWh (0.1 kW × 15 h).
- The cost of using electricity is the number of kilowatt-hours used × cost per kilowatt-hour.
  - For example, if each kilowatt hour costs 12p, the cost of using the light bulb was 18p (1.5 kWh × 12p/kWh).

Figure 1: A typical electricity bill

Switching to standby

B–A*

- Leaving equipment on standby for long periods of time increases our electricity usage.

Power stations

Energy changes

D–C

- In a power station, fossil fuels (coal, gas, oil) are burned, or nuclear fuels (uranium, plutonium) undergo fission. Water is heated, changing to steam. Steam drives turbines, which spin generators, generating electricity.

- Many power stations are about 35% efficient. Gas power stations are 60% efficient, using burning gases and steam to spin turbines. Power stations using waste energy directly for heating are 70-80% efficient.

How Science Works

- Using more efficient power stations reduces environmental damage because they use less fuel and produce fewer emissions for every kWh of electricity generated.

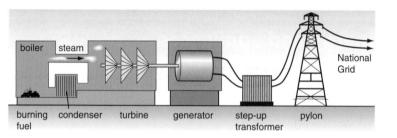

Figure 2: How a power station generates electricity

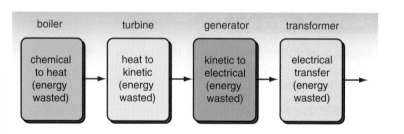

Figure 3: Energy changes taking place in a power station

> **Remember!**
> Fossil fuel and nuclear power stations generate electricity the same way, but use different energy sources to provide heat.

What should we burn?

B–A*

- Decisions on how to use inaccessible fuels must be made in future. There are advantages and disadvantages.

Improve your grade

Describe the energy changes taking place in these parts of a coal fired power station: the burning fuel; the boiler; the turbine; the generator. **AO1 (4 marks)**

Renewable energy

Four more renewable resources

- Renewable energy resources will not run out. These resources spin turbines directly:
  - When the wind blows, it spins blades on a wind turbine.
  - Water trapped behind hydroelectric dams falls through pipes, driving turbines.
  - Tidal barrages are walls built across river mouths. When the tide goes in or out, water flows though pipes in the barrage driving turbines.
  - Waves drive turbines in small wave generators.

- Steam drives turbines in biomass and geothermal power stations:
  - Biomass is organic waste that is burned.
  - Heat from rocks deep underground changes water to steam in geothermal power stations.

- Photovoltaic cells change sunlight to electricity directly. This is solar power.

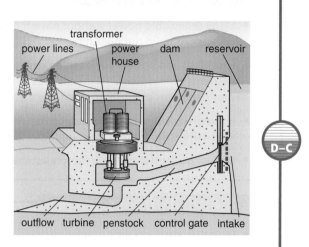

Figure 4: Hydroelectric power station

D–C

Storing energy

- Some renewable energy resources are weather dependent and unreliable. It may be possible to store some generated electricity in large underground batteries for later use.

Remember!
Renewable energy resources will not run out.

B–A*

Electricity and the environment

More effects of generating electricity

Type of energy source	Advantages	Disadvantages
Fossil fuels	Does not occupy large land areas.	Produces polluting gases and solid waste. Mining damages the environment.
Nuclear fuels	Does not produce greenhouse gases.	Creates radioactive waste that must be safely stored.
Hydroelectricity	Renewable.	Causes large scale flooding. Produces greenhouse gases.
Biomass	Renewable. Reduces need for landfill sites.	Produces greenhouse gases.
Wind power	Renewable. Does not produce polluting gases.	Can affect wildlife. Must be sited in windy areas.
Tidal power	Renewable. Does not produce polluting gases.	Causes large scale flooding and changes water flow.
Geothermal	Renewable.	Can release toxic gases from below the Earth's surface.
Solar	Renewable. Does not produce polluting gases.	Large scale use involves large land areas and can affect wildlife.

D–C

Not so green hydroelectric power

- Hydroelectricity can cause more global warming than fossil fuel power stations as vegetation rots at the bottom of the reservoir and at its edges when water levels change.

EXAM TIP

If you are asked about the environmental impact of an energy source, your answer must describe environmental damage (for example, flooding, disrupting river flow, emission of greenhouse gases) and not a general disadvantage (for example expensive or unreliable).

B–A*

Improve your grade

Explain whether a hydroelectric power station or a coal-fired power station is best for a city located near the coast.
AO3 (5 marks)

Making comparisons

Costs and reliability

D–C

- Many renewable power stations have high capital costs and low running costs. Fossil fuel power stations are cheaper to build but operating costs will increase as the supplies of fossil fuels fall.

- Fossil fuel and nuclear fuels can be stored, increasing their reliability. A single power station generates large amounts of electricity. Solar cells, wave and wind turbines are weather-dependent so less reliable, and each unit generates small amounts of electricity.

- Hydroelectric power stations generate electricity quickly when needed but the reservoir levels must be maintained, sometimes by pumping water back to the reservoir.

- Nuclear power stations have the longest start up and shut down times, followed by coal then gas power stations.

Electricity from sewage

B–A*

- Scientists are constantly looking for new ways to generate electricity.

- Experimental methods to generate electricity include oxidising sewage using bacteria. This creates charged particles which can be separated, setting up a voltage and allowing a current to flow.

How Science Works

- You should be able to compare the benefits and drawbacks of different energy sources in different situations.

Remember!
Make sure you know at least one advantage and disadvantage for each source of energy.

The National Grid

High voltages

D–C

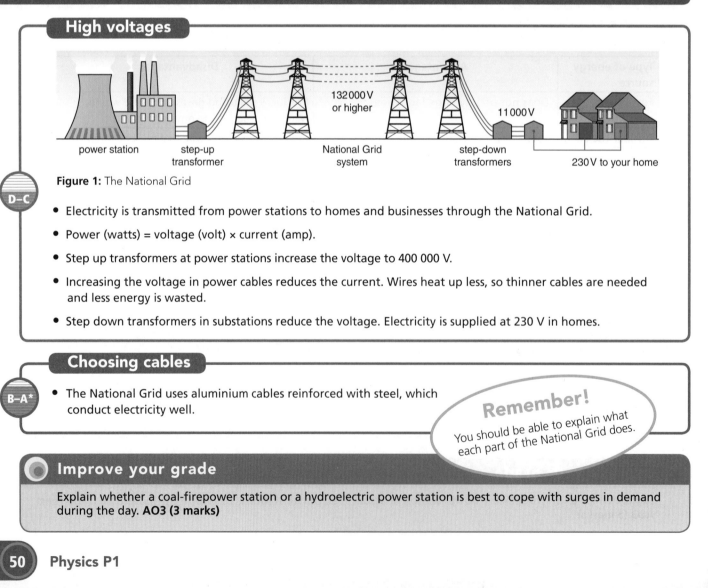

Figure 1: The National Grid

- Electricity is transmitted from power stations to homes and businesses through the National Grid.

- Power (watts) = voltage (volt) × current (amp).

- Step up transformers at power stations increase the voltage to 400 000 V.

- Increasing the voltage in power cables reduces the current. Wires heat up less, so thinner cables are needed and less energy is wasted.

- Step down transformers in substations reduce the voltage. Electricity is supplied at 230 V in homes.

Choosing cables

B–A*

- The National Grid uses aluminium cables reinforced with steel, which conduct electricity well.

Remember!
You should be able to explain what each part of the National Grid does.

Improve your grade

Explain whether a coal-firepower station or a hydroelectric power station is best to cope with surges in demand during the day. **AO3 (3 marks)**

What are waves?

Transferring energy

- Waves transfer energy from a source without transferring matter.

- Longitudinal waves oscillate in the same direction that the energy travels, and include sound waves.

- Transverse waves oscillate at right angles to the direction the energy travels, and include electromagnetic waves, water waves.

Figure 2: What is the direction of energy transfer for each of these waves?

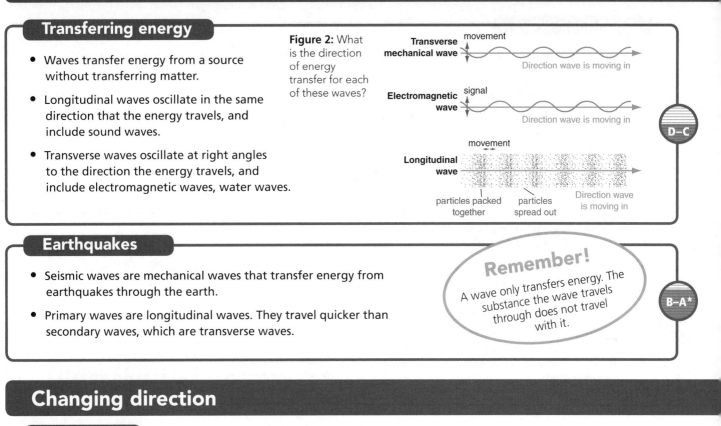

D–C

Earthquakes

- Seismic waves are mechanical waves that transfer energy from earthquakes through the earth.

- Primary waves are longitudinal waves. They travel quicker than secondary waves, which are transverse waves.

Remember!
A wave only transfers energy. The substance the wave travels through does not travel with it.

B–A*

Changing direction

Refraction

- The normal is an imaginary line drawn at right angles to a surface.

- Waves can be reflected (bounce off a surface). The angle between a reflected ray and the normal is the same as the angle between an incoming ray and the normal.

- Waves can be diffracted (spread through a gap or round an obstacle). Diffraction is greatest when the wavelength is about the same size as the gap or obstacle.

- Waves can be refracted (change direction at a boundary). Waves refract because they change speed in different materials. If a wave moves into a material where it travels slower, the wave refracts towards the normal.

Figure 4: Diffraction

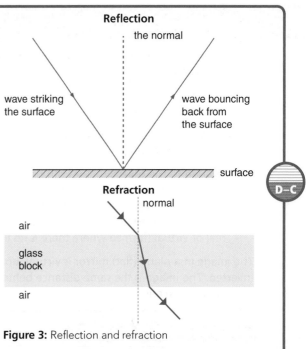

Figure 3: Reflection and refraction

D–C

Rainbows

- Raindrops refract sunlight. Blue light refracts more than red light. This is why we can see a rainbow if the Sun is behind us, shining into a rain storm.

Remember!
Make sure you know what each of the key terms means.

B–A*

Improve your grade

An echo is a reflected sound wave. Explain why you can hear echoes only in certain places, and why you may hear more than one echo. **AO3 (3 marks)**

Sound

Sound is a wave

D–C

- Vibrating objects create sound waves. Sound waves are longitudinal mechanical waves and cannot pass through a vacuum.

- All sound waves can be reflected (echoes), refracted or diffracted. Loud sound waves have a large amplitude; quiet sound waves have a small amplitude.

- Frequency measures the number of cycles per second and is measured in Hertz (Hz). Humans hear sounds between 20 Hz and 20 000 Hz.

- High pitched notes have a short wavelength and high frequency; low pitched notes have a long wavelength and low frequency. The higher the frequency of a wave, the more energy it can carry.

Using an oscilloscope to compare sounds

B–A*

- An oscilloscope shows a sound wave as a trace on a screen.
    - Amplitude is the height of the trace measured from the mid point to the highest (or lowest) point.
    - The wavelength is the distance from one peak to the next peak.
    - A grid is marked on the screen.
    The time base shows how many seconds each horizontal square represents so the frequency can be calculated.

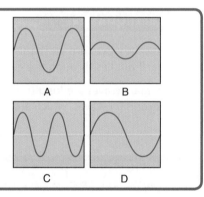

Figure 1: Sound B is quieter than sound A; sound C is higher pitched than sound D

Light and mirrors

What happens if a mirror is curved?

D–C

- Ray diagrams show how one or more rays of light travel when they are reflected or refracted.

- At least two rays are needed to show where the image of the object will be seen and its appearance.

- Compared to the object, images can be;
    - upright or inverted (upside down)
    - magnified (bigger), diminished (smaller) or the same size
    - laterally inverted (left and right sides reversed)
    - real or virtual (formed where there is no light from the object).

The image in a plane (flat) mirror is virtual, upright and laterally inverted. The image is the same distance behind the mirror as the object is in front of the mirror.

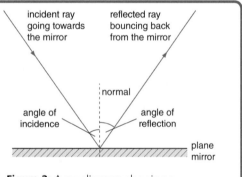

Figure 2: A ray diagram showing a reflected ray

Using curved mirrors

B–A*

- A flat mirror forms an image the same distance behind the mirror as the object is in front of it.

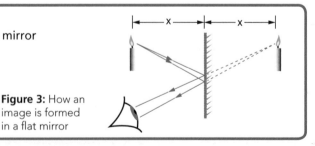

Figure 3: How an image is formed in a flat mirror

Improve your grade

Draw traces to show two sound waves. One sound is lower pitched and twice as loud as the other sound. Label the amplitude and wavelength on each trace. **AO1 (4 marks)**

Using waves

More about radiation and its uses

D–C

- Radio waves, microwaves, infrared and visible light are electromagnetic waves used in communication systems.

- Radio waves are used for radio and TV broadcasts. Local and national radio broadcasts use long wavelength radio waves. TV broadcasts use short wavelength radio waves.

- Microwaves are used in mobile phone networks and satellite communication, as the atmosphere does not absorb microwaves.

- Infrared radiation comes from anything warm. It is used in remote controls.

- Visible light allows us to see and is used in photography.

Diffraction

B–A*

- Diffraction affects how well electromagnetic waves are received. If waves cannot diffract (spread) around an obstacle, the signal they carry is not received clearly.

- TV and VHF (very high frequency) signals have wavelengths of several metres which is much smaller than hills or buildings. The waves cannot diffract around them. TV and VHF receivers must be in direct line-of-sight with the transmitter.

Figure 4: The signal from the transmitter comes in a straight line to these aerials

- Long-wave radio signals have wavelengths of several kilometres which is a similar size to hills and buildings. The waves diffract around them. The signals are received even where shorter waves cannot be received well.

The electromagnetic spectrum

Energy and wavelength

D–C

- Electromagnetic radiation is a continuous spectrum of transverse waves carrying energy.

- In a vacuum, all electromagnetic waves travel at the same speed, the speed of light. The speed (or velocity) of the wave is calculated using:
  - speed (m/s) = frequency (Hz) × wavelength (m)

- The shorter the wavelength of an electromagnetic wave, the higher its frequency.

- Electromagnetic waves with a short wavelength and high frequency carry most energy.

- Electromagnetic wavelengths range from about 10^{-15} m (gamma rays) to about 10^4 m (radio waves).

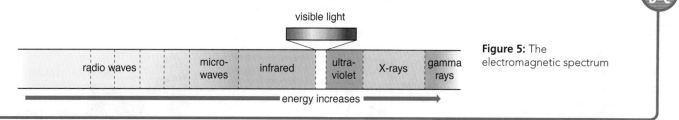

Figure 5: The electromagnetic spectrum

Light – wave or particle?

B–A*

- Electromagnetic waves can be reflected, refracted and diffracted, just like waves.

- They can also behave as particles, for example ultraviolet radiation can "knock" electrons off a metal surface. These particles of radiation are called photons. The idea that radiation can behave as waves or particles is called wave-particle duality.

Remember!
Although electromagnetic waves have some properties the same, waves that have a different frequency will carry different amounts of energy.

Improve your grade

Calculate the speed of radio waves with a wavelength of 10 000 m and frequency of 30 000 Hz. **AO2 (3 marks)**

Dangers of electromagnetic radiation

Protecting from the dangers of radiation

D–C

- Microwaves are strongly absorbed by water, passing through skin and heating up cells. Microwaves do not cause cancer, but may warm cells when they are absorbed. People are exposed to very low levels of microwave radiation over long periods of time from mobile phones. Many studies have looked for evidence that mobile phones cause harm but have not found evidence of any serious risk. The levels of microwaves that a user is exposed to can be reduced by improving the shielding of the phone, reducing the intensity of the signal or reducing the time of usage.

- Infrared radiation can cause burns.

- Very intense visible light can damage cells in the eye's retina.

- Ultraviolet radiation can cause sunburn in minutes, and prolonged exposure or repeated sunburn can cause skin cancer.

- X-rays and gamma rays can kill or damage cells, or cause burns.

Remember!
High frequency electromagnetic waves are more harmful than low frequency electromagnetic waves.

Mutations

B–A*

- Ultraviolet rays, X-rays and gamma rays are ionising. If they are absorbed in cells, they can ionise DNA molecules in the nucleus. The cell may die, or the genetic code may be damaged. The cell may mutate and could form a cancerous tumour.

- The risks increase if the radiation is very intense, or if exposure is over a long period of time or if high-energy waves are absorbed.

Telecommunications

The technological age

D–C

- Electromagnetic waves allow signals to travel long distances very rapidly.

- Microwaves communicate with satellites, as they are not absorbed by the atmosphere. Signals are transmitted to the satellite, then transmitted from the satellite to another place on Earth. Microwaves are used for satellite TV broadcasts, sat-nav and mobile phone networks.

- Bluetooth and WiFi networks use low energy radio waves to send signals short distances.

- Visible light and infrared signals travel long distances in optic fibres without being absorbed. Optic fibres form part of the telephone network, also used for Internet and email.

- Interactive TV and TV remote controls use infrared radiation.

Figure 1: How a satellite TV programme reaches you

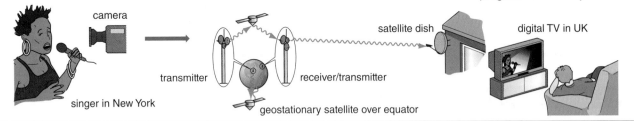

camera

satellite dish

digital TV in UK

transmitter

receiver/transmitter

singer in New York

geostationary satellite over equator

Communications satellites

B–A*

- Geostationary satellites orbit above the equator at the same rate that the Earth spins so they remain above the same place on Earth. Uses include telecommunication satellites and satellite TV transmission.

EXAM TIP

You should be able to choose the most suitable type of electromagnetic wave to use in different situations.

Improve your grade

Explain which type of electromagnetic wave is the best choice for satellite communications. **AO2 (3 marks)**

Cable and digital

Why use digital?

- Analogue signals can have any value, but digital signals are pulses that only have two values, on or off.

- Digital signals are higher quality than analogue signals because;
  - there is less interference between different digital signals
  - the digital signal quality is not affected by distance
  - digital signals can be made stronger without losing information.

- Fibre optic cables are thin flexible strands of very pure glass. They transmit information fast with good quality and can carry many signals.

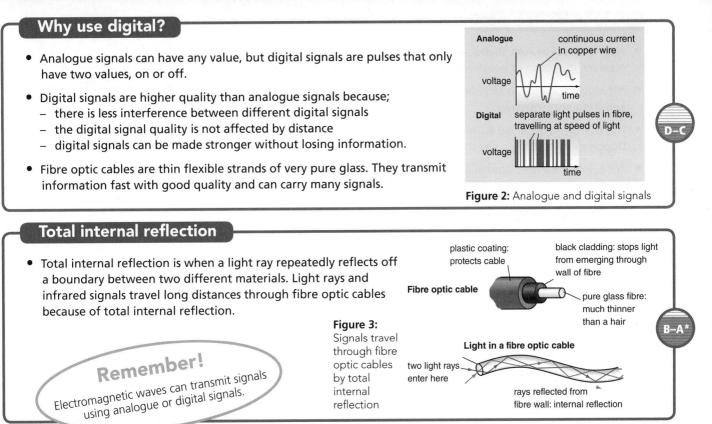

Figure 2: Analogue and digital signals

D–C

Total internal reflection

- Total internal reflection is when a light ray repeatedly reflects off a boundary between two different materials. Light rays and infrared signals travel long distances through fibre optic cables because of total internal reflection.

Figure 3: Signals travel through fibre optic cables by total internal reflection

Remember!
Electromagnetic waves can transmit signals using analogue or digital signals.

B–A*

Searching space

Looking further

- Telescopes can be based on Earth or in space.

- Advantages of earth-based telescopes compared to space-based telescopes are;
  - reduced costs to manufacture
  - easier to maintain, repair and update.

- Advantages of space-based telescopes compared with earth-based telescopes are;
  - clearer images as there is no atmospheric interference from moisture, dust, pollution or weather patterns
  - the full spectrum of electromagnetic waves from objects in space can be detected.

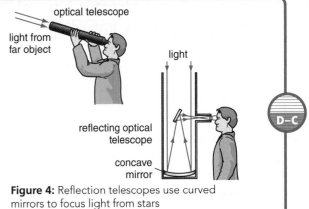

Figure 4: Reflection telescopes use curved mirrors to focus light from stars

D–C

Seeing the invisible

- Astronomers can detect a wide range of electromagnetic waves emitted from different objects in space.

Remember!
All types of electromagnetic waves travel at the same speed throughout space.

Radiation	Objects 'seen' in space
gamma ray	neutron stars
X-ray	neutron stars
ultraviolet	hot stars, quasars
visible	stars
infrared	red giants
far infrared	protostars, planets
radio	pulsars

B–A*

Improve your grade

Explain two advantages of using space based telescopes. **AO2 (4 marks)**

Waves and movement

Doppler and light

D–C

- When a source of light, sound or microwaves moves away from an observer, the observed wavelength increases and frequency decreases.

- If it moves towards an observer, the observed wavelength decreases and frequency increases.

- This change in observed wavelength and frequency is called the Doppler effect. It is greater if the source moves faster.

- Red light has a longer wavelength than blue light. Light from galaxies moving away from Earth appears redder (longer wavelength).

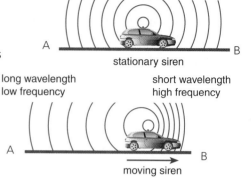

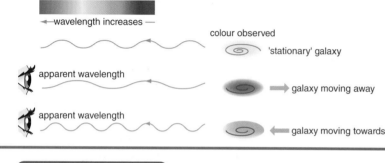

Figure 1: When the siren is stationary both people hear the same sound. When it moves, the person at B hears a higher pitched sound than at A.

Figure 2: Distant stars and galaxies move fast enough for the colour of their light to change.

Mystery object

B–A*

- The change in observed wavelength of light provides information on how distant objects move in relation to Earth.

> **Remember!**
> If the wavelength of light or sound increases, its frequency decreases.

Origins of the Universe

CMBR and the Big Bang theory

D–C

- Light from many galaxies is shifted towards the red end of the spectrum it appears to have a longer wavelength than light from our own galaxy. This is the red-shift.

- The red-shift is larger when galaxies move away faster. More distant galaxies have a larger red shift. They move away faster.

- The Big Bang theory states that the universe began from a very small initial point where all energy and matter were concentrated. About 14 billion years ago, matter and space expanded violently and rapidly from this point.

- The red-shift is evidence that the Universe is still expanding, supporting the Big Bang theory.

- Cosmic microwave background radiation (CMBR) is electromagnetic radiation filling the universe. It comes from radiation present shortly after the beginning of the universe.

- The Big Bang theory is currently the only theory that can explain the existence of CMBR.

When will the Universe stop expanding?

B–A*

- Galaxies move apart because of the energy they received at the Big Bang. They are also attracted to each other by gravity.

- If gravity within the universe is great enough, it may stop expanding and reach a fixed size, or collapse back in on itself.

> **Remember!**
> Evidence for the Big Bang includes evidence the Universe is expanding (red shift) and the radiation remaining from the Big Bang (CMBR).

Improve your grade

Explain the evidence we have that supports the Big Bang theory. **AO3 (6 marks)**

P1 Summary

The energy needed to warm 1 kg of a material by 1 degree Celsius is its specific heat capacity.

Solar panels use the Sun's radiation to warm water.

The transfer of energy by heating processes

Kinetic theory can be used to explain different states of matter.

U-values measure how effective materials are as insulators.

Convection transfers heat in liquids and gases; conduction transfers heat most effectively in solids.

The rate of heat transfer depends on surface area, volume, material and type of surface it is in contact with.

Factors that affect the rate at which energy is transferred

All objects absorb and emit infrared radiation. The temperature and colour of the surface affect how quickly this happens.

Conduction, convection, evaporation and condensation transfer energy and involve particles.

Energy can be transferred, stored or dissipated. It cannot be created or destroyed.

The efficiency of a device is the useful energy output / total energy input.

Energy transfers, efficiency and electrical energy

Electrical appliances carry out different energy transfers.

The amount of energy transferred depends on the equipment's power and time it is used for.

Electricity is generated in power stations where heat from a fuel or volcanic rocks changes water to steam. This spins turbines and generators.

Energy from water (hydroelectricity, tides and waves) and wind spins turbines directly. Energy from the sun produces electricity directly.

Generating and distributing electricity

The use of energy resources has an impact on the environment including greenhouse gas emissions, pollution, flooding and waste products.

The National Grid distributes electricity from power stations to consumers. High voltages in cables reduces energy losses.

Waves transfer energy and can be reflected, refracted and diffracted. Their speed is calculated using wavelength × frequency.

Electromagnetic waves are transverse waves that travel at the speed of light in a vacuum and form a continuous spectrum.

The Big Bang theory states that the Universe began as a very small dense and hot point and expanded rapidly. The red shift is evidence the Universe has continued expanding since then.

Waves, communication and the Universe

Radio waves, microwaves, infrared and visible light are used for communication.

Waves are reflected, and this is how an image in a mirror is formed.

Sound waves are longitudinal waves. The loudness and pitch of a note depends on the wave's amplitude and frequency.

Animal and plant cells

Cell organelles

D–C

- The different parts of a cell are called organelles. Each has a particular function.
- Figures 1 and 2 show the organelles you can see with a powerful light microscope.

EXAM TIP

Make sure you remember which organelles are found in both plant and animal cells and which are found only in plant cells. Figures 1 and 2 show you.

cell membrane, which controls the passage of substances in and out of the cell
ribosomes, where protein synthesis takes place
mitochondrion, where energy is released in respiration
cytoplasm, where metabolic reactions controlled by enzymes take place
nucleus, containing chromosomes made of genes. Controls the activities of the cell

Figure 1: An animal cell

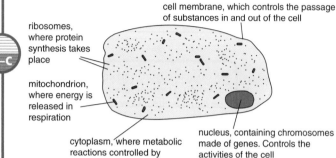

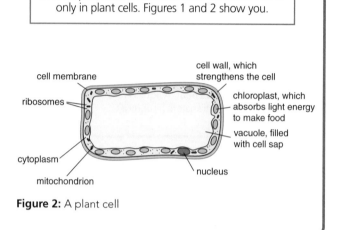

cell membrane
cell wall, which strengthens the cell
ribosomes
chloroplast, which absorbs light energy to make food
vacuole, filled with cell sap
cytoplasm
nucleus
mitochondrion

Figure 2: A plant cell

Electron microscopes

B–A*

- Research laboratories use electron microscopes.
- They can magnify an object two million times. (A light microscope can only magnify by two thousand times.)

Microbial cells

Yeast, algae and bacteria

D–C

- Yeast is a single-celled fungus. Its cells have cell walls but they are not made of cellulose like plant cell walls. Fungi cannot photosynthesise as they have no chloroplasts.
- Algae are simple, plant-like organisms. Their cells are similar to plant cells.
- Bacteria do not have a nucleus. Their genes are in the cytoplasm.

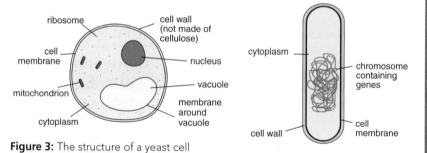

ribosome
cell wall (not made of cellulose)
cell membrane
nucleus
vacuole
mitochondrion
membrane around vacuole
cytoplasm

Figure 3: The structure of a yeast cell

cytoplasm
chromosome containing genes
cell wall
cell membrane

Figure 4: The structure of a bacterial cell

Viruses

B–A*

- Viruses are not made of cells. They do not have cell membranes, cytoplasm or a nucleus.
- Most viruses are made of a sphere of protein, with some DNA inside it.
- They are hundreds of times smaller than a cell.

Improve your grade

The image below was taken of some cells using a light microscope. State what type of cells they are and give reasons for your answer. **AO2 (2 marks)**

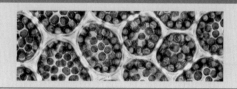

Diffusion

Cells and diffusion

- Diffusion is the spreading of the particles of a gas, or of any substance in solution, resulting in a net movement from a region where they are of a higher concentration, into a region where they are in a lower concentration.

- Most cells need oxygen so that they can respire. Oxygen diffuses into the cells from a higher concentration outside to a lower concentration inside.

- The concentration of oxygen inside a cell is kept low because the cell keeps on using it up.

Remember!

The speed of diffusion can be increased by increasing the difference in concentration (the concentration gradient) and increasing the temperature. (The faster that particles move around, the faster they will diffuse.)

D–C

A model cell

- Visking tubing contains millions of tiny holes which only let small molecules, like water, diffuse through. Large molecules, such as starch, cannot cross the membrane. We say it is partially permeable.

- This is similar to a cell membrane. Visking tubing can therefore be used as a model of a cell.

B–A*

Specialised cells

Examples of specialised cells

- The human body contains hundreds of different kinds of specialised cells.

- Red blood cells, goblet cells and ciliated cells are just three of them.

EXAM TIP

Cells are measured in a unit called a micrometre, symbol μm. There are 1000 micrometres in one millimetre.

Goblet cells are found in the lining of the alimentary canal, and in the tubes leading down to the lungs. They make mucus, which helps food to slide easily through the alimentary canal, and helps to stop bacteria getting down into your lungs.

mucus that has been made by the cell

cell membrane

cytoplasm

nucleus

20 μm

D–C

Figure 5: Goblet cells

Specialised plant cells

- Root-hair cells are specialised plant cells.

- They are fine hair-like extensions of a root.

- Their large surface area enables plants to maximise their absorption of water from the soil.

Remember!

All the different types of specialised cell are the result of differentiation during the growth of the organism.

B–A*

Improve your grade

Explain how a sperm cell is specialised.
AO2 (4 marks)

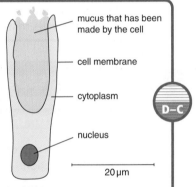

vesicle containing enzymes

nucleus

cell membrane

cytoplasm

mitochondria

30 μm

Tissues

Structure and function of tissues

- The cells that make up tissues are adapted for particular roles.
- Muscular tissue is specialised to produce movement.
- Glandular tissue is made up of cells which secrete (release) useful substances such as enzymes or hormones.

muscle cell

The cells are long and thin.

mitochondria – The cells use energy to make themselves get shorter (contraction). The many mitochondria in the cells provide the energy for contraction.

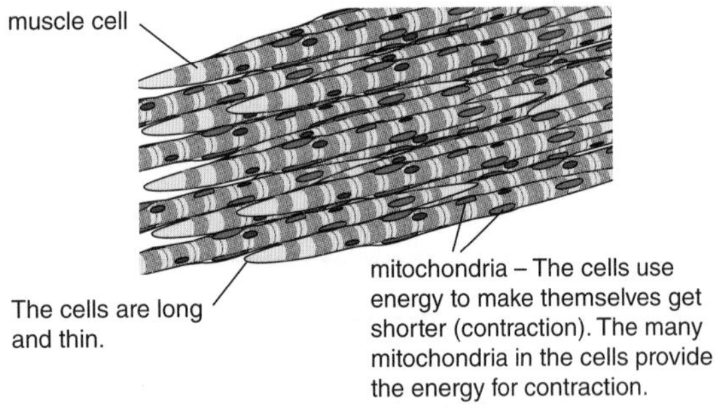

Figure 1: Muscular tissue

glandular cells

secreted substance

vesicles

The glandular cells contain many small vesicles of useful substances that the cell has made, such as enzymes or hormones. The substances are released outside the cell. This is called secretion.

Figure 2: Glandular cells

Single-celled and multicellular organisms

- Multicellular organisms have many different cells which are specialised for a particular function, an advantage over single-celled organisms.
- Some processes are easier for single-celled organisms, for example oxygen can easily diffuse into a single cell.
- Multicellular organisms need a transport system to bring oxygen to every cell in their body. They also have specialised tissues that are adapted for allowing things to move in and out of the body quickly.

Animal tissues and organs

Functions of the digestive system

- A system is a group of organs that performs a particular function.
- The function of the digestive system is to break down the food you eat so the food molecules can enter the blood.
- Each of the organs shown in Figure 3 has an important role in this.

Remember!

A group of cells is called a tissue. Organs are made up of different types of tissue. The heart contains muscle tissue that can contract and relax, nervous tissue to control the heart beat, and ligaments to hold the different tissues in place.

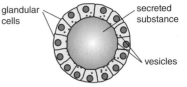

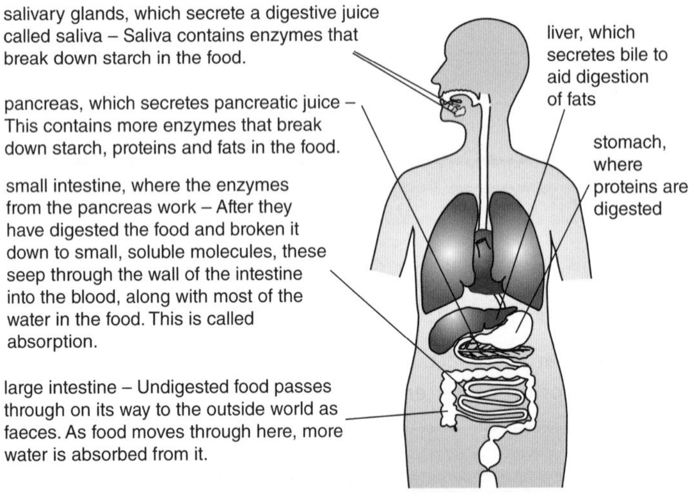

salivary glands, which secrete a digestive juice called saliva – Saliva contains enzymes that break down starch in the food.

pancreas, which secretes pancreatic juice – This contains more enzymes that break down starch, proteins and fats in the food.

small intestine, where the enzymes from the pancreas work – After they have digested the food and broken it down to small, soluble molecules, these seep through the wall of the intestine into the blood, along with most of the water in the food. This is called absorption.

large intestine – Undigested food passes through on its way to the outside world as faeces. As food moves through here, more water is absorbed from it.

liver, which secretes bile to aid digestion of fats

stomach, where proteins are digested

Figure 3: The functions of organs in the digestive system

Protecting the digestive system

- The whole of the digestive system is lined with epithelial tissue to protect the cells in the organs from digestive juices which could break them down.
- This tissue secretes large quantities of mucus, which forms a barrier over the inner surface of the digestive organs. The mucus also makes it easier for the food to slide through the digestive system.

Improve your grade

Describe why the stomach is classed as an organ. **AO2 (2 marks)**

Plant tissues and organs

Plant tissues

- The organs in a plant are made up of tissues.

- The whole plant is covered in a layer of epidermis. This helps to protect the underlying cells, stops the leaves from losing too much water and prevents pathogens from entering the plant.

- Most of the cells in a leaf are mesophyll cells. This is where photosynthesis takes place.

- Xylem and phloem tubes run through the entire plant.

- These are tubes which make up the plant's transport system: xylem carries water from the roots to the leaves; sugars are transported around the plant in phloem.

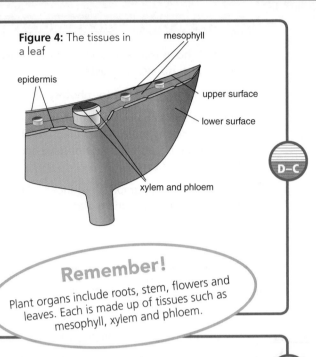

Figure 4: The tissues in a leaf

mesophyll

epidermis

upper surface

lower surface

xylem and phloem

D–C

Remember!

Plant organs include roots, stem, flowers and leaves. Each is made up of tissues such as mesophyll, xylem and phloem.

Leaf epidermis

- The epidermal tissue on the lower surface of the leaf has little holes called stomata.

- These allow gases to diffuse into and out of the leaf.

B–A*

Photosynthesis

Oxygen and energy for living things

- The word equation for photosynthesis is:

$$\text{carbon dioxide} + \text{water} \xrightarrow{\text{energy from sunlight}} \text{glucose} + \text{oxygen}$$

- Millions of years ago there was hardly any oxygen in the air. Oxygen was first made when bacteria evolved that could photosynthesise. Gradually, over millions of years, the amount of oxygen in the air built up. Now more than 20 per cent of the air is oxygen. It has all been made by bacteria, algae and plants.

- In photosynthesis, light energy is stored in glucose molecules.

- The energy is transferred to animals when they eat the plants.

- Glucose can be converted into starch and stored for later use. Starch molecules are big and, unlike glucose, cannot diffuse out of cells.

D–C

EXAM TIP

The equation for photosynthesis shows us that water and carbon dioxide are reactants. Glucose and oxygen are products and are made from the rearrangement of the atoms in the reactants.

Chlorophyll and light absorption

- White light is a continuous spectrum of colours, from red to violet.

- Different parts of the spectrum have their own characteristic wavelengths. Short wavelength light looks blue and long wavelength light looks red.

- Light energy is absorbed by chlorophyll found in the chloroplasts of plant cells.

- Chlorophyll looks green because it absorbs the blue and red parts of the spectrum and reflects the green part.

B–A*

Improve your grade

Without photosynthesis, humans would not survive. Explain why. **AO2 (2 marks)**

Limiting factors

Limits to the speed of photosynthesis

D–C

- Figure 1 shows that as the light intensity increases, the rate of photosynthesis also increases.

- Light is a limiting factor for photosynthesis.

- However, there comes a point when the rate of photosynthesis does not increase any more, even when the plant is getting more light. This may be because it does not have enough carbon dioxide or the temperature is too low.

- A similarly shaped graph is produced when carbon dioxide in the air around a plant is increased.

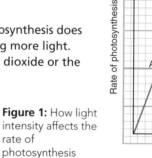

Figure 1: How light intensity affects the rate of photosynthesis

Greenhouses

B–A*

- Growing crops in a greenhouse gives the grower a lot of control over the conditions in which the plants live.

- A grower may be able to produce more tomatoes more quickly if they heat the greenhouse, but the cost of the fuel might outweigh the increase in what they are paid for the tomatoes.

Remember!

Light intensity, carbon dioxide concentration and temperature are all limiting factors for photosynthesis.

The products of photosynthesis

Mineral salts

D–C

- Glucose is made from carbon, hydrogen and oxygen atoms. Plants make other substances from glucose. These are shown in Table 1.

Table 1: Glucose products

Substance	Atoms the substance is made from			
	Carbon	Hydrogen	Oxygen	Nitrogen
carbohydrates	✓	✓	✓	
fats and oils	✓	✓	✓	
proteins	✓	✓	✓	✓

- To convert glucose into proteins, plants need nitrogen which they get in the form of mineral ions. They normally use nitrate ions, NO_3^-, and absorb them from the soil through their roots.

Choosing the storage product

B–A*

- Most plants store at least some energy as starch.
- Starch is a polymer made up of many glucose molecules linked together in a long chain.
- Unlike glucose, starch is insoluble. It forms grains inside cells, instead of dissolving and getting mixed up with everything else inside the cell.
- Fats and oils can store more energy per gram than starch but it is more difficult to make them and to break them down.

Improve your grade

Anna uses a fertiliser that is high in nitrates on her tomato plants. Explain why she does this. **AO2 (3 marks)**

Distribution of organisms

Factors affecting distribution

- Temperature: many species can live only in a particular temperature range.

- Availability of nutrients: very wet soils are usually short of nitrate ions, so only certain plants can live there. Animals can live only where their food is found.

- Amount of light: plants and algae must have light for photosynthesis. They cannot grow in really dark places.

- Availability of water: all organisms need water. Species that live in deserts have special adaptations that help them to obtain and conserve water. The seaweed egg wrack can only grow in places that are covered in water for most of the day or they get too hot and dry.

- Availability of oxygen and carbon dioxide: most organisms need oxygen, for respiration. They cannot live where oxygen is in short supply. Plants also need carbon dioxide, for photosynthesis.

D–C

Remember!
Organisms can only live in environments for which they are adapted.

Interaction of physical and biotic factors

- The distribution of organisms is also affected by biotic factors – those involving other organisms, including competition and predation.

- If egg wrack grows low down on the shore, it is covered by water for most of the time. This means that sea-dwelling herbivorous animals can graze on it for a much longer part of the day.

- The egg wrack also has to compete for space and light with other seaweeds that are better adapted for growing in deep water.

B–A*

Using quadrats to sample organisms

Using quadrats

- Quadrats can be used to measure the distribution of organisms in a habitat. It is best to place the quadrats randomly. Once you have placed your quadrat, you need to identify each species inside it.

- Then you can either:
  - count the numbers of each one
  - estimate the percentage of the area inside the quadrat that each species occupies.

- You should repeat this process many times. Work out the mean number of, or area covered by, each species.

- A transect is used to find out if the distribution of organisms changes as you move from one habitat to another.

(a) counting individuals within a quadrat

Results:
species 1	*
species 2	8
species 3	1

* too numerous to count

(b) estimating percentage cover in a quadrat

Results:
species 1	60%
species 2	20%
species 3	10%
bare ground	10%

species 1   species 2
species 3
bare ground

D–C

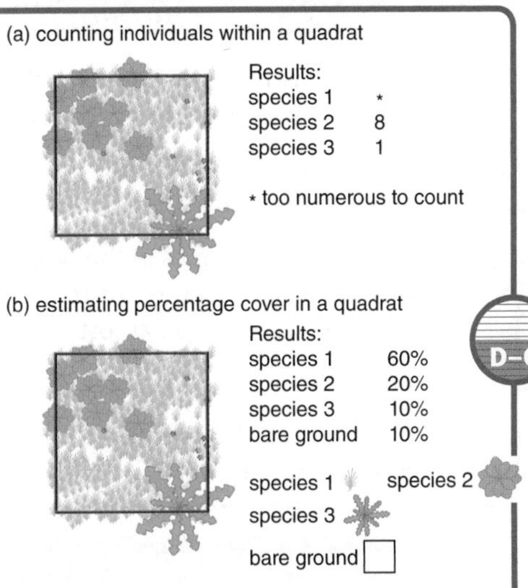

Figure 2: Two ways of collecting data from a quadrat

EXAM TIP

Placing the quadrat randomly ensures the results are valid. Placing the quadrat in at least 10 different places and calculating the mean will increase reliability.

Choosing quadrat size

- One way of finding the optimum size for a quadrat is to try out different sizes, and then count how many different species you find in each size.

- Choose the one that is able to contain at least one organism from each species that lives in the habitat you are sampling.

B–A*

Improve your grade

Simon went on holiday to Mexico. He noticed that the plants growing there were very different to the plants growing at home in the UK. Explain the reasons for this difference in distribution. **AO2 (3 marks)**

Proteins

Protein shapes

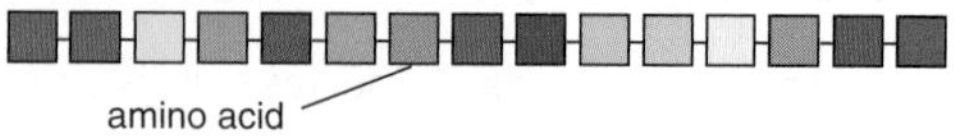

- Each protein molecule is made of a very long chain of hundreds of amino acids linked together.

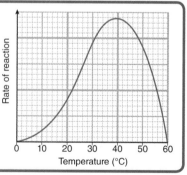

amino acid

Figure 1: The structure of part of a protein molecule

- The type and order of the amino acids determines the shape of the protein. The shape of a protein molecule affects the way it behaves.

- Some protein molecules, like those in muscles, make long, thin fibres.

- Enzymes, antibodies and hormones usually have a globular (ball-like) structure, which often has a dent in it that is a perfect fit for one other kind of molecule. For example, an antibody molecule might have a dent that perfectly fits a particular molecule on a particular bacterium.

D–C

Making proteins in the body

- Animals make proteins from the amino acids they get from food.

- During digestion, protein molecules are broken down into amino acids. These are then absorbed through the wall of the small intestine and are carried all over the body in the blood. Each cell takes up the amino acids that it needs from the blood. Inside the cell, on ribosomes, amino acids are linked together to make the particular kinds of proteins that the cell requires.

- The genes in the nucleus provide instructions about exactly which amino acids to link together, and in which order.

B–A*

Remember!
There are 20 different amino acids. The body can make all the different proteins it needs from these amino acids by arranging them in different orders.

Enzymes

Factors affecting enzyme activity

- An enzyme molecule is a long chain of amino acids, folded into a ball. There is a dent in the ball into which another molecule can fit. This dent is the active site of the enzyme.

- The molecule that fits into the enzyme is called its substrate. The enzyme makes the substrate react, changing it into a new substance.

- Most of the enzymes in the body work best at about 37 °C, which is normal body temperature. This is their optimum temperature. At temperatures above the optimum, the enzyme begins to uncurl and lose its shape. Once the active site has lost its shape, the substrate no longer fits.

- When the enzyme is permanently changed in this way, it is said to be denatured.

- Enzymes are also sensitive to pH. If the pH is a long way from the enzyme's optimum pH, then the enzyme denatures.

D–C

How temperature affects enzymes

- Figure 2 shows how temperature affects enzyme activity.

- When the temperature is low, the enzyme and its substrate are both moving slowly. They do not bump into each other very often. When they do collide, there is not very much energy involved, so the substrate may not react.

- As temperature increases, the molecules move more quickly so there is more chance that they will collide and react.

B–A*

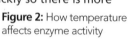

Figure 2: How temperature affects enzyme activity

Rate of reaction

Temperature (°C)

Improve your grade

Pepsin is an enzyme that helps break down proteins in the stomach. It has an optimum pH of 2. Use this information to explain why the stomach produces hydrochloric acid. **AO2 (2 marks)**

Enzymes and digestion

Types of enzymes

- Digestive enzymes pass out of cells in glandular tissue, go into the space inside the gut and become mixed up with the food.

- They cut large food molecules into smaller bits.

- There are three main groups of digestive enzymes (see Table 1).

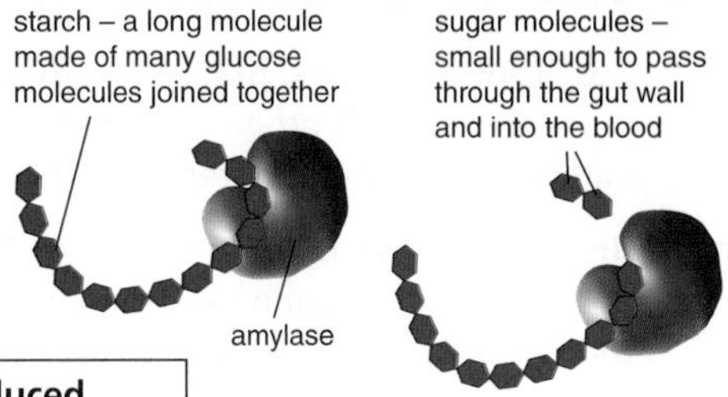

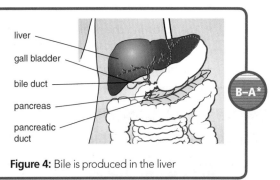

starch – a long molecule made of many glucose molecules joined together

sugar molecules – small enough to pass through the gut wall and into the blood

amylase

Figure 3: How amylase digests starch

	Substrate	Product	Where enzyme is produced
amylase	starch	sugars	salivary glands in the mouth and pancreas
protease	protein	amino acids	pancreas, stomach and small intestine
lipase	lipids (fats and oils)	fatty acids and glycerol	pancreas and small intestine

Table 1: The main groups of digestive enzymes

Acid and alkali

- The liver produces bile, which is stored in the gall bladder. When food arrives in the small intestine, bile flows along the bile duct and mixes with the food.

- Bile is an alkali. It neutralises the acid from the stomach. This provides the slightly alkaline pH which is the optimum pH for the enzymes in the small intestine.

liver

gall bladder

bile duct

pancreas

pancreatic duct

Figure 4: Bile is produced in the liver

Enzymes at home

Biological detergents

- Biological washing powders (detergents) contain lipase and protease enzymes.

- Some stains on clothes – for example blood stains – cannot be removed using ordinary detergents. The enzymes help to break down the stains into substances that dissolve in water. The stains can then wash away.

- The enzymes in detergents often work best at about 30 °C.

EXAM TIP

You should be able to apply what you know about how temperature affects enzymes to explain why biological washing detergents should be used at temperatures no higher than around 30 °C.

Skin complaints

- Some stains on clothes, like blood, are made up of proteins. The proteases in detergents change the proteins into small, soluble amino acid molecules. These can easily be washed away.

How Science Works

- In the 1970s, doctors began to notice many more patients with sore skin on their hands. The knowledge that the new biological detergents contained proteases that could break down skin protein (keratin) led many to link their use to the sore skin. Research has failed to confirm this idea. However, many people still report that their skin is sensitive to biological detergents.

Improve your grade

Dipesh mixed some starch with amylase in a beaker and left the mixture in a water bath at 37 °C. After 30 minutes, he tested the mixture to see if there was any starch present. Predict what he will find and give a reason for your answer. **AO2 (2 marks)**

Enzymes in industry

Baby foods, sugar syrup and slimming foods

- Baby foods: In young babies, the digestive system is not fully developed. Some baby food manufacturers add proteases to their products. These enzymes break down large protein molecules into amino acids. When the baby eats this pre-digested food, it can absorb the amino acids.

- Sugar syrup: This is used in making sweets and sports drinks. Starch solution is easy to make by cooking potatoes or maize and mixing them with water. The starch can then be changed into sugar syrup by adding carbohydrase enzymes such as amylase.

- Slimming foods: These often contain a very sweet sugar called fructose instead of glucose. Fructose is made from glucose using an enzyme called isomerase.

Making soft-centre chocolates

- A mixture of sucrose, flavouring, colouring and an enzyme called sucrase are mixed together to make a paste.

- The paste is moulded, left to set then liquid chocolate is poured over it.

- The chocolates are then warmed up a little, so that the enzyme begins to work on the sucrose inside them.

- The following chemical reaction happens inside the chocolate case:
    sucrose → glucose + fructose

- The mixture of glucose and fructose makes thick, soft syrup inside the chocolates.

Aerobic respiration

Releasing energy

- Most of the time, cells release energy by combining glucose molecules with oxygen. This is called aerobic respiration.

- The word equation is:
    glucose + oxygen → carbon dioxide + water (+ energy)

- Your body obtains oxygen from the air which enters your blood in your lungs. It is transported in the blood to all your body cells. The carbon dioxide that the cells make is carried back, in the blood, to the lungs.

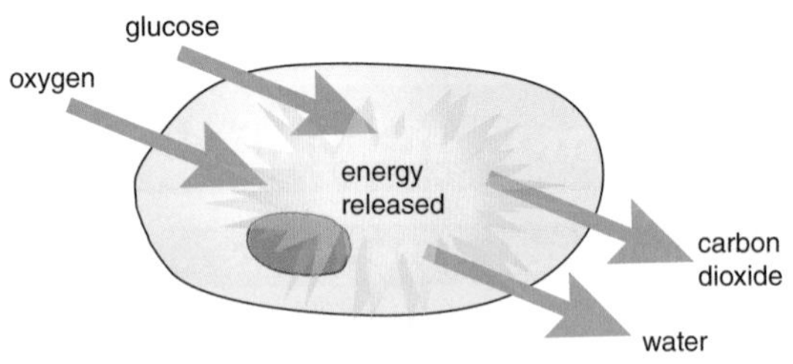

- Aerobic respiration takes place in the mitochondria of cells.

- They contain all the enzymes that are needed to make the reactions of respiration happen quickly.

Figure 1: Aerobic respiration

Respiration and photosynthesis

- Photosynthesis locks up energy from sunlight inside glucose molecules. Respiration releases this energy so that cells can use it.
- When there is sunlight, mesophyll cells in plant leaves are photosynthesising and respiring.
- Both reactions are happening at once – photosynthesis in their chloroplasts and respiration in their mitochondria.
- In bright light, photosynthesis happens a lot faster than respiration.
- At night, photosynthesis stops, but the cells carry on respiring.

EXAM TIP

Notice that the chemical equation for aerobic respiration is the same as the one for photosynthesis but in reverse. This means that if you only learn one for the exam, you will also know the other.

Improve your grade

Explain why breathing rate increases when you exercise. **AO2 (3 marks)**

Using energy

Using the energy

- Building large molecules: Cells use energy to make small molecules join together to make long chains.

- Muscle contraction: Muscles use energy to contract. They contain a store of glucose called glycogen, which is made up of many glucose molecules linked together. When a muscle needs energy, it breaks down the glycogen to produce glucose for use in respiration.

- Maintaining a steady body temperature: Mammals and birds keep their body temperature around 37 °C. If their environment is colder than they are, then heat is lost from their body. Respiration in cells releases energy, which increases body temperature.

- Making amino acids: Plants can make amino acids from sugar and nitrate ions. This requires energy.

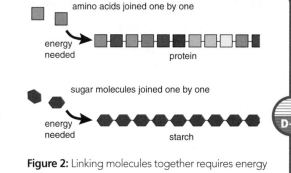

Figure 2: Linking molecules together requires energy

Energy for swimming

- Sperm cells are packed full of mitochondria. These produce energy to drive its tail movements so it can swim towards the egg (a distance of 15 cm).

- They do not have any food reserves such as glycogen. Instead, they use sugars from the fluid in which they were ejected from the male.

Anaerobic respiration

Lactic acid and oxygen debt

- Anaerobic respiration is incomplete breakdown of glucose to lactic acid: glucose → lactic acid (+ a little energy)

- It releases far less energy than aerobic respiration (aerobic releases 16.1 kJ per gram of glucose, anaerobic only releases 0.8 kJ).

- Lactic acid builds up in the muscles. It makes muscles feel tired and can cause cramps. They stop contracting efficiently.

- After you have stopped exercising you have to get rid of this lactic acid.

- You continue to breathe heavily. This takes in the extra oxygen that you need to get rid of the lactic acid (the oxygen debt).

Remember!
Anaerobic respiration is carried out whenever your body cannot get an adequate supply of oxygen to the cells.

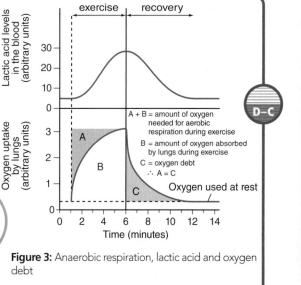

Figure 3: Anaerobic respiration, lactic acid and oxygen debt

Maximum oxygen uptake

- When you exercise, your muscles use more oxygen so you breathe more deeply and more quickly. Your heart beats faster so that oxygenated blood is pumped to your active muscles more quickly.

- The maximum volume of oxygen that your body is able to use per minute is called VO_2 max. The fitter you are, the higher the value of your VO_2 max.

Improve your grade

You are involved in a race. At first you sprint off feeling full of energy. However, halfway through, your legs start to ache and you have to stop. Explain why this happened. **AO2 (3 marks)**

Cell division – mitosis

Mitosis and chromosomes

D–C

- More cells are made when existing cells divide into two. This is mitosis.

- Mitosis is very important because it provides cells for growth and to replace dead or damaged cells.

- Normal body cells have 23 pairs of chromosomes.

- Before a cell divides by mitosis, it first copies each chromosome.

- When the cell divides by mitosis, the chromosomes are shared out equally between the two new cells.

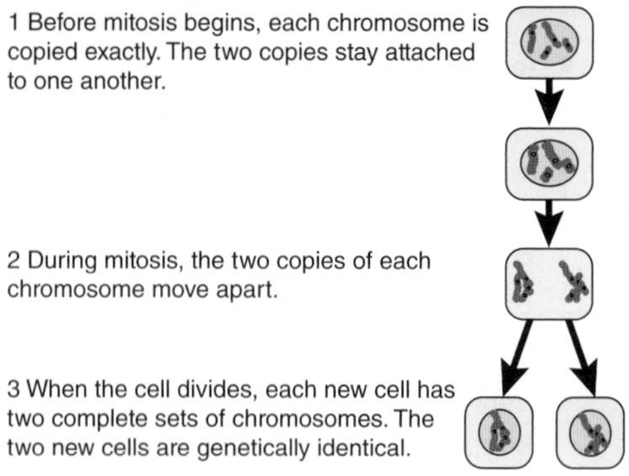

Figure 1: Before and during mitosis

1 Before mitosis begins, each chromosome is copied exactly. The two copies stay attached to one another.

2 During mitosis, the two copies of each chromosome move apart.

3 When the cell divides, each new cell has two complete sets of chromosomes. The two new cells are genetically identical.

How mitosis happens

B–A*

- Mitosis happens in the following stages.
    - Chromosomes are copied and line up along the middle of the cell.
    - The identical copies of each chromosome split apart and move to opposite ends of the cell.
    - The chromosomes now form into two nuclei.
    - New cell membranes form and two new cells have been made.

Cell division – meiosis

Meiosis and chromosomes

D–C

- Before meiosis happens, the chromosomes in the cell that is going to divide are copied. This is just the same as in mitosis.

- Then the cell divides. Unlike in mitosis, it divides twice. This means that four cells are produced – not two, as happens in mitosis.

- Each of the new cells gets only half the original number of chromosomes.

- Figure 2 shows what happens when a cell divides by meiosis.

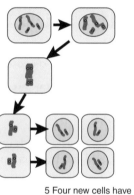

1 Before meiosis begins, each chromosome is copied exactly. The two copies stay attached to one another.

2 Each chromosome finds its partner from the other set.

3 The chromosomes separate from their partners, and the cell divides.

4 Each cell divides again, this time separating the two copies of each chromosome.

5 Four new cells have been produced, each with a single set of chromosomes.

Figure 2: Before and during meiosis

Comparing mitosis and meiosis

B–A*

- Table 1 shows the differences between meiosis and mitosis.

Table 1: Comparing mitosis and meiosis

Mitosis	Meiosis
The cell divides once.	The cell divides twice.
Two cells are made.	Four cells are made.
The new cells have the same number of chromosomes as the original cell.	The new cells have half the number of chromosomes as the original cell.
This is how new body cells are made.	This is how gametes are made.
It happens in all parts of the body.	It happens only in the testes and ovaries.

EXAM TIP

It is easy to get mitosis and meiosis mixed up. Remember that meiosis is the process that produces gametes (sex cells). An easy way to remember is that meiosis contains an e for egg and an s for sperm.

Improve your grade

Explain why mitosis is an essential process in the formation of a baby from a fertilised egg. **AO2 (2 marks)**

Stem cells

Embryo and adult stem cells

- In animals, the fertilised egg divides over and over again to produce an embryo. For the first few days, all of these cells stay as stem cells. Each one has the potential to develop into any kind of cell in the human body.

- Most cells differentiate and form specialised cells. A few remain as stem cells and can continue to divide and specialise throughout adult life.

- In plants, many cells can differentiate at any time during their lives.

- Stem cells from embryos and adults (particularly bone marrow stem cells) could be used to replace damaged tissues in the future.

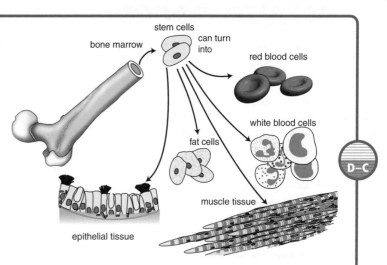

Figure 3: Bone marrow stem cells can produce many different kinds of specialised cells

Stem cells for the future

- 'Stem cell cures' are being sold over the internet. Most of these have never been tested, do not work and are potentially dangerous.

- It is important that patients are protected from these untested treatments but it is also important that trials are carried out as they could lead to success.

Genes, alleles and DNA

Alleles and DNA fingerprinting

- Most genes come in several different forms, called alleles. For example, a gene for hair colour might have one allele that gives black hair, another which gives brown hair, another which gives red hair and so on.

- Humans have two sets of chromosomes in each cell – one from each parent.

- There is a gene for the same characteristic at the same place on the same chromosome in each set. The two genes might be the same allele or different alleles.

- Everybody's DNA is slightly different.

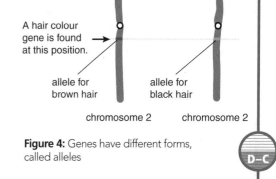

Figure 4: Genes have different forms, called alleles

- Forensic scientists often use DNA from cells in fluids, hair or bones to identify a body, or to identify a person who was at the scene of the crime. They make a DNA fingerprint by cutting up a sample of DNA into little pieces to form a series of stripes. Each person's pattern of stripes is unique.

Analysing your DNA

- Scientists are now able to link certain alleles with an increased risk of getting some diseases.

- There are now several commercial companies who will analyse your DNA and tell you which diseases you might develop later in your life.

Improve your grade

In the future, we may be able to use an individual's genome to calculate the likelihood of them developing certain diseases such as cancer. Evaluate this potential application. **AO2 (6 marks)**

Mendel

Mendel's experiments

D–C

- Mendel pollinated peas with purple flowers with pollen from white flowers. All the offspring had purple flowers. Mendel then tried breeding these together. He found that the offspring were a mixture of purple and white.

- He decided that:
  - each plant must have two 'factors' for flower colour
  - the purple factors were 'stronger' than the white factors
  - each factor must be separately inherited, that is, the factors did not 'blend' together.

- Mendel presented his findings to other scientists in 1865. At that time no one knew anything about how cells divided or that chromosomes and genes existed. It was not until after Mendel died in 1900 that other scientists rediscovered his work. Today, we know that Mendel's 'factors' are alleles of genes.

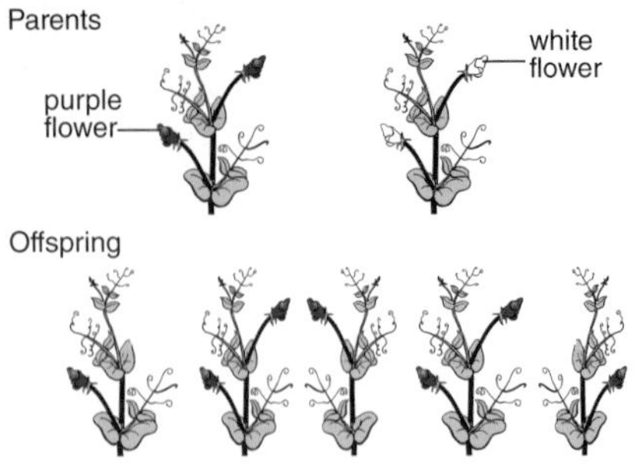

Figure 1: The result of crossing purple flowered pea plants with white flowered pea plants

Luck or good judgement?

B–A*

- Mendel was lucky with his experiments. By chance – or perhaps good judgement – he picked on characteristics that never do 'blend' together.

- There are many other features that do 'blend' together. For example, with some other kinds of flowers, if you cross red ones with white ones, you obtain pink ones.

How genes affect characteristics

Alleles come in pairs

D–C

- In rabbits, there might be a gene for fur colour: one allele of the gene may give black fur, another allele may give white fur.

- In a rabbit's cells, there are two complete sets of chromosomes. This means that there are two copies of each gene.

- If you call the allele for black fur B, and the allele for white fur b, then there are three possible combinations: BB, Bb or bb.

- Bb produces black fur. This is because the B allele is dominant. The b allele is recessive.

- When a dominant and a recessive allele are together, only the dominant one has an effect.

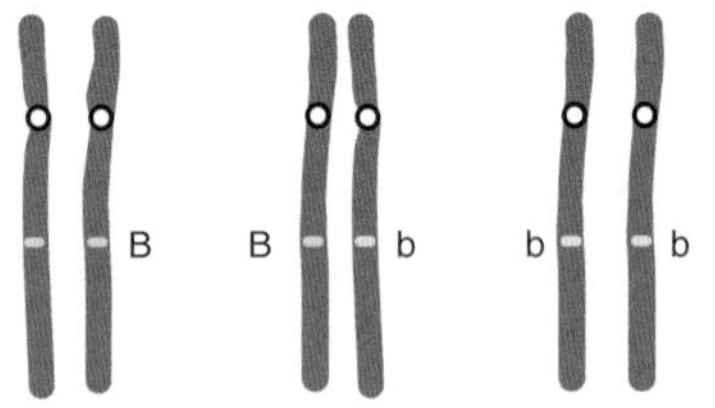

Figure 2: The three possible combinations of alleles for fur colour in a rabbit's cells

Figure 3: A rabbit with black fur could have the alleles BB or Bb. A rabbit with white fur can only have the alleles bb

BB   Bb   bb

Intersex conditions

B–A*

- Sex is determined by a pair of chromosomes called the sex chromosomes. Males are XY, females are XX.
- Around 1 in 100 people have some characteristics of both sexes (intersex conditions) caused by a zygote ending up with a combination such as XXY or XYY or even just a single X chromosome.
- Usually, if at least one Y chromosome is present, the person develops as a male, because it is the Y chromosome that determines maleness.

EXAM TIP

In the exam, you will need to be able to use some important genetic terms:

genotype – the combination of alleles, for example BB

phenotype – a characteristic, for example black fur

heterozygous – two different alleles, for example Bb

homozygous – two identical alleles, for example bb.

Improve your grade

Katy has red hair. Both her parents have brown hair. Explain how Katy inherited red hair when her parents do not have it. **AO2 (3 marks)**

Inheriting chromosomes and genes

Genetic diagrams for alleles of genes

- Genetic diagrams are used to show how alleles of genes are inherited.

- Figure 4 shows the outcome of a male rabbit with alleles Bb for fur colour breeding with a female with bb.

- Half of the sperm cells will have allele B and the other half will have allele b.

- All of the female's eggs will contain allele b.

- The genetic diagram shows you to expect about half of the baby rabbits to have the alleles Bb and have black fur. The other half would be bb and have white fur.

- It is important to remember that a genetic diagram only shows chances, not the actual results of the cross.

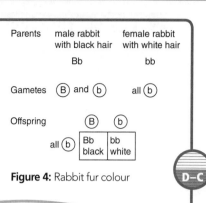

Figure 4: Rabbit fur colour

D–C

Remember!

It is usually a good idea to complete a genetic diagram by summarising the approximate chances of getting each of the different genotypes and phenotypes. In the above cross, the expected genotype ratio is 1Bb : 1bb.

Test crosses

- If an organism has a characteristic that is controlled by a dominant allele, then there are two possible combinations of alleles that it might have.

- The only way to find out which one you'll get is by doing a breeding experiment. This is a test cross.

EXAM TIP

If you are asked to draw a genetic diagram make sure you include the whole sequence of descriptions, not just the square showing the gametes and offspring.

B–A*

How genes work

DNA and protein synthesis

- A gene is a length of DNA that carries the code for making one protein.

- Each strand of a DNA molecule is made up of a sequence of four bases.

- The sequence of bases in the DNA determines the sequence of amino acids in the protein that is made.

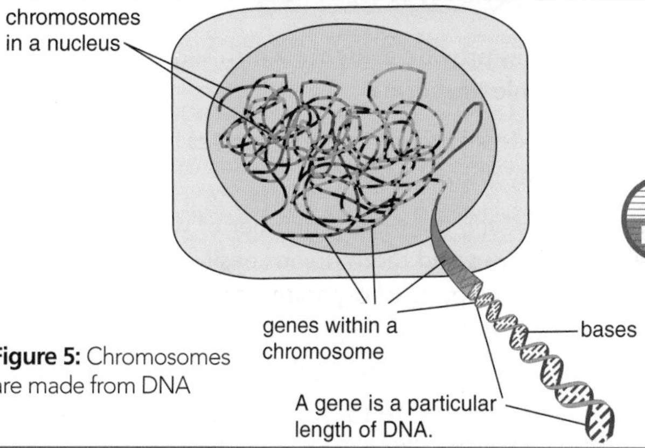

chromosomes in a nucleus

genes within a chromosome — bases

A gene is a particular length of DNA.

Figure 5: Chromosomes are made from DNA

D–C

Mutations

- Mutations in DNA form new alleles of genes.

- Many people have a slightly different version of the haemoglobin gene. It differs by only one base in the DNA but it is enough to cause a serious condition called sickle cell anaemia.

- In conditions where oxygen is in short supply (such as when doing a lot of exercise), it makes the red blood cells go so badly out of shape that they clump together and get stuck in blood capillaries.

B–A*

⊙ Improve your grade

The height of pea plants is controlled by a single gene. The tall allele is dominant. A tall pea plant was bred with a short pea plant. The offspring formed was in the ratio 1 tall : 1 short. Draw a genetic diagram to show the cross. **AO2 (4 marks)**

Genetic disorders

How they are inherited

- Polydactyly is a condition in which a person has more than five fingers on their hands, or more than five toes on their feet.

- It is caused by a dominant allele, so you only need to inherit one allele in order to have this condition.

- Figure 1 shows how polydactyly was inherited in one family.

- Cystic fibrosis is a disorder of cell membranes and affects the lungs and pancreas.

- It is caused by a recessive allele.

- A person with the disorder will have two recessive alleles. Someone with one recessive allele is called a carrier.

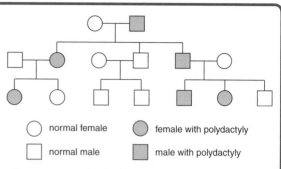

○ normal female   ● female with polydactyly

□ normal male   ■ male with polydactyly

Figure 1: Example of a family tree for polydactyly

Remember!

Carriers of cystic fibrosis will not have any symptoms of the illness. They may not even know they are a carrier unless they pass this allele on and have a child with cystic fibrosis.

D–C

Embryo screening

- Parents who know they are carriers for a serious genetic disease may choose to have a child by IVF so the embryos can be screened.

- This involves testing the embryos' DNA to find out if any have the alleles, then only implanting the unaffected ones into the mother's uterus.

B–A*

Fossils

The fossil record

- Fossils provide evidence which suggests that all species of living things that exist today have evolved from simple life-forms.

- The fossil record for most species is incomplete because most organisms do not form fossils when they die (they decay instead).

- However, there are a number of ways that the remains of plants and animals can be preserved as fossils.
  - Bones and teeth do not easily decay.
  - Some parts of organisms do not decay because conditions are not suitable for decay organisms.
  - Parts of an organism may be replaced by other materials such as hard minerals as they decay.
  - Traces of an organism, like footprints, may be preserved in rocks as prints.

- The fossil record for some organisms, such as the horse, is complete and shows how it has evolved over the last 50–60 million years.

D–C

How did life evolve?

- There is not enough valid or reliable evidence to be certain how life on Earth first evolved.
- Some scientists think that life first evolved in the seas, from the mix of chemicals that was present on the early Earth.
- All living things on Earth contain a genetic code in their DNA.
- This is a very strong piece of evidence that all life on Earth has evolved from the same common ancestors.

B–A*

Improve your grade

Polydactyly is caused by a dominant allele. Jessica has polydactyly. Ryan does not have it. Could their children have polydactyly? Explain your answer. **AO2 (3 marks)**

Extinction

Causes of extinction

- Life on Earth began about 3.8 billion years ago. As Earth slowly changed, many different plants and animal species became extinct (died out) and were replaced by new types.

- There are several causes of extinction:
  - a change in the environment – for example, the temperature might increase and so individuals that cannot adapt or move away will die out
  - new predators – may cause the extinction of species that are its prey and are not adapted to get away
  - new diseases – if a new, fatal, disease is introduced to a habitat then the species living there may not have immunity and so will die
  - a more successful competitor – if there is a limited amount of food in a habitat and two different species are competing for it, the less well-adapted one may become extinct.

Remember!
If a species is not able to quickly adapt to new situations it may become extinct.

D–C

Mass extinctions

- The fossil record suggests that there have been several periods in Earth's history when huge numbers of species became extinct over a relatively short period of time.

- It is thought that these were probably caused when a sudden, catastrophic event produced massive changes on Earth.

B–A*

New species

How new species arise

- Many new species arise through a series of steps. Here is an example.

1 Geographical isolation
A few lizards drift away from the mainland on a floating log and end up on an island. They are isolated from the rest of the species.

2 Genetic variation
In both the mainland lizards and the island lizards, there are many different alleles of genes which lead to variation.

3 Natural selection
The environment, predators, etc on the mainland are different from those on the island. Natural selection takes place and different features are selected for.

4 Speciation
Over time, more and more differences build up between the two populations of lizards until they are no longer the same species.

Remember!
Biologists define a species as a group of organisms that share similar characteristics, and that can breed together to produce fertile offspring.

D–C

Other kinds of isolation

- Geographic isolation is not the only way that two populations of the same species can become isolated.

- Any situation that forces members of the same species to start breeding within selected groups can cause a new species to arise.

- For example, one kind of fruit fly can breed on either apples or hawthorn fruits. It seems that those that hatched on apples always go back to apples to breed, while those that hatched on hawthorn fruits always go back to hawthorn fruits.

B–A*

Improve your grade

Around 300 years ago, the dodo lived on the island of Mauritius. It had no predators. People arrived on the island. They brought predators such as dogs, pigs and rats. Explain why the dodo became extinct. **AO2 (2 marks)**

Cells, tissues and organs

The parts of cells are called organelles. Each has an important part to play in the function of the cell.

Plant, animal and microbial cells have many similarities and some differences.

Oxygen and other dissolved substances move into and out of cells by diffusion.

In animals and plants, cells are grouped into tissues. Organs are made up of different tissues working together to carry out a function.

Different types of cells have different structures which enable them to carry out their specific functions. They are specialised.

Plants and the environment

Chlorophyll in plant cells absorbs energy from sunlight. This energy is used to convert carbon dioxide and water to glucose and oxygen in a reaction called photosynthesis.

The distribution of different species of organism in the environment is affected by physical factors. These include temperature, nutrients, light, water, oxygen and carbon dioxide.

Limiting factors, including light, temperature and carbon dioxide concentration, affect the rate of photosynthesis.

Glucose is used to supply energy; to make cellulose for cell walls; to make storage substances like starch and fats; and to make protein to build new cells and enzymes.

Quantitative data about the distribution of organisms can be collected using quadrats and transects.

Proteins and respiration

Protein molecules are long chains of amino acids. Hormones, antibodies and enzymes are all proteins.

Enzymes are biological catalysts which control all metabolic reactions, including digestion. They are also used in the home and in industry.

All living organisms release energy from glucose by respiration.

The rate at which enzymes work is affected by temperature and pH.

In aerobic respiration, energy is released when glucose is combined with oxygen. During exercise, when muscles are using a lot of energy, heart rate and breathing rate increase to provide muscles with extra oxygen.

In anaerobic respiration, a small amount of energy is released from glucose without using oxygen.

Cell division, inheritance and speciation

Cells normally divide by mitosis, which results in two genetically identical daughter cells. To produce gametes, cells divide by meiosis, which produces genetically different cells, each with half the normal number of chromosomes.

Clones can be created artificially by taking cuttings and carrying out tissue culture, embryo transplants and adult cell cloning.

New species can arise if two populations of a species become separated. Natural selection may result in them becoming so different that they can no longer interbreed.

Genes are passed from one generation to the next. A gene controlling a particular characteristic may have different forms, called alleles. Some diseases, such as cystic fibrosis, can be inherited.

Genetic diagrams can be used to predict the probable characteristics of the offspring of two parents.

Fossils provide evidence about some of the species that lived long ago.

Investigating atoms

How ideas about atoms have changed

- All matter consists of atoms, which cannot be divided up.

- All atoms of the same element are identical, but different from atoms of every other element.

- The Greeks suggested the idea of atoms in the 5th century BC, John Dalton revived the ideas in the 19th century.

- In 1897 the electron was discovered, 1909 the nucleus was discovered, 1911 Rutherford suggested that electrons orbit the nucleus, 1919 protons were discovered, and 1932 neutrons were discovered.

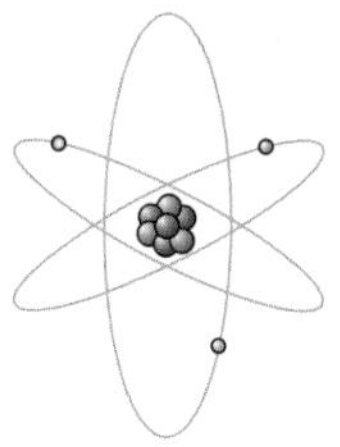

○ neutron
○ proton
○ electron

Figure 1: How can you tell that this atom is not drawn to scale?

D–C

Discovering the nucleus

- Rutherford's research team fired α-particles at gold foil. Flashes on a fluorescent screen showed where the α-particles hit.

- Most α-particles went straight through, showing that gold atoms are mostly empty space.

- Some were deflected and came out at an angle. To the team's surprise, a few came out backwards.

- Rutherford realised that the α-particles were bouncing off something inside the atoms. He called it the 'nucleus'.

B–A*

Mass number and isotopes

Properties of isotopes

- Not all atoms of an element have the same number of neutrons. Atoms of the same element with different masses are called isotopes.

- The proportion of each isotope of an element is fixed, and it is the same in both the pure element and in the compounds containing that element.

- Because isotopes have the same number of electrons, they all react in exactly the same way.

How Science Works

- The relative atomic mass (A_r) of an element is the average mass of an atom. You work it out using the percentage or number of atoms of each different isotope and its mass. Potassium has two isotopes, 90 per cent of the atoms are of potassium-39; the other 10 per cent are of potassium-40. The A_r of potassium is $\dfrac{(90 \times 39) + (10 \times 40)}{100} = 39.1$

D–C

Dating with isotopes

- Some elements have radioactive isotopes which decay over time.

- Knowing these and how quickly the isotopes decay, scientists can work out the age of archaeological remains such as Egyptian mummies.
  - Carbon-14 dating is used for remains suspected of being up to 60 000 years old.
  - Older samples need different isotopes. Uranium isotopes were used to calculate the age of the Earth.
  - Fossils are dated by dating the rocks in which they are found.

B–A*

Improve your grade

Work out the relative atomic mass (A_r) of chlorine. Chlorine has two isotopes, 25 per cent is chlorine-37, 75 per cent is chlorine-35. **AO2 (2 marks)**

Compounds and mixtures

Relative formula mass

D–C

- To work out the relative formula mass (M_r) of a compound you find the atomic mass of each element multiplied by the number of atoms of that element present in the formula, then add them up.

- calcium sulfate, $CaSO_4$

  atoms and A_r Ca = 40   S = 32   O = 16

  so   (40 × 1) + (32 × 1) + (16 × 4) = 136

- Atoms and molecules are far too small to be counted. To make counting atoms or molecules easy, chemists call the atomic or formula mass measured in grams a mole.

More complicated formulae

B–A*

- A group of atoms sometimes has a set of brackets round it, and a subscript number outside the bracket. You need to take this into account when working out the relative molecular mass.

- For example: aluminium sulfate, $Al_2(SO_4)_3$

  atoms and A_r Al = 27   S = 32   O = 16

  so   (27 × 2) + (32 × 3) + (16 × 12) = 342

Remember!

Atomic masses are either in the question or in the periodic table. In the periodic table, it is the larger of the two numbers in the box for the element!

Electronic structure

Electronic structures

D–C

- The electrons occupy the lowest available energy levels. They fill up each shell in turn. 2 electrons go in the first (except H which has only 1), up to 8 in the second, the third shell accepts 8 electrons and if the third shell fills up, then the remaining electrons go into the fourth shell – up to calcium.

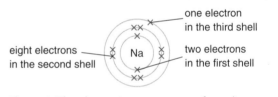

one electron in the third shell

eight electrons in the second shell

two electrons in the first shell

Figure 1: The electronic arrangement of a sodium atom, 2, 8, 1

- For elements beyond calcium the number of outer electrons is the same as the Group number. The number of outer electrons determines the reactivity of an element.

Noble gas structures

B–A*

- Noble gases (Group 0) are unreactive because they have a full outer shell of electrons.

- When atoms combine to make compounds, their outer electrons are shared, gained or lost until each atom has a Noble gas electronic configuration.

Improve your grade

Use the periodic table to:

a) Work out the electronic structure of these elements: magnesium, chlorine, nitrogen, aluminium, calcium. **AO2 (5 marks)**

b) Work out how many outer electrons each of these elements has: strontium, arsenic, gallium, astatine, germanium. **AO2 (5 marks)**

Ionic bonding

What is ionic bonding?

- When a metal reacts with a non-metal the metal atoms lose electrons and become positive ions, the non-metal atoms gain electrons and become negative ions.

- A sodium atom (Na) loses an electron to form a sodium ion (Na^+) with a 1+ charge. It has become a positive ion. A chlorine atom (Cl) gains an electron to form a chloride ion (Cl^-) with a 1− charge. It is now a negative ion.

- The oppositely charged ions are electrostatically attracted. This is ionic bonding.

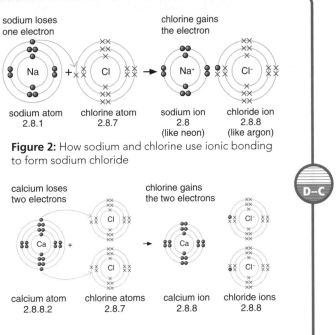

Figure 2: How sodium and chlorine use ionic bonding to form sodium chloride

Figure 3: How calcium forms ionic bonds

D–C

Conducting electricity

- Ionic compounds can conduct electricity, but only if the ions are free to carry the electric current. Solid sodium chloride does not conduct as the ions cannot move; dissolved in water or molten, it can conduct electricity. This suggests the bonding is ionic.

B–A*

Alkali metals

Similarities and differences in reactions

- Group 1 metals react in similar ways because their atoms all have just one electron in their outer shell.
  - lithium just floats on water and fizzes gently
  - potassium bursts into flame, zooms across the water and spits
  - caesium explodes on contact with water.
- The reactivity increases down the Group.

How Science Works

- Trends or patterns, like the reactivity of Group 1 metals, are important. For example, rubidium is below potassium and above caesium in Group 1. So it will probably have a more violent reaction with water than potassium, but less than caesium.

D–C

Discovering alkali metals

- Group 1 metals, (alkali metals) were discovered in the 19th century. They were too reactive to be separated from their compounds until electrolysis was invented.

Remember!
Reactivity of metals increases as you go down the group, but in non-metals the reactivity increases going up the group.

B–A*

Improve your grade

Draw diagrams to show how calcium and fluorine lose and gain electrons to make the compound calcium fluoride (CaF_2). **AO1 (3 marks)**

Halogens

The chemical reactivity of halogens

D–C

- This table shows information about Group 7 elements, the halogens.

Table 1: The halogens

name	symbol	description	molecule
fluorine	F	pale yellow poisonous gas	F_2
chlorine	Cl	green poisonous gas	Cl_2
bromine	Br	dark orange–red poisonous liquid that easily vaporises	Br_2
iodine	I	a dark grey solid which on heating becomes a purple gas	I_2

- When halogen elements form compounds fluorine becomes fluoride, bromine becomes bromide, and the ion always has a charge of 1–.

- Fluorine is the most reactive Group 7 element, then chlorine, bromine, and finally iodine which is least reactive.

Reactive elements, unreactive compounds

B–A*

- Chlorine is a highly toxic gas. We use it to kill bacteria to make water safe to drink.

- Compounds from chlorine are often very safe and useful:
  - sodium chloride is great on chips, polyvinyl chloride (PVC) is used to make window frames and guttering, not to mention clothes.

Remember!
When a compound is made from two different elements, the compound will have different properties from the original elements.

Ionic lattices

Electronic structures

D–C

- Solid sodium chloride forms an ionic lattice that is very strong. The oppositely charged ions attract each other in all directions. As a result, each sodium ion is attracted equally to six chloride ions. Each chloride ion is attracted to six sodium ions.

- Melting and boiling points are high, as it is hard to separate the ions from each other. Only when the ions are free to move by melting or dissolving in water can electricity be conducted. Ionic lattices have melting points greater than 500 °C.

- When ionic compounds conduct electricity (electrolysis) they decompose back to their elements. The metal element is formed at the negative electrode or cathode, the non-metal element is produced at the positive electrode or anode.

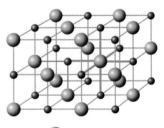

● Na^+ ○ Cl^-

Figure 1: The sodium chloride lattice

Multi-atom ions

B–A*

- Many negative ions are not single atoms that have lost electrons to become ions, but several non-metal atoms that are covalently bonded together, and have gained electrons as well. For example, the nitrate ion NO_3^-, sulfate ion SO_4^{2-}, and carbonate ion CO_3^{2-}.

- One non-metal ion has a positive charge: this is the ammonium ion, NH_4^+.

Remember!
A compound is likely to have ionic bonding if it has high melting and boiling points, and can only conduct electricity when molten or when in solution.

○ Improve your grade

Explain in detail why sodium chloride:
a has a high melting point
b conducts electricity when in solution. **AO1 (3 marks)**

Covalent bonding

Sharing by numbers

- When atoms of two non-metals combine, they share electrons to achieve noble gas electronic structures, which are stable.

- The outer electron shells of atoms overlap. This sharing forms a covalent bond which holds the two atoms together. Covalent bonds are very strong.

- A chlorine atom has seven outer electrons, so it shares one more from another atom, see Figure 2. It can share from another chlorine atom, or any non-metal atom.

- An oxygen atom has six outer electrons, so needs two more to make eight. In water, the oxygen shares two of its electrons, one with each hydrogen to make H_2O.

- To make an oxygen molecule, oxygen atoms share two electrons with another oxygen atom to make a double bond; two shared pairs of electrons.

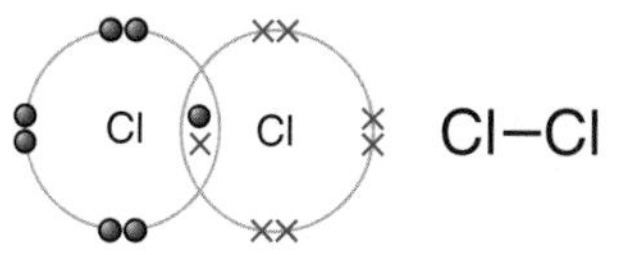

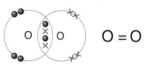

Figure 2: A chlorine molecule

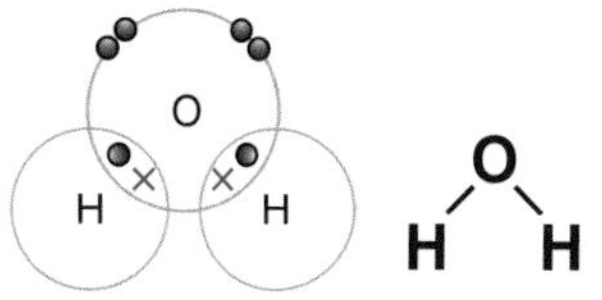

Figure 3: A water molecule

Figure 4: An oxygen molecule

> ### EXAM TIP
> When drawing electronic structures to show bonding, to save time and to make it clearer, you can miss out the inner filled electron shells.

D–C

How do covalent bonds work?

- The nuclei of both atoms involved in a covalent bond are positively charged.

- The electrons involved in the bond are attracted to both nuclei, and so remain a fixed distance apart from each nucleus. This distance is called the bond length.

B–A*

Covalent molecules

Properties

- Covalent molecules have no charged particles that are free to move. So they cannot conduct electricity.

- Whilst ionic compounds can dissolve in water, covalent compounds do not.

- Covalent bonds are very strong, but they only exist between the atoms in the molecule. There are weak attractions between the molecules, intermolecular forces. These weak forces are easily broken, so covalent molecules have low melting and boiling points.

D–C

Water is odd

- Water has a very low relative molecular mass, and should be a gas at room temperature, but water molecules have strong intermolecular forces and so water has high melting and boiling points.

> **Remember!**
> Covalent molecules have strong bonds inside the molecule but weak bonds between the molecules. An easy-to-melt, poor conductor of electricity is probably a covalent compound.

B–A*

Improve your grade

Draw diagrams to show the covalent bonds present in these compounds: HCl, H_2, CO_2, NH_3, and CH_4. **AO1 (5 marks)**

Covalent lattices

Bonding structure and properties

D–C

- Some covalent compounds have giant structures. These structures are called lattices.

- Examples of giant structures include diamond and graphite and silicon dioxide (sand).

- They have very strong structures so the melting and boiling points are very high.

name	structure	hardness	electrical conductor
diamond (carbon)		hardest natural substance	no
graphite (carbon)		soft, used as a lubricant	yes
silicon dioxide (SO_2)		very hard, used as sandpaper	no

Explaining graphite's properties

B–A*

- Graphite has layers of atoms that can slide over each other. Three of the four outer carbon electrons form covalent bonds. The fourth is free to move, or delocalised, and can carry an electric current through the structure.

- Each layer is only attracted to the one above and below by weak intermolecular attractions. These allow graphite to be soft and slippery.

Remember!
Pencils leave a mark on paper because the bonds between the layers of graphite are weaker than the bonds between the paper fibres.

Polymer chains

Polymers vary

D–C

- Polymers are thermosoftening (soften on heating) or thermosetting (harden on heating). Thermosoftening polymers are easy to recycle, thermosetting ones are harder.

- A polymer's use depends on both the monomer and the process used to make it.
  - High density polyethene (HDPE) is used for plastic bottles, and water pipes.
  - Low density polyethene (LDPE) is used to make film and plastic bags.

- HDPE and LDPE are made from the same monomer, but by different catalysts, temperatures and pressures, making different polyethenes with different properties.

- A thermosoftening polymer is a tangle of smooth chains. It is easy for the molecules to slide past each other and melt. A thermosetting polymer has more side chains and links to other chains. It is very hard for the chains to slide at all, so it cannot melt.

Intermolecular forces

B–A*

- Polymer molecules are long covalent chains made from carbon atoms, with various side groups attached.

- Intermolecular forces between molecules are weak, but because the molecules are so large the effect is greater, giving them higher melting points than expected.

- Thermosetting polymers soften and melt over a range of temperatures.

Improve your grade

Draw diagrams to show the difference between the structure of a thermosetting and a thermosoftening polymer.
AO1 (2 marks)

Metallic properties

Alloys

- Alloys are mixtures of two or more metal elements.

- In a pure metal, all the atoms are the same size. They fit into perfectly regular layers which can slide easily. Larger or smaller atoms in alloys disrupt the regular lattice. The layers slide less easily, so alloys are harder and less malleable than pure metals.

- Most metals in common use are alloys, because they are stronger than pure metals and have lower melting points.

- Special alloys have been developed that can remember their shape.

- These shape memory alloys such as nitinol are useful, as when warmed they return to their original shape. Uses include dental braces; and plates to hold broken bones together.

- In metals, the outer electrons of each atom are delocalised or free to move.

- As the sea of electrons can move, the metal is able to conduct electricity.

- The structure allows metals to bend, as the positive ions can move around each other.

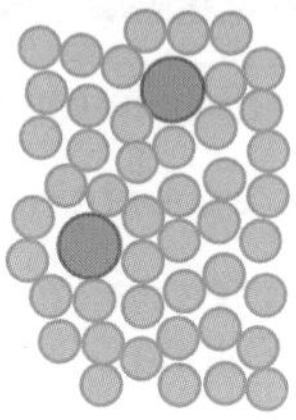

Figure 1: An alloy lattice

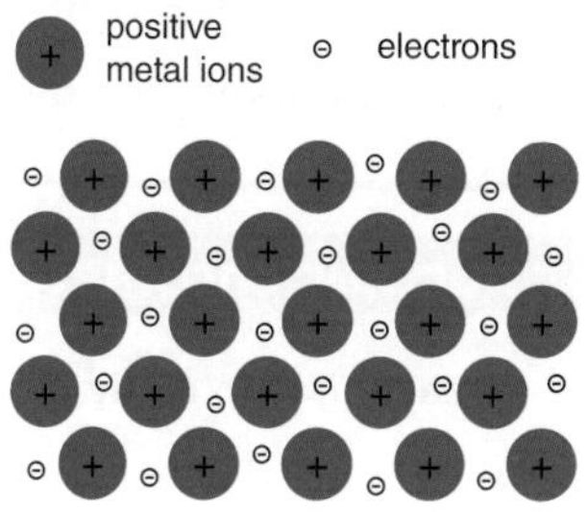

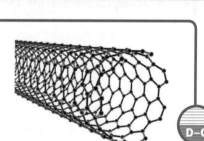

positive metal ions   electrons

Figure 2: Metallic bonding

D–C

Conduction

- When a potential difference is applied to a metal, the delocalised electrons flow from the negative connection towards the positive connection creating an electrical current.

- Heating a metal makes the particles in it vibrate more. The vibrations pass along the metal lattice, 'warming' the metal so metals are good conductors of heat.

B–A*

Modern materials

Applications of modern materials

- Modern materials include 'Smart' materials, such as photochromic, thermochromic, and shape memory alloys. There are also very small materials known as nanoparticles.

- Photochromic materials change colour according to the intensity of light.

- Thermochromic materials change colour according to the temperature.

- Nanoparticles are very small, being from 1–100 nm in size, with up to 300 atoms.

- Carbon has several nanoparticles, such as nanotubes; and Buckminsterfullerene (C_{60}).

- Uses of smart materials include self-cleaning and shading glass.

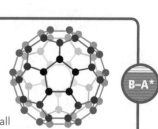

Figure 3: A carbon nanotube

D–C

Fullerenes

- Buckminsterfullerene (C_{60}) is a fullerene. They are based around five or six carbon-atom rings joined together in a geodesic sphere. It is possible to trap a molecule inside the sphere, so allowing highly toxic drugs to be carried to specific sites in the body to kill infections or cancerous cells.

Remember!
Nanoparticles have an enormous surface-area-to-volume ratio, so they make excellent catalysts.

Figure 4: A buckminsterfullerene ball

B–A*

Improve your grade

Explain how the metal lattice allows for metals to be stretched without breaking. **AO1 (2 marks)**

Identifying food additives

Chromatography

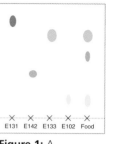

Figure 1: A chromatogram of food colours

- Paper chromatography shows that the colourings in a food are those listed on the label.

- When a solvent rises up the paper, it carries the colours with it and separates them.

- Each chemical in a mixture travels up the paper at the same speed as it does when alone.

- The chromatogram in Figure 1 shows that the food contains E133 and E102, as the food has spots at the same height as the samples. It does not have any E131 and E142, but it has a third chemical that is not identified.

- Chromatography is a separating technique also used with drugs and medicines.

D–C

Further chromatography

B–A*

- To save time in analysing chromatograms, chemists use the retention factor (R_f).

- Any chemical will always travel the same distance up a standard paper in the same time in the same solvent. The retention factor can be calculated like this:

$$R_f = \frac{\text{distance moved by the substance}}{\text{distance moved by the solvent front}}$$

- Thin-layer chromatography uses a glass plate coated with a thin layer of absorbent materials. It is quicker than paper chromatography, but also more expensive.

Instrumental methods

Retention times and mass spectrometry

D–C

- Analytical chemists use a technique called gas chromatography. They send a gas solvent with the substances to be analysed, through a tube packed with a solid material.

- The components travel through the tube at different speeds and are detected at the end. The results are shown as a graph, with each peak representing a component and the height of the peak indicating how much was present.

- The position of the peak is used to work out the retention time for each component.

- Often a mass spectrometer is attached to the gas chromatography equipment, a GC-MS method. This measures the relative atomic or molecular mass of each component.

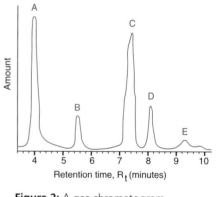

Figure 2: A gas chromatogram retention–time graph

How does mass spectrometry identify substances?

B–A*

- The mass spectrometer knocks an electron of each atom or molecule, leaving a positively charged particle or molecular ion. It can then measure the mass of the particle.

- It also breaks up some of the molecules into smaller pieces, and this helps to distinguish between different molecules with the same relative molecular mass.

⊙ Improve your grade

Look at Figure 2. It shows the retention–time graph for a gas chromatogram sample.
  a   How many different components were in the sample?
  b   List the components in order of the quantity of each present. Start with the most plentiful. **AO3 (2 marks)**

Making chemicals

Routes and conditions

- There are several types of chemical reaction that can be used to make a chemical:
  - oxidation or reduction, losing or gaining oxygen
  - neutralisation, reacting an acid with an alkali or a base
  - precipitation, reacting two soluble compounds to make an insoluble product
  - electrolysis, using electricity to split a liquid or solution.

- Whichever type of reaction is used, the reaction conditions are important in terms of speed, costs, and safety:
  - temperature, higher temperatures make reactions go quicker
  - concentration, more concentrated solutions react faster
  - pressure, higher pressure compresses gas molecules, giving more molecules in the same volume and leading to faster reactions
  - particle size, smaller pieces react faster than larger pieces
  - catalysts, these can speed up the reaction rate, and are unchanged by the reaction.

D–C

Chemical calculations

- Chemists calculate the correct masses of reactants to use to make products, so that only the smallest amounts needed are used to make the products and waste products are kept to the minimum.

- This keeps costs to the minimum and reduces waste.

B–A*

Chemical composition

Empirical formulae

The relative quantities of each element present in a compound can be worked out using the percentage composition method:

- find the relative formula mass (M_r), for example $FeSO_4$ = 152

- find the mass of the element in the compound, Fe = 56 × 1, = 56

- divide 56 by M_r and multiply by 100 to get the percentage of iron in iron sulfate = 39.4%.

Working out a formula from experimental data

To calculate the formula of copper oxide from an experiment, you need to know: how much copper was used, how much copper oxide was produced, how much oxygen was needed.

If 1.27 g of copper was burnt, and made 1.59 g of copper oxide, what is the formula of copper oxide?

To answer this, you need to:

- calculate the mass of oxygen used as 1.59 g – 1.27g = 0.32 g

- divide the mass of copper by its A_r (1.27/63.5 = 0.02) and oxygen by its A_r (0.32/16 = 0.02)

- divide each result by the smallest one, i.e. 0.02, to give the ratio of atoms in the formula, so Cu = 0.02/0.02 = 1, O = 0.02/0.02 = 1, so the formula is CuO.

D–C

Analysing hydrocarbons

You can analyse hydrocarbons using the same method. Burning 1.68 g of hydrocarbon produced 5.28 g of carbon dioxide, and 2.16 g of water. So:

- mass of carbon present (12/44 × 5.28 = 1.44 g), mass of hydrogen present (2/18 × 2.16 = 0.24 g)

- divide each mass by its A_r C = 1.44/12 = 0.12, H = 0.24/1 = 0.24

- ratio of atoms is 0.12 : 0.24, or 1:2, so the empirical formula is CH_2, so it is an alkene.

B–A*

| A_r, H = 1, C = 12, O =16 |
| M_r, CO_2 = 44, H_2O = 18 |

Improve your grade

Calculate the formula of lithium oxide made from 5.6 g of lithium and 6.4 g of oxygen. **AO2 (2 marks)**

Quantities

Quantities from equations

D–C

- Look at this equation for the production of ammonia from nitrogen and hydrogen:
  $N_2(g) + 3H_2(g) \rightleftharpoons 2NH_3(g)$. The equation shows that one nitrogen molecule reacts with three hydrogen molecules to produce two molecules of ammonia gas.

- We can then work out the **relative reaction masses** of each reactant or product by multiplying the A_r or M_r by the number of reacting molecules for each gas.
  $N_2 = 28 \times 1 = 28$, $H_2 = 2 \times 3 = 6$, and $NH_3 = 17 \times 2 = 34$.

- 28 units of nitrogen will react with 6 units of hydrogen to make 34 units of ammonia. To work out how much ammonia can be made if 56 tonnes of nitrogen is used, you divide 56 by 28 to find the reacting amount (= 2); so 56 tonnes of nitrogen will make 34×2 tonnes of ammonia = 68.

> ### EXAM TIP
>
> Whenever there is an important or difficult calculation required that needs a balanced equation, the balanced equation will be given in the question.

Sir Terry Pratchett's sword

B–A*

- If a good copper ore contains 5% by mass of copper carbonate. How much copper could be made from 25 kg of ore?

- $2CuCO_3(s) + C(s) \rightarrow 2Cu(s) + 3CO_2(g)$

- the M_r for $CuCO_3$ is $63.5 + 12 + (16 \times 3) = 123.5$
  - reacting mass for $CuCO_3$ is $123.5 \times 2 = 247$
  - reacting mass for copper is $63.5 \times 2 = 127$
  - Assuming the 25 kg of ore is used and contains $(5/100) \times 25 = 1.25$ kg of copper carbonate, the possible yield will be $(1.25 \times 127)/247 = 0.64$ kg

A_r
Cu = 63.5
O = 16
C = 12

How much product?

Calculating percentage yield

D–C

- Other factors affect the amount of product produced in a chemical reaction.

- Reactants are rarely 100% pure. The mass of actual reactant is less than the mass weighed out, so forms less product: the same reactants can form different products: if the reaction is reversible, some of the products turn back into reactants.

- To compare the effectiveness of making a chemical in different ways, we calculate the percentage yield

$$\text{percentage yield} = \frac{\text{actual yield}}{\text{theoretical yield}} \times 100$$

> **Remember!**
> When calculating percentage yields, it doesn't matter what units are used for the actual yield or the theoretical yield so long as they are both measured in the same units.

Percentage yields, economics and the environment

B–A*

- Percentage yield indicates how effectively reactants are converted into the product.

- It is usually more economic to use a reaction with a high percentage yield.

- Maximising percentage yields makes economic sense. Less raw materials are needed and less are wasted in making unwanted by-products that have to be disposed of.

Improve your grade

Calculate the mass of calcium oxide that can be made from 100 tonnes of calcium carbonate.
The balanced equation for the reaction is:
$CaCO_3(s) \rightarrow CaO(s) + CO_2(g)$ **AO2 (3 marks)**

Reactions that go both ways

Examples of reversible reactions

- Heating solid ammonium chloride causes it to decompose into two gases, hydrogen chloride and ammonia. When the two gases cool, the ammonium chloride is reformed. This is a reversible reaction.

- Copper sulfate crystals are blue, but heat them and they turn white, as the water of crystallisation is removed. Add a little water to the white anhydrous copper sulfate and the white powder turns blue again.

$$CuSO_4.5H_2O(s) \rightleftharpoons CuSO_4(s) + 5H_2O(\ell)$$

- The double-headed arrow shows that the reaction can go both ways, left to right or forward, right to left or backward.

D–C

Reversible reactions used in industry

- Ammonia is made using a reversible reaction:

$$N_2(g) + 3H_2(g) \rightleftharpoons 2NH_3(g)$$

- Sulfuric acid is needed to make many other chemicals. It too is manufactured using two reversible reactions.

$$2SO_2(g) + O_2(g) \rightleftharpoons 2SO_3(g) \quad \text{then,}$$

$$SO_3(g) + H_2O(\ell) \rightleftharpoons H_2SO_4(\ell)$$

How Science Works

- When investigating reversible reactions and the effect of different conditions upon them, it is really important to be clear what the independent variable and control variables are in the investigation.

B–A*

Rates of reaction

Measuring rates of reaction

- When chemists talk about the rate of reaction they mean how much chemical reacts, or is formed, in a given time.

- rate of reaction = $\dfrac{\text{amount of reactant used up}}{\text{time taken}}$ or $\dfrac{\text{amount of product formed}}{\text{time taken}}$

- You can alter the rate of a chemical reaction by doing one or more of these: increasing the temperature, increasing the concentration of a reactant, using smaller pieces of a reactant, or adding a catalyst.

- Often a gas is produced and so it is easy to measure the quantity produced. Figure 1 shows ways to measure a gas.

D–C

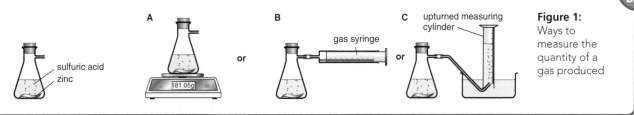

A   sulfuric acid zinc   or   B   gas syringe   or   C   upturned measuring cylinder

Figure 1: Ways to measure the quantity of a gas produced

Calculating reaction rates

- You can either calculate the average reaction rate for a reaction; or the reaction rate at a given time.
  - To calculate the average reaction rate from the graph, you can see that in 4 minutes the reaction produced 12.8 cm³ of hydrogen gas. So 12.8/4 = 3.2 cm³ per minute.
  - To calculate the rate at 2–2.5 minutes, use the graph to find the change in volume from 2 to 2.5 minutes. This is 0.8 cm³, the time for this change to happen is 0.5 minutes, so 0.8/0.5 = 1.6 cm³ per minute.

Figure 2: Measurements from a magnesium–acid reaction

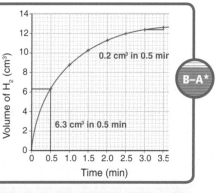

0.2 cm³ in 0.5 min

6.3 cm³ in 0.5 min

Volume of H₂ (cm³)

Time (min)

B–A*

Improve your grade

Use Figure 2 to calculate the rate of the reaction at the following times:
a   0 to 0.5 minutes
b   1.0 to 2.0 minutes
c   2.5 to 3.0 minutes. **AO3 (3 marks)**

Collision theory

Explaining collision theory

- For a chemical reaction to occur, the reactants must collide with each other. Any change that increases the number of collisions will increase the rate of a reaction. Not all collisions cause a reaction. The collision has to be hard enough to cause the reactants to react. This can be done by increasing the temperature, increasing the concentration, increasing the pressure, using smaller particles of a solid, adding a catalyst.

D–C

• reactant particles * successful collision

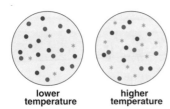

lower temperature   higher temperature

Figure 1: Effect of temperature on the number of successful collisions per second in a liquid or gas

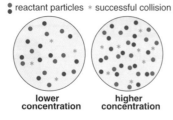

lower concentration   higher concentration

Figure 2: Effect of concentration on the number of successful collisions per second in a liquid or gas

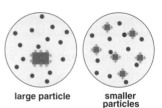

large particle   smaller particles

Figure 3: Effect of surface area of a solid on the number of successful collisions per second

How much change?

B–A*

- Doubling the concentration of one reactant often doubles the rate.

- Cutting a cube of solid reactant into eight, doubles its surface area and doubles the rate.

- Increasing the temperature always increases the rate.

- Different reactions need different catalysts. Not all reactions can be catalysed.

Remember!
Not all collisions cause reactions.

Adding energy

Investigating the effects of temperature

D–C

- Magnesium reacts with sulfuric acid, giving off hydrogen gas.

$$Mg(s) + H_2SO_4(aq) \rightarrow MgSO_4(aq) + H_2(g)$$

- Figure 4 shows results from using the same amounts of magnesium and acid at three different temperatures.

- The steeper the graph's curve, the faster the rate. All three reactions produce the same final volume of hydrogen gas, as they have the same quantities of starting reactants.

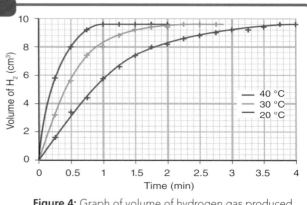

Figure 4: Graph of volume of hydrogen gas produced at different temperatures

Activation energy

B–A*

- To cause a reaction, particles must collide with sufficient energy to break bonds.

- The minimum energy needed for this to happen is called the activation energy.

◉ Improve your grade

A student was investigating a chemical reaction. She decided to increase the temperature of the reaction, whilst decreasing the concentration of one reactant. Use collision theory to explain why the rate of reaction remained unchanged. **AO3 (2 marks)**

Concentration

The effect of concentration

- Higher concentration means that there are more reactant particles in the same volume. There are more collisions per second, increasing the reaction rate. Increasing the concentration does not affect the energy of collisions.

- Reactants are used up during a reaction and their concentrations decrease, so the reaction slows down. That is why the graphs are curves.

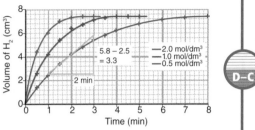

Figure 5: Rate-of-reaction graphs for magnesium reacting with sulfuric acid at different concentrations

- To find the rate at any point during the reaction find the gradient of the graph at that point.

- Look at Figure 5. What is the rate at 0.5 mol/dm³ after 2 minutes?
  - Draw a tangent at this point, and construct a triangle as shown.
  - Find the scale lengths on each axis: x-axis = 2 min, y-axis = 3.3 cm³
  - Rate = amount of product ÷ time = 3.3/2 = 1.65 cm³ of H₂/min

Two solutions

- Doubling the concentration of one reactant will double the rate of the reaction. If you double both reactants then the reaction rate will double twice, that is go four times as fast and be complete in a quarter of the time.

How Science Works

- A chemical reaction is complete when there are no more reactants available to react. It may stop because all the reactants have been used up or, more likely, when just one of the reactants has been used up. The remaining reactant is said to be in **excess**.

Size matters

Size and surface area

Solids only react at their surface. Breaking a solid into smaller pieces exposes more surface area. That provides more reactant particles to be available for collisions: the reaction rate increases.

The ratio between the two surface areas shows the change in rate of reaction. For example, if you reduce particle size from 1 cm³ to 1 mm³:

- for 1 cm cube surface area:
  - each face = 10 mm × 10 mm = 100 mm²
  - total = 6 × 100 mm² = 600 mm²

- for 1 mm cube surface area:
  - each face = 1 mm × 1 mm = 1 mm²
  - total = 6 × 1 mm² = 6 mm²

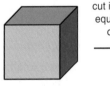

cut into eight equal-sized cubes

Figure 6: Smaller particles, larger surface area

- A 1 cm cube contains 1000 cubes of 1 mm side length, so:
  - surface area of 1000 cubes = 6000 mm²
  - 6000/600 = 10, so the reaction will be ten times faster.

Liquids and gases

- Liquids react with gases much faster if sprayed into tiny droplets first.

- Oil-fired boilers and furnaces spray in the fuel oil, so that it burns rapidly.

Improve your grade

Use Figure 5 to find the rate of the reaction after 1 minute when the concentration is 2 mol/dm³. **AO2 (2 marks)**

Clever catalysis

Using catalysts

- Catalysts are chemicals that speed up reactions, but are not used up in the process. They can be used over and over again.

- Car exhaust fumes contain poisonous carbon monoxide and oxides of nitrogen. In a catalytic converter platinum, rhodium and palladium, catalyse reactions between the gases producing less harmful ones, for example carbon monoxide becomes carbon dioxide, nitrogen oxides are converted to nitrogen. Many catalysts are transition metals or their compounds.

Table 1: Some catalysts and their uses

catalyst	process	product
iron	Haber Process	ammonia
platinum	oxidising ammonia	nitric acid
nickel	hydrogenating vegetable oils	vegetable spreads such as margarine

How catalysts work

- A catalyst provides a way for a reaction to occur with lower activation energy. so more collisions now result in reactions, and the rate is increased.

- The reactants do not react directly. One combines with the catalyst. This combination reacts with the second reactant, making the product and regenerating the catalyst.

- When another chemical reacts with the catalyst it can be poisoned, and stop working.

- Enzymes are biological catalysts. Like other catalysts, they are specific to reactions and we each have hundreds of different enzymes working all the time in our bodies to keep us alive. They are also used in biological washing powders, brewing and baking.

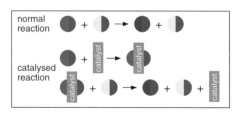

Figure 1: Each circle represents a reactant. The catalyst reacts and changes, but is regenerated again

Controlling important reactions

Economics and safety

- When planning an industrial process, chemists need to consider all the variables that affect rates of reaction. Making ammonia is a good example:

- $N_2(g) + 3H_2(g) \rightleftharpoons 2NH_3(g)$
  - the temperature used is 450 °C
  - the pressure used is 200 atmospheres (this has the effect of increasing the concentration)
  - an iron catalyst is used
  - the catalyst has a large surface area.

- The choice of conditions produces a small yield of ammonia (15–20%), but quickly and cheaply. The unreacted nitrogen and hydrogen are fed back in, so eventually all of the reactants become ammonia.

- Increasing temperature speeds up the reaction, but reduces the amount of ammonia formed, the high pressure produces more ammonia, but is very expensive, the iron catalyst speeds up the reaction, whilst being cheap.

- Many chemical reactions are exothermic, they produce heat. This can alter the rate of the reaction so the chemical plant needs to be able to keep the temperature constant.

Catalysts in nature

- Animals and plants use catalysts. Photosynthesis relies upon chlorophyll as a catalyst to convert carbon dioxide and water quickly to sugars.

- For plants and animals to respire, special biological catalysts – enzymes – are needed to change the sugar back to carbon dioxide, water and the energy we need.

Improve your grade

What is activation energy? Explain how a catalyst lowers the activation energy for a particular reaction. **AO1 (3 marks)**

The ins and outs of energy

Energy stores

- All substances store energy, but different substances store different amounts.

- If reaction products store less energy than the reactants did, the reaction is exothermic. The extra energy is transferred to the surroundings, heating them up.

- If the products store more energy (an endothermic reaction), they must have absorbed energy from somewhere. The surroundings provide the energy by cooling down.

- These changes can be shown by an energy-level diagram.

- Exothermic and endothermic reactions are useful. Calcium oxide reacts exothermically with water to provide a heat source for self-heating cans. Ammonium nitrate is used to provide an endothermic reaction to make sports injury packs, to cool muscles.

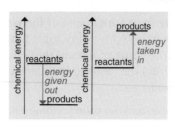

Figure 2: Exothermic and endothermic energy-level diagram

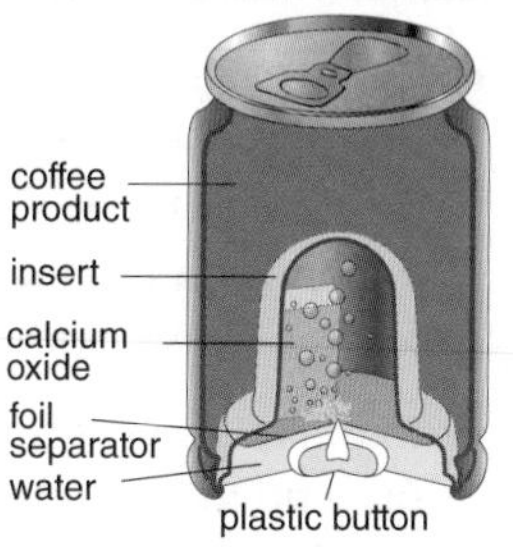

Figure 3: Self-heating coffee can

D–C

Energy and reaction rates

- Energy changes affect temperature, which affects reaction rates.

- Endothermic reactions cool, so to maintain the rate they need to be supplied with heat.

- Exothermic reactions produce heat, so need to be cooled down if the rate is not going to increase.

- With reversible reactions, if the forward reaction is endothermic, then the backward reaction will be exothermic.

B–A*

Acid–base chemistry

Acid–base reactions

- Most bases are insoluble in water. Bases that do dissolve in water are alkalis, because their solutions contain hydroxide ions in the solution. Alkalis are usually hydroxides of Group 1 and 2 metals in the periodic table.

- A base reacts with an acid to form a salt and water, for example zinc oxide and sulfuric acid.

$$ZnO(s) + H_2SO_4(aq) \rightarrow ZnSO_4(aq) + H_2O(\ell)$$

- Metal carbonates also produce carbon dioxide.

$$CaCO_3(s) + 2HCl(aq) \rightarrow CaCl_2(aq) + H_2O(\ell) + CO_2(g)$$

- Alkalis react with acids to give a salt and water only.

- Group 1 carbonates and hydrogencarbonates are bases because they react with acids to give salt and water, and carbon dioxide.

- Neutralisation reactions are exothermic and can be represented by this ionic equation:

$$H^+(aq) + OH^-(aq) \rightarrow H_2O(\ell)$$

D–C

Ammonia

- Ammonia solution is alkaline because when it dissolves in water it becomes ammonium hydroxide solution.

- If ammonia reacts directly with an acid, such as hydrochloric acid, it does not produce water. Instead, it produces the salt.

$$NH_3(g) + HCl(aq) \rightarrow NH_4Cl(aq)$$

B–A*

Improve your grade

Sketch an energy-level diagram for the exothermic reaction that takes place when nitric acid reacts with sodium hydroxide. Write on the line for 'reactants', the names of the reactants; and on the line for 'products', the products. **AO1 (4 marks)**

Making soluble salts

Practical methods

D–C

- Soluble salts can be made from acids in one of three ways: reacting with a metal, reacting with an insoluble base, reacting with an alkali. The name of the salt that's made depends on the metal, or on the metal in the name of the alkali or base, and the acid.

- To make a salt from a solid: add the solid to some dilute acid until no more will dissolve, if the reaction is slow, then warm the mixture gently, filter the solution to remove unreacted solid, evaporate the filtered solution until it is nearly all gone, then allow it to cool and the salt crystals will appear.

- To make a salt using an alkali: place 25 cm³ of alkali in a conical flask, and add some universal indicator, add the acid from a burette, syringe or measuring cylinder until the acid is neutralised – note the volume of acid required, then either repeat the experiment omitting the universal indicator; or add a little powdered charcoal to the coloured solution and heat, then filter to remove the charcoal, evaporate the solution until it is nearly all gone.

Table 1: Some acids and their salts

acid		salt made
sulfuric	H_2SO_4	sulfate
hydrochloric	HCl	chloride
nitric	HNO_3	nitrate

Remember!

The higher a metal is in the reactivity series, the easier it will react with the acid. At the top of the reactivity series the reaction will be extremely violent – so Group 1 salts are made by other methods.

Neutralisation may not work

B–A*

- Not all salts are soluble. Some metals either have a surface layer of oxide that prevents them from reacting; or the reaction with the acid coats them with the salt, preventing the rest of the metal atoms reacting with the acid.

EXAM TIP

If asked to make a salt, first decide on the acid you need, then on a suitable source of the metal.

Insoluble salts

Useful precipitations

D–C

- An insoluble salt is one that will not dissolve in water. Insoluble salts are made by mixing together two solutions, each containing one part. The metals swap partners. The solid product formed by mixing two solutions is called a precipitate. Filtering removes the precipitate, which is washed before drying in an oven.

- Salts are ionic compounds. A mixture of two salts in solution contains positive ions of two metals and negative ions of two non-metal groups. If any pair combines to form an insoluble salt, they produce a solid precipitate.

- Precipitation reactions are useful to remove unwanted ions from water; calcium ions cause hardness in water and are removed using sodium carbonate solution, phosphate ions in washing powder are removed by adding aluminium ions to the waste water. Calcium carbonate is produced for toothpaste this way.

Chemical analysis

B–A*

- Sulfates can be identified by adding barium chloride solution. A white precipitate of barium sulfate appears.

- Halogen compounds can be identified by adding silver nitrate.

- Sodium hydroxide can identify metal ions in solution.

Improve your grade

Describe the method to make some solid chromium chloride from chromium metal. AO3 (4 marks)

Ionic liquids

Conduction in liquids

- Pure water does not conduct electricity as it is made of molecules. Tap water is not pure as it contains many dissolved ions, so it does conduct electricity.

- When a liquid conducts, the electric current must enter and leave the liquid through solid conductors called electrodes. This is electrolysis.

- The ions in a solution or molten compound move around randomly. In electrolysis, positive ions are attracted to the negative cathode and negative ions are attracted to the positive anode.

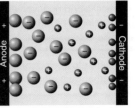

Figure 1: Ions in a solution that is conducting electricity

- As the ions move, they carry electric charge through the liquid from one electrode to the other.
  - At the anode, the negative ions release their electrons; whilst at the cathode, the positive ions pick up electrons at the same rate as the negative ions are releasing their electrons.
  - Metal ions gain electrons at the cathode, and become atoms that coat the electrode or react with the liquid. Non-metal ions at the anode lose electrons, and may be released as a gas or react with the liquid.

- Electroplating is another use of electrolysis (see Figure 2). An object is coated with a thin layer of metal, as shown. The silver anode dissolves into the solution, maintaining the concentration of silver ions.

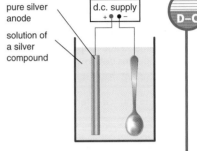

Figure 2: Electroplating a spoon

D–C

Direct and alternating current

- Electrolysis only takes place with direct current (d.c.), mains electricity is alternating current (a.c.), so does not allow electrolysis; but the water still conducts electricity.

Remember!
As electrolysis of a solution continues, the concentration of the compound in the solution (electrolyte) is reduced, so more of the compound needs to be added. In electroplating, the anode is often made from the metal to be plated; it dissolves into the electrolyte so maintaining the metal ion's concentration.

B–A*

Electrolysis

More applications of electrolysis

- Electrolysis of molten sodium chloride decomposes the compound producing sodium metal at the cathode and chlorine gas at the anode.
  - Electrolysis of a solution of sodium chloride (brine) is slightly different. The water makes the brine a mixture of ions (Na^+, H^+, OH^- and Cl^-). Chlorine gas is produced at the anode, at the cathode, the lower of sodium and hydrogen (the positive ions) in the reactivity series is made, which is hydrogen; and the sodium ions react with the hydroxide ions to make a solution of sodium hydroxide.

- Aluminium metal is obtained by electrolysis from aluminium oxide (Al_2O_3) dissolved in cryolite to reduce the melting point of the oxide.

- Copper is purified by electrolysis.

D–C

Half-equations

- Two reactions happen during electrolysis, one at each electrode. These can be represented as **half-equations**.

- In sodium chloride electrolysis:
  - at the cathode, $Na^+ + e^- \rightarrow Na$ (this is reduction)
  - at the anode, $2Cl^- \rightarrow Cl_2 + 2e^-$ (this is oxidation)

B–A*

Improve your grade

Write half-equations to show the reactions, at both the anode and cathode, for the electrolysis of copper chloride ($CuCl_2$).
AO2 (2 marks)

Structure and bonding

Chemical bonding involves either transferring or sharing electrons in the outer shells of atoms.

Ionic compounds are held together by strong electrostatic forces of attraction.

Atoms that lose electrons become positively charged ions. Atoms that gain electrons become negatively charged ions.

When atoms share pairs of electrons, they form covalent bonds.

Structure, properties and uses

When melted or dissolved in water, ionic compounds conduct electricity. Covalent compounds do not.

Ionic compounds have regular structures (giant ionic lattices), high melting and boiling points.

Diamond and graphite have different properties determined by their structures.

Metals consist of giant structures of atoms arranged in a regular pattern.

The properties of thermosoftening and thermosetting polymers depend on what they are made from and how they are made.

Nanoscience refers to structures that are 1–100 nm in size and of the order of a few hundred atoms.

Atomic structure, analysis and quantitative chemistry

Atoms of the same element can have different numbers of neutrons.

Elements and compounds can be identified using instrumental methods.

The relative atomic mass (A_r) and the relative formula mass (M_r) allow numbers of particles to be compared.

The masses of reactants and products can be calculated from balanced symbol equations.

The amount of a product obtained is known as the yield.

A reversible reaction is one where the products of the reaction can react to produce the original reactants.

Rates of reaction

Catalysts change the rate of chemical reactions but are not used up during the reaction.

The minimum amount of energy that particles must have to react is called the activation energy.

The rate of reaction = $\dfrac{\text{amount of reactant used}}{\text{time}}$ or = $\dfrac{\text{amount of product formed}}{\text{time}}$

Chemical reactions can only occur when reacting particles collide. Collision theory explains why changes to conditions affect rates.

Exothermic and endothermic reactions

An exothermic reaction transfers energy to the surroundings. An endothermic reaction takes in energy from the surroundings.

If a reversible reaction is exothermic in one direction, it is endothermic in the opposite direction.

Acids, bases and salts

Metal oxides and hydroxides are bases. Soluble hydroxides are called alkalis. Ammonia makes an alkaline solution to produce ammonium salts.

Soluble salts can be made from acids by reacting them with metals, insoluble bases, or alkalis. Salt solutions can be crystallised to produce solid salt. Insoluble salts can be made by mixing solutions of ions so that a precipitate is formed.

In neutralisation reactions, hydrogen ions react with hydroxide ions to produce water.

Hydrogen ions, $H^+(aq)$, make solutions acidic and hydroxide ions, $OH^-(aq)$, make solutions alkaline.

Electrolysis

When an ionic substance is melted or dissolved in water, the ions are free to move about within the liquid or solution.

During electrolysis, positively charged ions move to the negative electrode and are reduced, and negatively charged ions move to the positive electrode and are oxidised.

Aluminium is manufactured by the electrolysis of a molten mixture of aluminium oxide and cryolite.

Reactions at electrodes can be represented by half-equations, for example:
$2Cl^- \rightarrow Cl_2 + 2e^-$ or $2Na^+ + 2e^- \rightarrow 2Na$

The electrolysis of sodium chloride solution produces hydrogen, sodium hydroxide and chlorine.

See how it moves

Speed and average speed

- Average speed is calculated in m/s using speed = total distance travelled (m)/time taken (s).

Speed from the graph

- Distance time graphs show the distance an object moves in a period of time.

- A line sloping up shows the object moving away from its starting point. If it slopes down, the object is moving back.

- The gradient (slope) at any point gives the speed at that time.

- A steeper gradient shows a faster speed.

- A flat gradient shows the object is stopped.

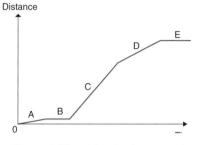

Figure 1: The object is always moving forward, but is stopped at B and E

EXAM TIP

The unit of speed depends on units used for distance and time, for example if distance is measured in km and time is measured in hours, speed is measured in km per hour.

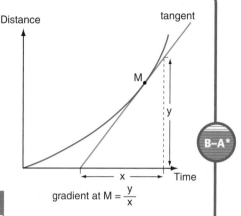

gradient at M = $\frac{y}{x}$

Figure 2: Use the gradient of a d-t graph to calculate speed. If speed changes, the line curves and you must draw a tangent.

Speed is not everything

Velocity and acceleration

- Velocity is an object's speed in a given direction, for example a skydiver falls downwards. A change in velocity (or speed) is called acceleration.

- Acceleration (m/s²) = $\frac{\text{change in velocity (m/s)}}{\text{time taken (s)}}$.

- The change in velocity is final velocity – original velocity.

- A negative acceleration (deceleration) means the object slows down.

Using velocity time graphs

- Velocity time graphs show how an object's velocity changes with time. The gradient of the graph gives the object's acceleration.

- A steeper gradient means greater acceleration.

- A flat line shows a steady speed.

- The area under a velocity time graph shows the distance travelled. To calculate the area, think of the shape as a combination of triangles and rectangles.

EXAM TIP

It is very easy to confuse velocity time graphs with distance time graphs. Check carefully before writing your answer.

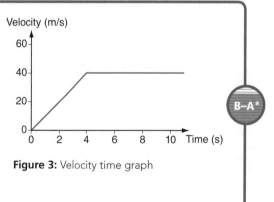

Figure 3: Velocity time graph

Improve your grade

Sam travels 5 m in 10 s, before stopping for 2 s. He then takes 6 s to return to the start. Draw a distance time graph that shows Sam's journey. **AO2 (3 marks)**

Forcing it

How things move

D–C

- The resultant force on an object is the single force that would make an object move in exactly the same way as all the original forces acting together.

- Forces acting in the same direction add, and forces acting in opposite directions subtract.

- If there is no resultant force, forces are balanced. The object does not change speed, so remains stationary.

- If there is a resultant force, stationary objects start moving and moving objects accelerate in the direction of the force.

- If the resultant force is in the opposite direction to motion, the object decelerates.

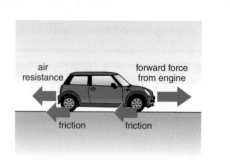

Figure 1: Unbalanced forces on an accelerating car

Using resultant forces

B–A*

- In a vehicle, fuel consumption increases if drag forces increase at a certain speed. A larger forward force is needed to overcome drag forces. More energy is needed from the engine to provide a larger forward force and keep going at the same speed.

Remember!

The resultant force when a car accelerates is positive (forwards) – drag forces are smaller than the force from the engine. There is no resultant force when the car drives at a steady speed. When a car brakes, the resultant force is negative so the car slows down.

Forces and acceleration

Investigating acceleration

D–C

- If there is a resultant force, the object accelerates in the direction of the unbalanced force.

- Acceleration is calculated using: $\text{acceleration (m/s}^2) = \dfrac{\text{force (N)}}{\text{mass (kg)}}$.

- Acceleration increases as the force applied increases.

- Acceleration decreases if the mass of the object increases.

Mass in space

B–A*

- The weight of an object is a force that depends on the mass of the object and the gravity acting on it. To find out the mass of an object in space, scientists apply a known force to it and measure its acceleration.

How Science Works

- Streamlined objects have small cross-sectional areas and smooth surfaces. This reduces drag forces so the objects can move faster for the same forward force. Many cars, boats and submarines are streamlined.

Improve your grade

Calculate the total distance travelled by the stone in the first 10 seconds in Figure 3 on page 93. **AO2 (3 marks)**

Balanced forces

Forces on moving objects

- For every force, there is an equal sized force acting in the opposite direction.

- When you sit on a chair, it pushes up on you as hard as you push down on it.

- The weight of a skydiver pulls down on a parachute. This is matched by the upwards tension in the parachute strings.

- When a gun is fired, the bullet feels a forwards force. The bullet exerts an equal sized backwards force on the gun, which recoils.

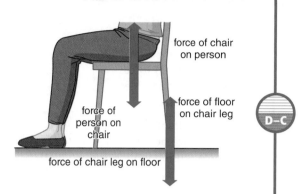

force of chair on person

force of floor on chair leg

force of person on chair

force of chair leg on floor

Figure 2: Upwards reaction forces balance the weight acting down

D–C

That lifting feeling

- When a lift accelerates upwards, the force from the floor pushing upwards is greater than the weight of the person in the lift. The person feels heavier than normal.

B–A*

Stop!

Thinking and braking

- Total stopping distance is thinking distance + braking distance.

- Reaction time is the time taken for a driver to react to a hazard.

- Thinking distance = speed × reaction time, and is the distance travelled before the driver reacts and brakes. It increases if the driver:
  - is distracted or tired
  - has taken alcohol or drugs (including some medicines).

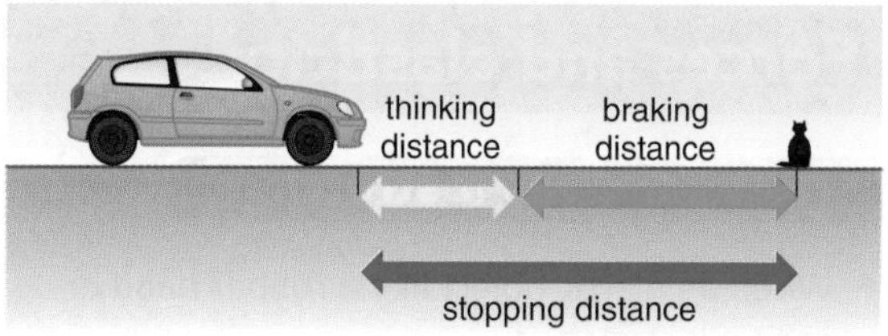

- Braking distance is the distance travelled while the brakes are applied and the car is slowing down. It increases if:
  - the road is wet or icy
  - the tyres are worn down
  - the brakes are in bad condition.

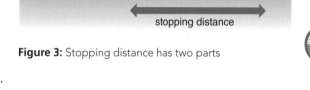

thinking distance

braking distance

stopping distance

Figure 3: Stopping distance has two parts

D–C

Aquaplaning

- Water from the surface of a wet road is channelled out through the treads on a tyre. If the tyre tread is worn, water cannot be removed so the car slides over a layer of water on the road and cannot be controlled. This is called aquaplaning.

How Science Works

- Factors that increase stopping distance also increase the risk of accidents. Drivers who fall asleep at the wheel can be prosecuted for dangerous driving.

B–A*

Improve your grade

Describe how drag forces compare with forces from the engine for a car that accelerates, then reaches a steady speed then decelerates. **AO2 (4 marks)**

Terminal velocity

Terminal velocity

- Moving objects experience drag forces in the opposite direction to motion. Drag forces increase as an object moves faster.

- The resultant force is the difference between the force causing motion and drag forces.

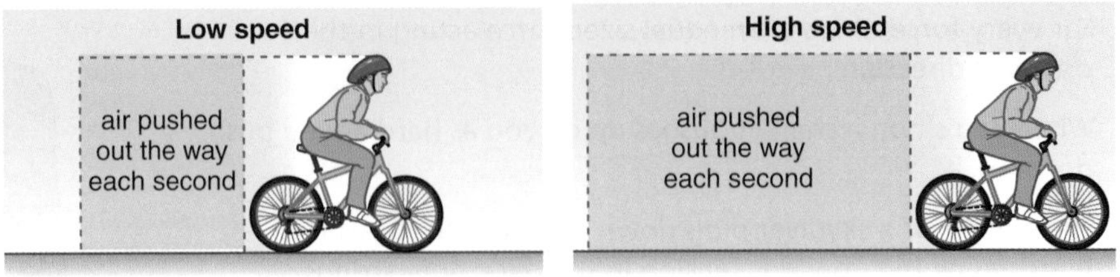

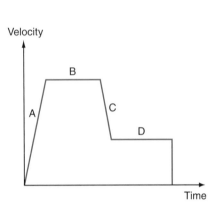

Figure 1: More air is pushed out of the way at high speed

- A skydiver feels a constant downwards force, weight. Weight is calculated using weight (N) = mass (kg) × gravity (N/kg).

- As the skydiver accelerates, drag forces increase until they match his weight and he reaches a top speed (terminal velocity).

- When the parachute is opened, drag forces increase suddenly. The resultant force is upwards so he decelerates and drag forces decrease.

- Finally he reaches a slower terminal velocity.

Remember!
Skydivers always fall downwards. When the resultant force is upwards, the skydiver decelerates.

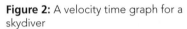

Figure 2: A velocity time graph for a skydiver

D–C

Forces and elasticity

How far does it stretch?

- When an elastic object like a rubber band or spring is stretched:
  - the extension is proportional to the force applied up to a limit
  - above the limit, extension is not proportional to the force.

Remember!
Extension is the difference between the original length and the stretched length.

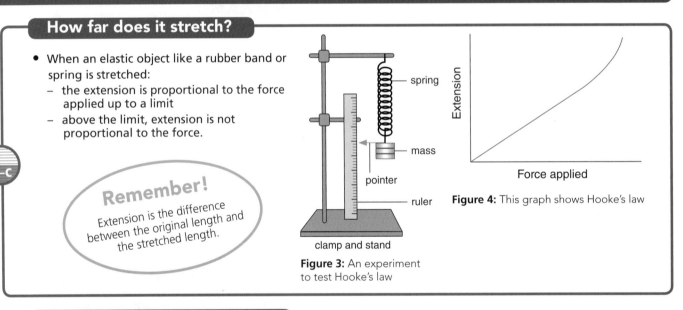

Figure 3: An experiment to test Hooke's law

Figure 4: This graph shows Hooke's law

D–C

Using elastic potential energy

- Whenever an elastic object is stretched or squashed, work is done to change its shape. The work done is stored as elastic potential energy. This energy can be usefully transferred for example in a catapult or bow, clockwork toy, or small electrical generator.

B–A*

Improve your grade

Sketch the velocity time graph for a marble that is released into a measuring cylinder of water. The marble reaches its terminal velocity. **AO2 (3 marks)**

Energy to move

Kinetic energy transfers

- A moving object has kinetic energy. This energy has been transferred:
  - from chemical energy in food a person eats
  - from chemical energy in fuel used in an engine.

- The kinetic energy from a moving object is transferred to the surroundings as a result of frictional forces. These include:
  - friction between car tyres and the road surface
  - air resistance felt by aircraft and other moving objects.

- The energy transfer can be useful.
  - Regenerative brakes slow down the car using the engine. The car's kinetic energy is used to charge the car's battery as it slows down.
  - In a car crash, car crumple zones are designed to distort, absorbing kinetic energy.

D–C

Storing kinetic energy

- Flywheels are heavy, fast-spinning wheels that store kinetic energy for short periods of time.

B–A*

Working hard

How much and how far?

- Work is done, and energy is transferred whenever a force moves.

- The amount of work done is calculated using: work done (J) = force (N) × distance moved in the direction of the force (m).

- A person lifts a 60 N parcel by 1.5 m. Work done is 60 N × 1.5 m = 90 J.

- A smaller force is needed to drag something up a ramp compared with lifting it directly. The force along the ramp is working over a longer distance, and also needs to work against friction.

Remember!
The force must move to do work. A person lifting a box does some work but when they stand still just holding a box they are not doing any work.

D–C

The clock's stopped!

- When the battery runs out, clocks stop with their moving hands in the "quarter to" position. Most work must be done to lift the hand upwards per tick in this position.

EXAM TIP

Check diagrams so you use the correct measurements for distance and force when calculating work done.

B–A*

Improve your grade

a Calculate the work done lifting a piano 0.8 m into a lorry. The piano weighs 1850 N. **AO2 (3 marks)**
b Calculate the force needed to drag the piano into the lorry if a ramp 2 m long is used. **AO2 (2 marks)**

Energy in quantity

Calculating gravitational potential energy

- Gravitational energy is energy an object has because of its position. The object gains more energy when it is lifted higher above the ground.

- Gravitational potential energy (joules) = mass (kg) × gravity (N/kg) × height (m).

Remember!
Gravity on Earth is 9.8 N/kg (usually rounded to 10 N/kg).

For example, the gravitational potential energy gained by a 3 kg rabbit lifted 0.8 m by its owner is
3 × 10 × 0.8 = 24 J

Remember!
Use the vertical height above the ground when calculating gravitational potential energy.

Investigating kinetic energy

- All moving objects have kinetic energy. Kinetic energy is calculated using:
  - Kinetic energy (J) = ½ × mass (kg) × velocity^2 (m/s)^2
  - for example the kinetic energy of a person (mass 100 kg) running at 3 m/s is ½ × 100 × 3^2 = 450 J
  - The kinetic energy of an object:
    » doubles if its mass doubles
    » increases fourfold if the speed doubles.

- At its highest point, a pendulum has gravitational potential energy and no kinetic energy. As it swings, some gravitational energy changes to kinetic energy and back again.

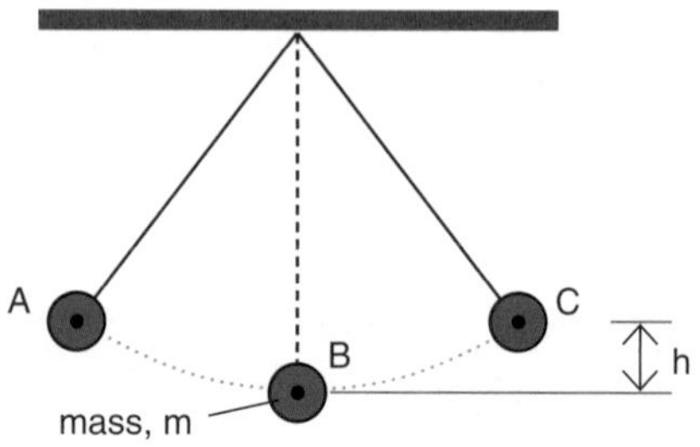

Figure 1: A swinging pendulum changes gravitational energy into kinetic energy and back again

Remember!
½ × mass × velocity^2 has a different format to many other equations. Common mistakes include forgetting to multiply by ½, and forgetting that velocity is squared.

How Science Works

- For a moving vehicle, doubling the speed quadruples the kinetic energy and also quadruples the braking distance.

Energy, work and power

Finding the power

- Power measures how quickly work is done, or energy is transferred. It is measured in watts (W) or kilowatts (kW).

- Power is calculated using: power (W) = $\dfrac{\text{energy transferred (J)}}{\text{time (s)}}$ and $\dfrac{\text{force (N)} \times \text{distance (m)}}{\text{time (s)}}$

- The power of a motor that does 300 J of work in 10 seconds is 300 J / 10 s = 30 W.

The same work but easier

- People need not pull as hard to drag an object up a ramp compared with lifting it. They only work against a component of gravity, not the direct force.

Remember!
One kilowatt is 1000 W.

Improve your grade

Calculate the power of a motor that can lift 600 N in 1 minute. **AO2 (3 marks)**

Momentum

Collisions and explosions

- All moving objects have momentum.

- Momentum (kgm/s) = mass (kg) × velocity (m/s)

- Momentum has a direction because velocity has magnitude and direction.

- Momentum is conserved. If no external forces act, momentum before and after a collision or explosion is the same.

- In a collision, find the total momentum of moving objects before the collision. The total momentum before and after the collision is the same.

- If a stationary object explodes into two parts, the momentum before and after is zero. Each part after the explosion has the same momentum in opposite directions.

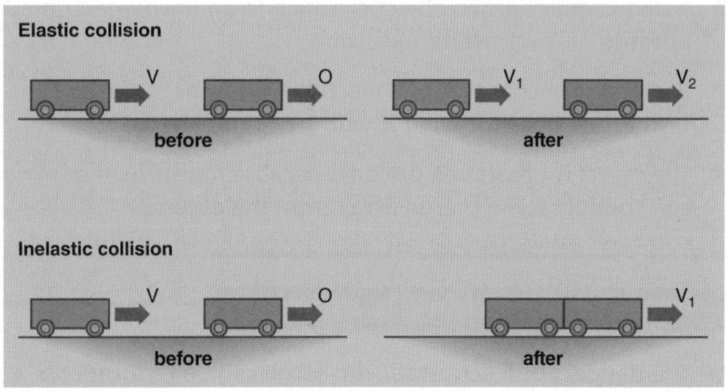

Elastic collision

before

after

Inelastic collision

before

after

Figure 2: Elastic and inelastic collisions

D–C

Remember!

You can choose which velocity direction is positive, but always use a diagram to help you see what is happening in a question.

Jet engines

- A jet plane moves because of fast moving compressed air rushing out of the back of a jet engine. The plane gains momentum, moving in the opposite direction to the air.

Figure 3: Air rushing out of the balloon makes it move

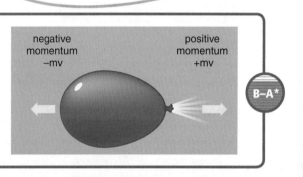

B–A*

Static electricity

Investigating charges

- Atoms contain electrons with a negative charge. When different materials are rubbed together, electrons move from one material to the other. Some materials tend to gain electrons and others lose electrons.

- Objects with the same charge repel, objects with opposite charges attract.

Remember!

Electrons are the only particles that move between different materials. Losing electrons makes a material positively charged.

D–C

Identifying unknown charges

- A gold leaf electroscope with a negative charge can be used to investigate the charge on the other objects. If a negatively charged object comes close, the gold leaf rises more. If a positively charged object comes close, the gold leaf rises less.

How Science Works

- Materials like cling film gain or lose electrons so easily that they just need to touch other materials for an electrostatic charge to build up.

B–A*

Improve your grade

Two trolleys each with a mass of 1 kg roll towards each other. One trolley travels at 2 m/s to the left and one travels at 3 m/s to the right. They collide and stick together. What is the velocity and direction of the trolleys after the collision? **AO2 (4 marks)**

Moving charges

Electrostatic induction

D–C

- A negatively charged object can stick to an uncharged object because of electrostatic induction.

- The negative charge repels electrons from the surface of the uncharged object leaving it with a positive charge.

- The effect is greater on dry days because moisture in damp air can conduct some charge away from the objects.

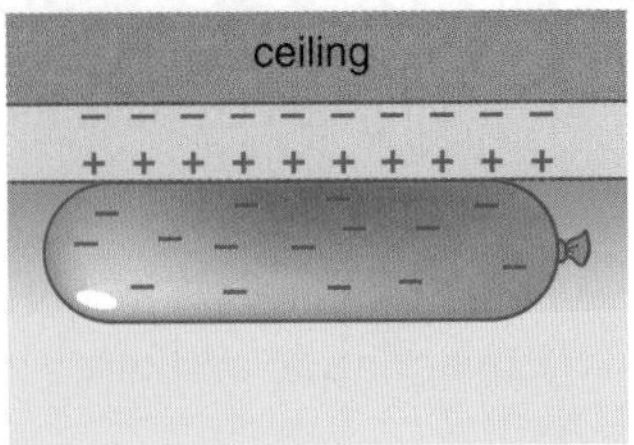

Figure 1: The balloon sticks because of electrostatic induction

Investigating kinetic energy

B–A*

- The dangers of electrostatic discharge can be reduced.

- A lightning conductor attracts lightning strikes, which travel down the metal rod to flow safely away.

Circuit diagrams

Series and parallel

D–C

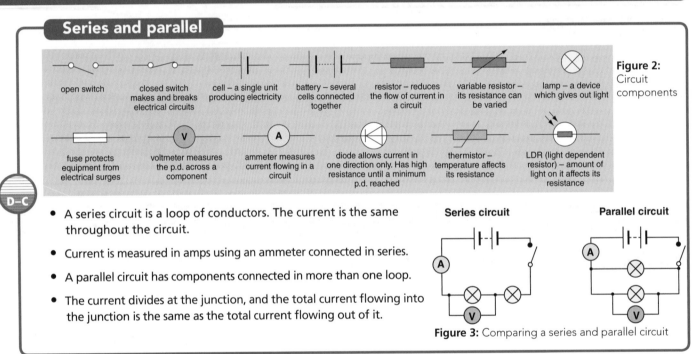

Figure 2: Circuit components

- A series circuit is a loop of conductors. The current is the same throughout the circuit.

- Current is measured in amps using an ammeter connected in series.

- A parallel circuit has components connected in more than one loop.

- The current divides at the junction, and the total current flowing into the junction is the same as the total current flowing out of it.

Figure 3: Comparing a series and parallel circuit

Supplying or using energy

B–A*

- Cells supply energy to the circuit from stored chemical energy in the battery.

- Other components transfer this energy to the surroundings.

- Potential difference is measured in volts using a voltmeter connected in parallel. It measures the energy transferred per unit of charge.

- In series circuits, the potential difference is shared between components.

- In parallel circuits, the potential difference across each loop is the same as the potential difference from the cell.

Improve your grade

Max rubbed a plastic ruler and a metal ruler with a piece of cloth.
a Which ruler became charged? AO1 **(1 mark)**
b Max held the charged ruler near a charged balloon. The balloon moved away from the ruler.
   Explain why AO2 **(2 marks)**

Ohm's law

Factors affecting resistance

- Electric current is a flow of moving charges. The charges flow more easily through materials such as copper, because copper has low resistance. If the resistance of a component or wire is high, a small current flows when a given potential difference is applied.

- Resistance is calculated in ohms (Ω) using: resistance (Ω) = voltage (V) ÷ current (A).

- A wire's resistance at a certain temperature is greater if the wire is longer or thinner. This is because electrons are more likely to collide with the nuclei in the material.

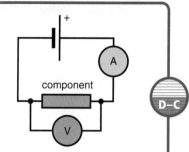

Figure 4: A circuit used to find resistance of a component

D–C

Current and potential difference for a resistor

- For some materials, voltage is proportional to current, and resistance is constant. This is called Ohm's law, and applies to ohmic materials.

- A voltage–current graph can be used to find the current in wire at a certain voltage. These values are used to calculate resistance.

> **EXAM TIP**
> You may be asked to draw the circuit used to measure resistance.

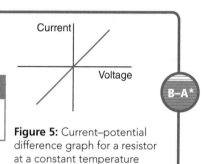

Figure 5: Current–potential difference graph for a resistor at a constant temperature

B–A*

Non-ohmic devices

Direction, light and temperature

- A filament bulb does not obey Ohm's law. As the current in a filament bulb increases, its resistance increases.

- A diode only allows current to flow in one direction (forward). The arrow on the symbol shows the forward direction. Resistance is very high in the reverse direction.

- A light dependent resistor (LDR) has a smaller resistance when light shines on it.
  - LDRs are used in automatic light systems.

- A thermistor has a smaller resistance when it is heated.
  - Thermistors are used in fire alarms and electronic thermometers.

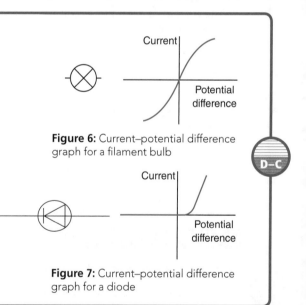

Figure 6: Current–potential difference graph for a filament bulb

Figure 7: Current–potential difference graph for a diode

D–C

How do they work?

- Diodes, LDRs, thermistors and light emitting diodes are made from semiconductor materials. Electrons gain enough energy when the material is heated or light shines on to be able to flow through it.

> **Remember!**
> Potential difference is also called voltage, and is measured in volts.

B–A*

Improve your grade

When the potential difference across a bulb is 6V, the current through it is 0.1A. What is the resistance of the bulb?
AO2 (3 marks)

Components in series

Potential difference and resistance in series

- In a series circuit the following applies:
  - the current is the same in all places
  - the total resistance of several components is each component's resistance added together
  - the supplied potential difference is shared across components in the same proportion as their resistance. A component with a high resistance has a larger potential difference than a component with a small resistance.

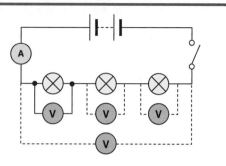

Figure 1: Potential difference in a series circuit

Cells in series

- When you add several cells together, their potential difference adds together.

- If one cell is connected the opposite way to the others, its potential difference is negative.

> **Remember!**
> Use voltage = current × resistance to find the voltage across a single component in a circuit.

Components in parallel

Potential difference and resistance in parallel

- The current in a parallel circuit has more than one path. It divides at junctions and joins up again.

- The current in each loop depends on the total resistance in that loop. A smaller current flows through loops with a high resistance compared to loops with a small resistance.

- The total current leaving and returning to the battery is the sum of currents in each loop.

- All bulbs stay as bright as if one bulb was in the circuit.

- The potential difference across each component in a parallel circuit is the same.

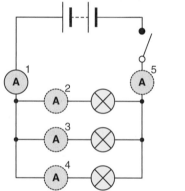

Figure 2: Current in a parallel circuit

Series and parallel

In a loop in a parallel circuit with more than one component:

- the potential difference is shared across the components

- the current is the same through each component.

EXAM TIP
You must be able to use the rules to calculate current and voltage in both series and parallel circuits.

Improve your grade

A series circuit is set up, using four 1.5 V cells in series.
a   What is the total voltage supplied to the circuit?
The circuit includes a motor (resistance 10 ohms) and a bulb (resistance 5 ohms).
b   What is the total resistance in the circuit?
c   Calculate the current in the circuit.
d   What is the potential difference across each component? **AO2 (5 marks)**

Household electricity

Using a cathode ray oscilloscope

- Cells and batteries produce direct current (d.c.), which flows in one direction.

- Alternating current (a.c.) repeatedly changes direction. Mains electricity changes direction 50 times per second (its frequency is 50 hertz) and is about 230 V.

- An oscilloscope can display a.c. or d.c. voltage. Dials set the number of volts per vertical square, and the time per horizontal square.

- The amplitude (maximum height) of the trace shows the maximum potential difference supplied (in volts).

- The number of squares between the peaks shows time per cycle (in seconds).

Figure 3: An oscilloscope trace of a.c. current

D–C

Calculating period and frequency

- The frequency of the supplied voltage is calculated using:
  frequency (Hertz) = 1/time per cycle (seconds).

  For example, if the frequency is 50 Hz, the time per cycle is $\frac{1}{50}$ = 0.02 s.

Remember!
Amplitude is measured from the centre of a trace to the highest (or lowest) point.

B–A*

Plugs and cables

Cables and fuses

- In a three-pin plug, wires are covered in colour-coded plastic:
  - the neutral wire is blue
  - the live wire is brown
  - the earth wire is green and yellow.

- Appliances with a plastic outer case and no touchable metal parts are double insulated. They use two-core cable (live and neutral wires).

- Other appliances use three-core wire (earth, live and neutral wires). If there is a fault and the equipment becomes live, the current flows through the earth wire and blows the fuse.

- The fuse is connected in series with the live wire. It protects the appliance and flex from overheating. If the current is too large, the fuse melts and breaks the circuit.

Figure 4: Inside a three pin plug

D–C

EXAM TIP

Make sure you can describe dangerous habits when using electricity and spot mistakes when wiring plugs.

Figure 5: Three-core cable

Shocking

- Electric shocks can kill. Wet skin has lower resistance than dry skin so electrocution is more likely.

- Mistakes when wiring a plug include:
  - connecting wires to the wrong pins
  - not gripping the cable tightly under the cable grip
  - not attaching wires firmly to the pins
  - having too much bare wire exposed.

EXAM TIP

You should be able to explain why certain materials are used for different parts of a plug and cable.

B–A*

Improve your grade

Explain why a double insulated appliance does not need an earth wire. **AO1 (2 marks)**

Electrical safety

Residual current circuit breakers (RCCBs)

- RCCBs protect users from electrocution. The current in the live and neutral wires should always be the same. Residual current circuit breakers (RCCBs) break the circuit in less than 0.05 seconds if there is a difference in the current in these wires. The RCCB can be reset once the fault is repaired.

- Many homes now have RCCBs in them.

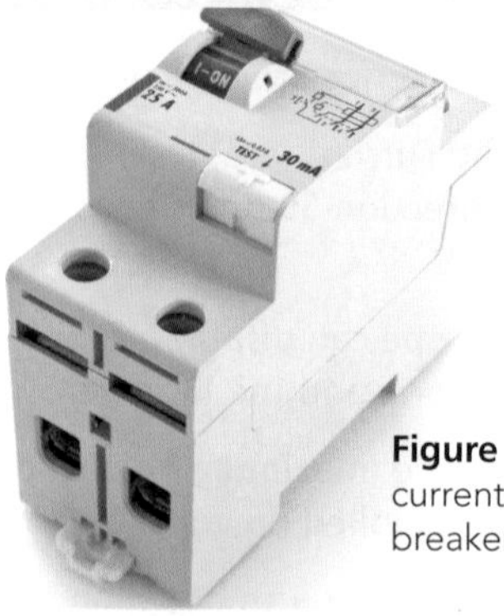

Figure 1: Residual current circuit breaker

Choosing the right fuse

- Fuses are rated in amps, for example 3 A, 13 A.

- If the fuse rating is too low, the fuse melts even if the appliance works correctly. If it is too high, it won't melt if there is a fault.

- Use current (A) = $\dfrac{\text{power (W)}}{\text{voltage (V)}}$ to calculate which fuse to use.

How Science Works

- Mains voltage is supplied at 230 V so a 3 A fuse only works if the power of the equipment is 690 W or less. Otherwise, a 13 A fuse is needed.

Current, charge and power

Current charge and work done

- An electric current transfers energy.

- The energy transferred is calculated using: energy transferred (J) = power (W) × time (s).

- The electric current is a flow of charge (electrons). Charge is measured in coulombs, and has the symbol Q.

- Current is the rate that charge flows. It is calculated using: current (A) = charge (C)/time (s).

Calculating the energy transferred

- A battery supplies charge with energy. One volt supplies each coulomb of charge with one joule of energy.

- Calculate the energy supplied using: energy (J) = potential difference (V) × charge (C)

Remember!
You may be asked to rearrange equations. Practise doing this before the exam.

EXAM TIP

When choosing the equation to use, check what you are asked for and what you are given in the question.

Improve your grade

Fuses rated at 3 A, 5 A and 13 A are available. Which fuse should be used in each case?
a   a lawn mower (power 2000 W)
b   a lamp (power 60 W)
c   a toaster (power 800 W) **AO2 (3 marks)**

Structure of atoms

Electrons, protons and neutrons

- Atoms are made from particles called electrons, protons and neutrons, which have different masses and charges. The atomic number is the number of protons in the atom. The mass number is the number of neutrons and protons.

Particle	Charge	Relative mass
neutron	0	1
proton	1	1
electron	−1	negligible

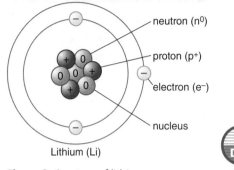

Lithium (Li)

Figure 2: An atom of lithium

- Atoms have no overall charge as they have equal numbers of electrons and protons. An atom that loses electrons becomes a positive ion. An atom that gains electrons becomes a negative ion.

- All atoms of the same element have the same number of protons, but can have different numbers of neutrons. Atoms of the same element with different numbers of neutrons are called isotopes.

- Unstable atoms have different numbers of protons and neutrons. Some will decay emitting ionising radiation from the nucleus.

D–C

Subatomic particles

- The model of the atoms is evolving. Scientists are looking for experimental evidence of unknown particles, and the forces that hold them together.

Remember!
There are a lot of new terms in this topic. Learn their definitions carefully.

B–A*

Radioactivity

Explaining the properties

- Radioactive atoms emit ionising radiation from the nucleus. Atoms decay at random times. An atom that absorbs ionising radiation may change into an ion. Types of ionising radiation include:
  - Alpha particles which consist of 2 neutrons and 2 protons (a helium nucleus).
  - Beta particles which are electrons emitted when a neutron changes to a proton.
  - Gamma radiation which is high energy electromagnetic radiation.

D–C

Nuclear equations

- The mass number is the number of protons and neutrons in an atom. The proton number is the number of protons in an atom.

- Nuclear equations show how isotopes decay by emitting alpha and beta radiation. The mass number and the proton number on each side of the equation must balance.

- This equation shows alpha decay

$$^{224}_{88}\text{Ra} \rightarrow {}^{220}_{86}\text{Rn} + {}^{4}_{2}\text{He}$$

- This equation shows beta decay

$$^{212}_{83}\text{Bi} \rightarrow {}^{212}_{84}\text{Po} + {}^{0}_{-1}\text{e}$$

B–A*

EXAM TIP

Make sure you know the properties of nuclear radiation and can use them to explain how the radiation affects nearby atoms and molecules.

Improve your grade

Why are the risks from handling an alpha source safely different from the risks from handling a gamma source?
AO3 (3 marks)

More about nuclear radiation

Explaining the scattering experiment

- The plum pudding model was an early model of the atom. It was not a successful model so a scientist called Rutherford carried out an experiment to explain the structure of an atom.

- He fired alpha particles at gold foil and observed the scattering of the particles.

- Most alpha particles passed straight through as atoms were mainly empty space.

- Positively charged alpha particles were deflected by positive charge concentrated in the nucleus.

- A few massive alpha particles were reflected by the very dense nucleus.

- The pattern helped to explain the structure of atoms. Rutherford's nuclear model replaced the plum pudding model.

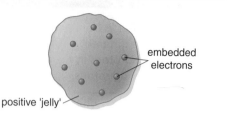

Figure 1: The plum pudding model.

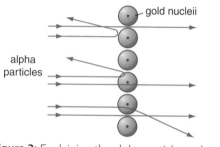

Figure 2: Explaining the alpha particle tracks.

More deflections

- Alpha and beta particles have opposite charges so they are deflected in opposite directions by electric and magnetic fields.

- Alpha particles are more massive than beta particles so they are deflected less.

Background radiation

Natural and lifestyle

- Background radiation is radiation that is all around us and comes mainly from natural sources.

- Natural sources include rocks, soil and food.

- Sources are affected by lifestyle choices, for example, on a long flight you increase your exposure to cosmic radiation.

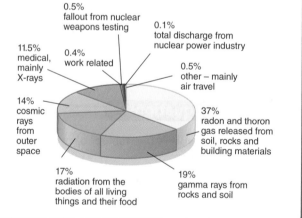

Figure 3: Sources of background radiation.

Dangerous levels – or not?

- It is hard to assess the risks from background radiation as many different factors affect your lifetime cancer risk.

How Science Works

- Radon gas is radioactive and seeps from rocks, collecting in buildings. The increased risk of lung cancer from radon varies in the UK depending on rock types and exposure to them. There is a very high risk of contracting lung cancer if you smoke.

Half-life

Measuring half-life

- We cannot predict if a specific radioactive atom will decay, but we can predict how many atoms will decay in a sample in a given time.

- Half-life is the time taken for the original radioactivity or count rate of a sample to halve.

- After a time equal to two half lives, the activity falls to a quarter of its original value.

- Each radioactive material has its own half-life, which varies from millions of years to milliseconds. The half-life does not change for a given material.

D–C

Graphs to identity radioactive isotopes

- The graph shows how the activity of a sample varies with time.

To find the half-life;

- work out the value of half the original count rate

- use the graph to read the time taken for the activity to reach this level.

Remember!
Whenever you start counting and whatever the original count rate, the count rate or activity falls to a half of the original value after one half-life.

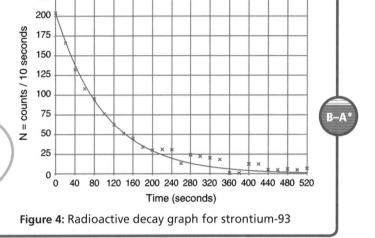

Figure 4: Radioactive decay graph for strontium-93

B–A*

Using nuclear radiation

Using radioisotopes

- Radioisotopes are radioactive isotopes. They have different uses.

- Gamma rays kill living cells and penetrate materials so are used for sterilisation and cancer treatments.

- Beta radiation penetrates thin sheets of metal and cardboard so is used to monitor the thickness of materials.

- Medical tracers are radioactive materials injected, inhaled, or eaten by a patient. The position of the radioactivity is monitored to check blood flow, or identify blockages or tumours.

- Alpha radiation is used in smoke detectors. The particles ionise air so a current can flow. If smoke absorbs the radiation, the current cannot flow switching the alarm on.

Remember!
Radioactive isotopes with a short half-life are used with living organisms to reduce harm. In monitoring equipment, isotopes with a long half-life are used to maintain consistent readings.

D–C

Plant growth

- Radioactive water emits beta radiation so it can be used to monitor plant growth.

B–A*

Improve your grade

The half-life of one material is 20 days, and its original activity is 100 000 counts per minute. Another sample has a half-life of 15 days, and an original count rate of 200 000 counts per minute. What is the count rate of each sample after 60 days? **AO2 (3 marks)**

Nuclear fission

Critical mass and chain reaction

Nuclear fission is when a large nucleus splits into two or more parts. In nuclear power stations:

- uranium-235 and plutionium-239 are used as fuels,

- atoms of uranium or plutonium absorb an extra neutron and become unstable,

- energy is released when the nucleus splits into two parts plus two or three neutrons.

- more uranium or plutonium atoms may absorb the extra neutrons. This can start a chain reaction involving more atoms at each stage.

- Control rods in nuclear power stations control the size of the chain reaction by absorbing surplus neutrons.

- The critical mass of a nuclear fuel is the minimum amount needed to keep the chain reaction going.

D–C

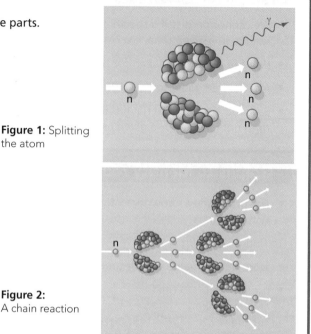

Figure 1: Splitting the atom

Figure 2: A chain reaction

Why is so much energy released?

B–A*

- Strong forces in the nucleus mean that large amounts of energy are released when the atom is split. Some of the mass of the nucleus is changed into energy during nuclear fission.

> **EXAM TIP**
>
> You should be able to describe the stages taking place during nuclear fission.

Nuclear fusion

Using nuclear fusion

D–C

- Nuclear fusion is when two small nuclei join together forming a nucleus of a different element and releasing energy. This is how energy is released in stars.

- Very high temperatures (above 15 million degrees Celsius) and very high pressures are needed for nuclear fusion because positively charged protons in nuclei repel each other.

- Nuclear fusion in stars produces elements lighter than iron.

- Elements heavier than iron are produced during supernova (when massive stars explode releasing huge amounts of energy).

The proton-proton chain

B–A*

- Nuclear fusion of hydrogen takes place in three stages in stars.

> **EXAM TIP**
>
> Be very careful not to confuse nuclear fission and nuclear fusion, and learn the correct spellings of each.

Improve your grade

Explain why the presence of iron in the Sun is evidence that the Sun formed from the remains of older stars.
AO2 (3 marks)

Life cycle of stars

Why does the main sequence last so long?

- All stars have a life cycle. When gravity pulls dust and gas together, a protostar forms. Smaller masses (planets) may also form. The life cycle of a star is determined by its size.

- The star's main sequence lasts billions of years. Nuclear fusion changes hydrogen to helium, and heavier nuclei. The star is stable because forces are balanced within the star (inward gravitational forces match outward forces due to the heat).

- As hydrogen fuel is used up, nuclear fusion reactions change. The star expands becoming a red giant.

- When nuclear fusion reactions change again, the inner core collapses. The star becomes a white dwarf. Nuclear reactions form elements up to iron.

- When its nuclear fuel runs out, the star cools and becomes a black dwarf.

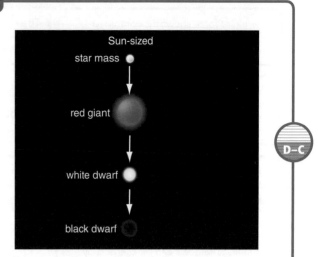

Figure 3: The life cycle of a star like the Sun

D–C

A matter of size

- Stars much bigger than the Sun become a red super giant.

- When its fuel is used up, the core collapses quickly and the star explodes as a supernova. Elements heavier than iron form in a supernova.

- After the supernova, stars up to 3 times heavier than the sun become neutron stars. Heavier stars create a black hole.

> **Remember!**
> You should be able to explain how changes in the forces acting on a star make it move between different stages.

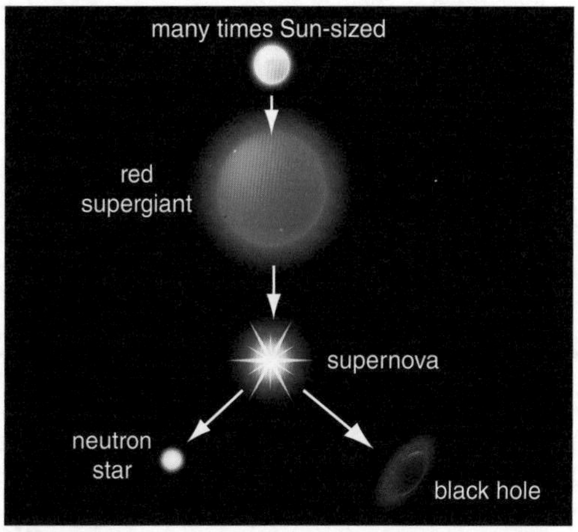

Figure 4: The life cycle of a star much bigger than the Sun

B–A*

Improve your grade

Explain how the stages in a star's life cycle are controlled by its mass. **AO2 (3 marks)**

P2 Summary

A resultant force shows the combined effect of several forces. It makes an object accelerate (change speed or direction).

The stopping distance of a vehicle is thinking distance plus braking distance.

Moving objects reach a terminal velocity (top speed) when drag forces matches driving forces.

The speed of an object can be found using a distance time graph. The acceleration and distance traveled can be found using a velocity time graph.

Forces and energy

Momentum is mass × velocity. It is conserved when objects collide or explode apart if no external forces act.

A force transfers energy when it does some work on an object.

Objects gain gravitational potential energy if they are lifted up. Moving objects have kinetic energy. Stretched or squashed elastic objects store elastic energy.

Insulating materials can be charged by rubbing. They gain or lose electrons.

Static electricity

Objects carrying the same charge repel. Objects carrying the opposite charge attract.

Electric current is a flow of charge.

Electrical circuits

Resistance measures how easily a current flows through a component.
Resistance = potential difference ÷ current

In a series circuit: the current is the same throughout, the potential difference is shared between components and the resistance of components adds.

Circuit diagrams use standard symbols.

In a parallel circuit, the current is shared between branches and the potential difference is the same for each branch.

Batteries supply direct current (d.c.); mains supply is alternating current (a.c.) at about 230V 50Hz.

Using mains electricity safely and the power of electrical appliances

Earthing, fuses and RCCBs protect the user and appliance.

Electric cable can be two-core or three-core. A three pin plug must be correctly wired.

Power is the potential difference supplied × current, or energy transferred ÷ time.

Energy transferred is the potential difference supplied × charge.

An atomic nucleus contains protons and neutrons. Electrons orbit the nucleus. Radioactive materials emit ionising radiation from their nucleus.

Atomic structure and radioactivity

Background radiation comes from natural sources (for example rocks,) and man-made sources (for example medical uses).

Half-life is the time taken for the count rate of a radioactive sample to halve.

Alpha, beta and gamma radiation have different properties, uses and dangers.

Ions are atoms that have lost or gained electrons. Isotopes have the same number of protons, but different numbers of neutrons.

Nuclear fission is the splitting of an atomic nucleus.

Nuclear fission and nuclear fusion

Nuclear fusion is when two atomic nuclei join to form a larger one.

Fission of uranium-235 or plutonium-239 during a chain reaction releases energy in nuclear reactors.

All stars have a life cycle. Nuclear fusion in stars releases energy and produces all naturally occurring elements.

Improve your grade

B1 Improve your grade

Page 6

Harry's diet is very high in saturated fats. Suggest two ways that this could affect his health. **AO2 (3 marks)**

It will make him put on weight.

This answer got 1 mark out of 3 (grade EF). To improve the answer, the candidate should have stated another way that Harry's diet could affect his health. A diet high in saturated fat will increase the risk of developing heart disease. Also, it is scientifically correct to use the term 'mass' instead of weight in this case.

Page 7

Use the graph to explain the change in the number of pathogens. **AO2 (4 marks)**

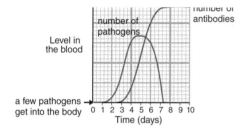

The number of pathogens started to increase, then after 5 days they started to decrease. This happened because the number of antibodies had increased which started to destroy the pathogens.

This answer got 3 marks out of 4 (grade B). The candidate described the rise in the number of pathogens but did not explain why this happened (the pathogens would start to reproduce). Including the time when the numbers of pathogens went down was good.

Page 8

Explain why a doctor will not prescribe you antibiotics for a bacterial throat infection. **AO2 (3 marks)**

Because we need to stop using so many antibiotics.

This answer achieved 1 mark out of a possible 3 (grade CD), as it stated a fact but did not go on to explain why doctors should not prescribe antibiotics for this infection. A simple throat infection, although uncomfortable, will soon be cleared up by the person's own white blood cells. Using antibiotics gives an advantage to any antibiotic-resistant bacteria that may have formed in the throat by only killing off the non-resistant bacteria and allowing the resistant ones to reproduce.

Page 9

You wish to grow some harmless E. coli bacteria on some agar jelly. You use an inoculating loop to transfer the bacteria to the jelly. Explain why it is important to hold the inoculating loop in a flame first. **AO1 (2 marks)**

To make it sterile

This answer gets 1 of the marks (grade D) as the candidate forgot to mention why making the loop sterile is important. It is to prevent the agar becoming contaminated with any other type of microorganism.

Page 10

An injury that results in breaking of the spine may result in the person being paralysed. Explain why. **AO2 (3 marks)**

The nerves in the spine might be broken.

This answer only makes one suggestion and thus only achieves one of the three possible marks (grade EF). To gain full marks, explain that the spinal cord is part of the central nervous system and contains many neurones that send messages from the brain to all parts of the body. When these impulses meet the muscles, they contract. If the nerves in the spinal cord are broken then impulses cannot cross them, so muscles will not respond.

Page 11

James is dancing in a nightclub. He starts to sweat. Explain how sweating helps to cool him down. **AO2 (2 marks)**

The sweat evaporates from his skin. It needs energy to change from a liquid into a gas and so uses the heat from the skin as a source of the energy. This cools down the skin and reduces his body temperature.

This answer gets full marks and so achieves an A* grade. The candidate has shown that they have a deep understanding of the science by adding the information about why the skin needs heat to evaporate.

Page 12

FSH is found in fertility drugs. Explain how taking FSH will increase a woman's fertility. **AO2 (2 marks)**

It will increase the amount of eggs that mature in her ovaries.

This answer achieves one of the two possible marks (grade D). The candidate should have continued to explain that increasing the amount of mature eggs means that more than one will be released at a time, which will increase the chance of one being fertilised.

Page 13

Explain how the hormone auxin brings about a response to light called phototropism. **AO2 (4 marks)**

Light shines on the plant from one direction. There is more auxin on the shaded side so there are more cells here. This bends the shoot towards the light.

This answer gets two marks (grade D). The candidate has misunderstood how auxins affect plant growth. They do not increase the number of cells but increase the length of them.

Page 14

Some doctors believe that everyone over the age of 50 should be prescribed statins on the NHS for free. Evaluate this decision. **AO2 (4 marks)**

It is a good decision because trials have shown that statins reduce blood cholesterol levels, so it will reduce their risk of dying from a heart attack.

The candidate has achieved two marks out of a possible four (grade C). For a higher grade, they need to give any disadvantages of this decision. For example, some people over 50 do not have high cholesterol so do not need statins. For them, they would be taking a drug that they do not need which could waste the NHS's money. Also, there is now evidence that statins have side-effects.

Page 15

Explain how cacti are adapted to living in the dry desert. **AO2 (3 marks)**

Cacti have long roots to grow deep into the soil and spread out as well as to absorb maximum amounts of water. They can also store water in its tissues to use when they can't absorb any water from the ground. They have no leaves, which reduces water lost by evaporation. Their spikes stop animals from eating them.

This answer gets full marks (grade A*). The candidate has described all of its features and explained how they enable the cactus to have a supply of water at all times, which is the main problem for a plant living in the desert. They have shown extra knowledge by explaining the function of the spikes.

Page 16

Sewage contains a lot of microorganisms. Explain why mayfly larva cannot live in water polluted with sewage. **AO2 (2 marks)**

The microorganisms are pathogens and kill the mayfly larva.

The candidate has not achieved any marks for this question as they have not understood how microorganisms will kill the larva. Increasing the amount of microorganisms in the water will decrease the amount of dissolved oxygen. The oxygen-loving mayfly larvae would not be able to survive in water with low levels of dissolved oxygen.

Page 17

Explain why biomass decreases as you move along a food chain. **AO2 (3 marks)**

The amount of energy decreases as you go along a food chain. Less energy means less biomass.

This answer gets two marks (grade B). The candidate should have gone on to explain why the energy decreases. A suitable reason might be that energy is wasted as heat, and in waste such as urine.

Page 18

Write a simple plan for an investigation to prove that the decay of bread by mould requires moisture. **AO2 (2 marks)**

Take 2 slices of bread. Add a little water to one. Place both in a plastic bag and leave in a warm place.

This answer has achieved 3 marks out of 4 (grade C). The candidate has not explained how they would compare the decay of each one. They could just look at the growth of mould on each slice. An even better method is to measure the area of each slice that is covered with mould.

Page 19

Emma and Alice are identical twins. Emma has blonde hair and Alice has brown hair. Is this variation due to genetics or the environment? Explain your answer. **AO2 (2 marks)**

It is due to the environment because they are identical twins so have the same genes.

This answer gets full marks (grade A). To further show knowledge, the candidate could have explained why they have the same genes: identical twins are formed when a fertilised egg splits in half.

Page 20

Buttercup the cow produces the most milk in her herd. Her farmer is considering using her eggs for embryo transplants. Explain to him why cloning her might be an even better idea. **AO2 (3 marks)**

If you clone her then you will get another cow exactly like her who will produce lots of milk.

This answer gained two marks out of three (grade B). To get full marks, the answer should explain why in this case cloning is better than embryo transplants. Embryo transplants involve using the eggs from the animal with desired characteristics and fertilising them with sperm from a male animal. The resulting embryos will have a mix of genes from both parents, and not necessarily the desired ones that you were trying to achieve. Cloning animals will produce offspring that are identical to the parent.

Page 21

In 1859, Charles Darwin published a book containing his ideas about natural selection and evolution. Explain why many people at the time did not believe what it said. **AO2 (3 marks)**

Most people thought that God had created all the living things on Earth in the form that they are in now.

The candidate correctly identified one of the reasons why people did not accept Darwin's theory, so achieves one mark out of three (grade D). They should also have explained that nobody understood how characteristics of the parent could be passed to its offspring, as genes had not yet been discovered. Also, at that time, there was not much scientific evidence to support the theory.

Page 22

The cheetah is the fastest land animal on Earth. It uses its speed to catch its prey. Use natural selection to explain how cheetahs have got faster over time. **AO2 (5 marks)**

The cheetahs that were the quickest caught prey and survived.

This answer gained two marks out of five (grade ED). The candidate should have put much more information into their answer. When tackling questions about natural selection, apply these stages: Some cheetahs were faster than others (variation). The fastest cheetahs caught more prey than the slower ones (competition). They survived (survival) to reproduce and pass on the genes for speed onto their offspring (reproduction). If this keeps happening, then over many generations the cheetah will get faster.

C1 Improve your grade

Page 24

Draw the electronic structures of the following atoms with proton numbers 3, 9, 11, 16, and 20.
AO2 (5 Marks)

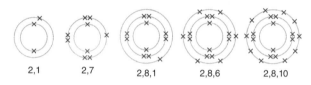

| 2,1 | 2,7 | 2,8,1 | 2,8,6 | 2,8,10 |

The answer gets 4 marks (grade C) as the first four are correct. In the last one, where the proton number is 20, the answer has the last 10 electrons in the third orbit rather than six in the third and two in the fourth orbits. For full marks the structure should be 2,8,8,2.

Page 25

Explain why oxygen has a molecule with a double covalent bond and fluorine has only a single covalent bond. You should refer to the number of outer electrons in both atoms in your answer. **AO2 (4 Marks)**

Fluorine has electronic structure 2,7, so needs one electron to fill the shell. So it makes a single bond. Oxygen is 2,6, and needs two electrons to fill the shell, so it makes two bonds.

This answer gets 2 marks (grade D). It correctly gives the electronic structure of the two atoms, but fails to mention sharing electrons with the other atom, or the need to achieve a Noble gas electronic structure.

Page 26

Balance these two chemical equations:

a $Mg(s) + O_2(g) \rightarrow MgO(s)$
b $C_3H_8(g) + O_2(g) \rightarrow CO_2(g) + H_2O(\ell)$
AO2 (2 Marks)

a $2Mg(s) + O_2(g) \rightarrow 2MgO(s)$
b $C_3H_8(g) + O_2(g) \rightarrow 3CO_2(g) + 4H_2O(\ell)$

This answer gets 1 mark (grade C). Part a) is correctly balanced, but in part b) the hydrogen and carbons are correct, but the oxygens have not been adjusted. It should read $5O_2$.

Page 27

Draw a table to show which of these metals is extracted by carbon and which by electrolysis. AO1 (2 Marks)

iron, magnesium, aluminium, copper, zinc, lead, potassium, and calcium

carbon	electrolysis
iron	magnesium
zinc	aluminium
lead	potassium
copper	calcium

This answer gets 2 marks (grade C). All the metals are correctly placed. You should remember that copper is extracted by carbon, but is often purified by electrolysis.

Page 28

Explain the difference between oxidation and reduction. **AO1 (2 Marks)**

Oxidation is gaining oxygen and reduction is the loss of oxygen.

This answer gets 2 marks (grade C). Oxidation and reduction are the opposite of each other. In a reaction, if one substance is oxidised the other is reduced.

Page 29

Explain why the waste heaps of an old copper mine would be a suitable site for phytomining or bioleaching. **AO2 (2 Marks)**

Copper mines produce waste which still has copper ore in it.

This answer gets 1 mark (grade C). It recognises that the waste heaps contain copper ore, but fails to explain that phytomining or bioleaching would be a cheap and effective way of obtaining the copper in the waste.

Page 30

Explain why titanium and aluminium metals are both reactive and corrosion free. **AO1 (2 Marks)**

Both metals get coated with an oxide layer very quickly in the air. This layer seals the surface, and prevents further corrosion. The layer is only two or three molecules thick.

This answer gets 2 marks (grade B). It clearly makes the two points needed, that an oxide layer is formed, and that this prevents further corrosion.

Page 31

Suggest why using biofuels may create more problems than it solves. **AO2 (3 Marks)**

Biofuels are grown on land that can be used to grow food. This may reduce the amount of food the world produces, and lead to starvation. It may also cause food prices to rise beyond what people can afford.

This answer gets 3 marks (grade A*). It clearly makes the point about land use, food costs, and food production. It doesn't mention that rainforests may be felled in order to grow profitable crops to make biofuel from.

Page 32

Draw out the structure of the straight chained alkanes with 5 carbon atoms and 7 carbon atoms. **AO2 (2 Marks)**

$$H-\underset{\underset{H}{|}}{\overset{\overset{H}{|}}{C}}-\underset{\underset{H}{|}}{\overset{\overset{H}{|}}{C}}-\underset{\underset{H}{|}}{\overset{\overset{H}{|}}{C}}-\underset{\underset{H}{|}}{\overset{\overset{H}{|}}{C}}-\underset{\underset{H}{|}}{\overset{\overset{H}{|}}{C}}-H$$

$$H-\underset{\underset{H}{|}}{\overset{\overset{H}{|}}{C}}-\underset{\underset{H}{|}}{\overset{\overset{H}{|}}{C}}-\underset{\underset{H}{|}}{\overset{\overset{H}{|}}{C}}-\underset{\underset{H}{|}}{\overset{\overset{H}{|}}{C}}-\underset{\underset{H}{|}}{\overset{\overset{H}{|}}{C}}-\underset{\underset{H}{|}}{\overset{\overset{H}{|}}{C}}-H$$

This answer gets 1 mark (grade C). The second structure should have seven carbon atoms, not six. Check your diagrams carefully.

Page 33

Draw structural diagrams to show how $C_{12}H_{26}$ can be converted to C_3H_6, and another molecule. State which of the two products is unsaturated, and why. **AO1 and 2 (3 Marks)**

It is the C_3H_6 that is unsaturated because it has a double bond.

This answer gets 2 marks (grade B). The diagram for the conversion is good, but on the C_3H_6 diagram there is a missing hydrogen on the left. Always count that each carbon atom has 4 bonds. The answer recognises propene as the unsaturated molecule and correctly gives the reason.

Page 34

PVC (polyvinylchloride) is a commonly used polymer. Its monomer has the structure:

Show, using three monomer molecules, the structure and bonding of a PVC polymer chain. **AO2 (2 Marks)**

Each of the lines represents a single covalent bond or a pair of shared electrons.

This answer gets 2 marks (grade B). The diagram shows the structure and bonding correctly. The additional statement makes it clear that the candidate knows that this molecule has covalent bonding, and what covalent bonding is.

Page 35

Describe the difference between recycling polymer waste and re-using it by incineration. Give two reasons why recycling is more environmentally friendly than incineration. **AO1 (3 Marks)**

Recycling polymer waste means re-using it as a polymer after some processing. Re-using it by incineration is when the polymer is burnt. The polymer is changed to heat and carbon dioxide.

This answer gets 1 mark (grade C). There is a clear description of the two processes but the second part of the question is ignored in the answer. Recycling is less damaging as it needs less energy for recycling than for making new polymers, saves on using fossil fuels, and stops polluting gases caused by incineration; it also reduces waste sent to landfill.

Page 36

Explain why biofuels are considered to be carbon neutral. **AO1 (2 Marks)**

Biofuels are carbon neutral because they only put into the atmosphere the same carbon dioxide as they took out by growing.

This answer gets 2 marks (grade C). The answer scores full marks, but could be improved by adding that photosynthesis takes the carbon dioxide out of the air, rather than using the word 'growing'.

Page 37

Draw a diagram to show how an emulsifier would surround a water droplet to make a water in oil emulsion. **AO2 (2 Marks)**

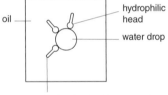

This answer gets 2 marks (grade A/B). The diagram is clear and shows both hydrophobic and hydrophilic ends of the emulsifier. It clearly shows that it is a drop of water in oil, and not the other way round. It doesn't label the emulsifier.

Page 38

Explain how the Atlantic Ocean is getting wider each year. Suggest another part of the world where a similar process may be taking place. **AO1 (3 Marks)**

America is drifting away from Britain by convection currents in the sea. This is also happening with Europe.

This answer gets 0 marks (grade D). The candidate knows about convection currents, but has placed them in the sea, rather than in the mantle. The increasing gap is caused by the tectonic plates drifting on the mantle-driven convection currents and, where the two plates are moving apart, fresh rock is being formed. The middle of the Pacific Ocean is another place where this is happening.

Page 39

Sketch a timeline showing how the proportions of water vapour, carbon dioxide, oxygen, and nitrogen have changed since the Earth was formed. **AO2 (4 Marks)**

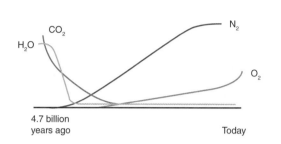

This answer gets 4 marks (grade A). It clearly shows how each of the four gases' concentration in the air has changed over time. It would have been helpful, but not necessary, to add in when life began (3.5 billion years ago), and also when life began on the land.

Page 40

Explain why the Miller–Urey experiment does not prove how life evolved. **AO3 (2 Marks)**

It shows how amino acids, which are the building blocks of life, were formed.

This answer gets 1 mark (grade C). The answer is not an explanation, it simply states the outcome of the experiment. It should have mentioned that amino acid formation was necessary for life to develop, but that this is only a theory, as none of the amino acids formed proteins or were alive at all.

P1 Improve your grade

Page 42

Explain whether a kettle of hot water cools down quicker if its outer surface is coloured white or dark green. **AO2 (3 marks)**

It cools quicker if it is dark green more heat is lost by radiation. (2)

This answer gets 2 out of 3 marks (grade B). The candidate correctly said that heat loss by radiation is affected by the colour of the kettle. For full marks, explain that the heat is lost more quickly from dark surfaces.

Page 43

Explain why convection can take place in a liquid but not in a solid. **AO2 (3 marks)**

In a solid particles are in fixed positions. During convection heat is carried by particles that move.

This answer gets 2 out of 3 marks (grade B). The candidate correctly described the arrangement of particles in a solid, but could also have said that in a liquid they are free to change places. For full marks, explain that particles move from a hotter to a cooler place.

Page 44

Explain why a towel dries quicker on a windy summer day. **AO2 (3 marks)**

Water evaporates from the towel quicker. (1)

This answer gets 1 out of 3 marks (grade C/D). The candidate did not describe how the wind or sun increases evaporation. For full marks, either say wind moves air away from the towel so the air just above the towel does not become saturated or that on a summer's day the air is warmer so particles have more energy so can break bonds more easily.

Page 45

The specific heat capacity of copper is 390 J/Kg °C. Explain whether copper heats up quicker than the same mass of water when they are put in a hot place. The specific heat capacity of water is 4200 J/ Kg °C. **AO2 (3 marks)**

Copper gets hotter than the water.

This answer gets 0 out of 3 marks (grade E/F). Copper heats up quicker but if the water and copper are left for long enough, their final temperature is the same. Remember to include 3 points for a 3 mark question, and to use the information given. For full marks, explain that copper needs less energy than water to heat up by 1 degree. If they absorb the same energy, the copper has a larger increase in temperature.

Page 46

Explain what this Sankey diagram shows in as much detail as possible. **AO2 (3 marks)**

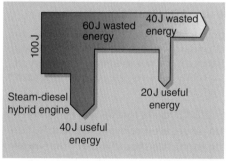

The efficiency is 20%. 60 J of useful energy are transferred and 100 J of wasted energy.

This answer gets 0 marks (grade E/F). The energy is transferred in two stages, which the candidate has confused. For full marks, explain what is meant by efficiency, show calculations and discuss both stages. In the first stage, efficiency is 40%. During the second stage, efficiency is 33%. Overall efficiency is 60%.

Page 47

Two different bulbs are switched on for 10 minutes. Calculate the energy transferred by each one i) a 60 W filament bulb ii) a 10 W energy efficient bulb. **AO2 (4 marks)**

Energy transferred = power x time. The energy transferred is 600 for bulb 1 and 100 for bulb 2.

This answer gets 1 mark (grade D) as the candidate forgot to convert minutes to seconds, and did not show the unit for energy. For full marks, show working. For example the 1st bulb transferred 60 x 10 x 60 = 36 000 Joules.

Page 48

Describe the energy changes taking place in these parts of a coal-fired power station: the burning fuel; the boiler; the turbine; the generator. **AO1 (4 marks)**

burning fuel: chemical → heat

boiler: heat → kinetic

turbine: kinetic (steam) → kinetic (turbine)

generator: kinetic → electrical

This answer gets 4 marks (grade A). The initial energy and final energy had been correctly identified in each case.

Page 49

Explain whether a hydroelectric power station or a coal-fired power station is best for a city located near the coast. **AO3 (5 marks)**

Coal is best (1);

Near the coast rivers flow slowly (1) they cannot be easily dammed (1);

Coal can be transported by water easily (1); it will produce enough electricity for a city (1).

This answer gains 5 marks (grade A)

Page 50

Explain whether a coal-fired power station or a hydroelectric power station is best to cope with surges in demand during the day. **AO3 (3 marks)**

The hydroelectric power station can be started in minutes. However, it causes flooding and can only be used in some parts of the UK. Water can be pumped into the reservoir during quiet periods ready for later use. Coal fired power stations produce greenhouse gases.

This answer gets 2 out of 3 marks (grade B). The comments about environmental issues are irrelevant to the question. To get full marks the candidate should explain that more power stations need to produce electricity if there is a surge in demand for electricity.

Page 51

An echo is a reflected sound wave. Explain why you can hear echoes only in certain places, and why you may hear more than one echo. **AO3 (3 marks)**

Sound waves must bounce off a surface to hear an echo. More than one echo is heard if the sound bounces off several objects.

This answer gets 2 out of 3 (grade B). To get full marks give examples of hard surfaces that sound can echo off e.g. cliffs or buildings.

Page 52

Draw traces to show two sound waves. One sound is lower pitched and twice as loud as the other sound.

Label the amplitude and wavelength on each trace. **AO1 (4 marks)**

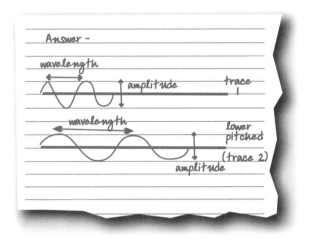

This answer gets 2 out of 4 (grade C/D). The amplitude is wrongly labelled, and the candidate has not shown the trace of a louder note. To get full marks correctly label amplitude as height from mid point to peak, and for a note twice as loud, show a wave with double the amplitude (twice as high).

Page 53

Calculate the speed of radio waves with a wavelength of 10 000 m and frequency of 30 000 Hz. **AO2 (3 marks)**

speed = 30,000÷10,000 = 3 m/s

This answer gets 1 out of 3 marks (grade D). The candidate did not write down the correct equation, or substitute correctly into it. They only get marks for showing the correct unit.

Page 54

Explain which type of electromagnetic wave is the best choice for satellite communications. **AO2 (3 marks)**

Microwaves are best as they are not absorbed by the atmosphere.

This answer gets 2 out of 3 (grade B). To get full marks it is important to show you know why microwaves must travel through the atmosphere, which is that satellites orbit above the earth's atmosphere.

Page 55

Explain two advantages of using space based telescopes. **AO2 (4 marks)**

They can see more as they are closer to the stars, and they can see in all directions.

This answer gets 0 out of 4 marks (grade E/F). The candidate does not realise that the atmosphere has a big impact on the signals reaching the Earth's surface. To get full marks explain that the radiation is not absorbed by the atmosphere so fainter objects can be seen. Another advantage is that objects in space emitting all types of electromagnetic radiation can be detected as the telescope is outside the atmosphere.

Page 56

Explain the evidence we have that supports the Big Bang theory. **AO3 (6 marks)**

The Big Bang was a massive explosion billions of years ago as we can still hear the echo. Galaxies are still moving away from us.

This answer gets 2 out of 6 (grade D/E). Answers like these often allow you to include several points, but expect you to include explanations. The candidate correctly stated the Big Bang theory, but wrongly said the "echo" was heard – it is detected as electromagnetic radiation. To get full marks choose points like these: the red shift shows distant galaxies are moving away from Earth, and further away galaxies move away faster so they used to be closer together.

Background microwave radiation caused by the Big Bang is detected in all directions.

B2 Improve your grade

Page 58

The image below was taken of some cells using a light microscope. State what type of cells they are and give reasons for your answer. **AO2 (2 marks)**

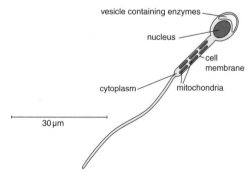

Plant cells because they are green

This answer gains 1 mark out of 2 (grade B). The candidate identified the correct type of cell but the reason given was not sufficient. To gain full marks the answer should have talked about organelles: it could have mentioned that these cells contain chloroplasts or that they have a cell wall. Full marks would also be given if the candidate said they were algae cells.

Page 59

Explain how a sperm cell is specialised. **AO2 (4 marks)**

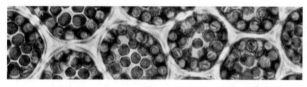

It has a tail to help it swim.

This answer gains 1 mark out of 4 (grade D). The candidate correctly identified that it has a tail for swimming but did not explain why this is required (to reach the egg for

fertilisation). To gain full marks, the answer needs to include at least two of the adaptations explained. For example, the candidate could have mentioned that the sperm has a vesicle containing enzymes, which are used to digest the outside of the egg to allow the nucleus of the sperm to fuse with the nucleus of the egg during fertilisation.

Page 60

Describe why the stomach is classed as an organ. **AO2 (2 marks)**

It contains lots of cells.

This answer did not achieve any marks (grade U), as the candidate did not understand what we mean by an organ. To gain full marks the answer should explain that the stomach is an organ because it is made up of lots of different tissues, which work together to carry out a function.
(It would not be necessary, although interesting, to write that the tissues are glandular tissue to secretes juices, muscular tissue to contract the stomach walls, and epithelial tissue to protect the stomach from the juices it makes.)

Page 61

Without photosynthesis, humans would not survive. Explain why. **AO2 (2 marks)**

Photosynthesis produces oxygen which we need to live.

This answer gains 1 mark out of 2 (grade B). To achieve maximum marks, the candidate should have also explained that photosynthesis produces glucose, which is a store of energy. Animals eat the plant and release the energy which is also essential for life.

Page 62

Anna uses a fertiliser that is high in nitrates on her tomato plants. Explain why she does this. **AO2 (3 marks)**

So the plants have nitrates which they can use to make protein.

This answer gains 2 marks out of a possible 3 (grade B). To achieve the missing mark, the candidate should have gone on to explain why proteins are essential for the healthy growth of the plant: they are used to build new cells. (It would not be necessary, although interesting, to write that the proteins are also used to build enzymes. Also, that without a supply of nitrate, the tomato plants would not grow to their full potential nor produce many tomatoes. Plants absorb minerals such as nitrates through their roots, so this is why fertilisers are added to the soil.)

Page 63

Simon went on holiday to Mexico. He noticed that the plants growing there were very different to the plants growing at home in the UK. Explain the reasons for this difference in distribution. **AO2 (3 marks)**

The temperature is much warmer in Mexico, so the plants that grow there are adapted to living at higher temperatures. The rainfall is also lower in Mexico, so the plants there have adaptations to help them cope with lack of water.

This answer gains full marks (grade A) as the candidate mentioned two differences in physical factors between Mexico and the UK, and explained how these affect the distribution of the plants. The plants growing in Mexico may not survive in the UK because of the low temperatures and high rainfall.

Page 64

Pepsin is an enzyme that helps break down proteins in the stomach. It has an optimum pH of 2. Use this information to explain why the stomach produces hydrochloric acid. **AO2 (2 marks)**

Pepsin has an optimum pH of 2 so it works best in acidic conditions.

This answer gained 1 mark out of a possible 2 (grade B). The answer did not really explain what is meant by optimum pH and why this is important. The optimum pH is the pH at which an enzyme works best at. The acid produced by the stomach means that pepsin can break down protein in the stomach at a fast rate.

Page 65

Dipesh mixed some starch with amylase in a beaker and left the mixture in a water bath at 37 °C. After 30 minutes, he tested the mixture to see if there was any starch present. Predict what he will find and give a reason for your answer. **AO2 (2 marks)**

There would be no starch left because the amylase breaks down starch.

This answer achieves full marks (grade A) but it would be more complete, and show the candidate's knowledge better, if it included the fact that the starch would have been broken down into sugars. Remember that enzymes are only catalysts in the breaking down of food. The starch would eventually break down by itself, it would just take a lot longer.

Page 66

Explain why breathing rate increases when you exercise. **AO2 (3 marks)**

Your muscles are contracting quickly so are using up energy at a fast rate. Breathing rate increases to get oxygen to the muscles quickly.

This answer gains 2 marks out of a possible 3 (grade B). For full marks, the candidate needs to go on to explain why the muscles need a good supply of oxygen: to carry out aerobic respiration in order to release energy from glucose. The energy is used to enable the muscles to contract. If you are aiming for an A* grade it is a good idea to include word equations wherever possible: glucose + oxygen → carbon dioxide + water (+ energy).

Page 67

You are involved in a race. At first you sprint off feeling full of energy. However, halfway through, your legs start to ache and you have to stop. Explain why this happened. **AO2 (3 marks)**

My heart could not beat fast enough and my lungs could not breathe fast enough to get oxygen to my leg muscles, so they could not carry out respiration and make energy for my leg muscles to work.

This answer gains 1 mark (grade C/D). The candidate was correct in stating that the lungs and heart could not work fast enough to get the required levels of oxygen to the leg muscles but respiration would not stop – anaerobic respiration would take over. For full marks, the answer should explain this, then go on to explain that the legs would start to ache and stop working efficiently because of the build-up of lactic acid in the muscle cells.

Page 68

Explain why mitosis is an essential process in the formation of a baby from a fertilised egg. **AO2 (2 marks)**

Mitosis is needed to produce the egg and sperm cells that join to form the fertilised egg.

This answer is incorrect and will score no marks (grade U). The candidate has mixed up the terms mitosis and meiosis. For full marks the answer should state that the fertilised egg is one cell (called a zygote), which will divide by mitosis to form the many cells that make up a baby.

Page 69

In the future, we may be able to use an individual's genome to calculate the likelihood of them developing certain diseases such as cancer. Evaluate this potential application. **AO2 (6 marks)**

If a person's DNA shows they might get a disease, they might not be able to get a job because the employer is worried about them getting ill and taking time off work. But it might be useful to know this, as you can be told about what symptoms to look out for and how to change your lifestyle to reduce the risk of the illness developing. However, you are not told if you will definitely develop the disease, only if you are more slightly at risk. This might cause people to worry and some people would rather not know.

The candidate has fully evaluated this new technology by giving detailed benefits and drawbacks (grade A).

Page 70

Katy has red hair. Both her parents have brown hair. Explain how Katy inherited red hair when her parents do not have it. **AO2 (3 marks)**

Katy got the gene for red hair from her parents.

This answer gains 1 mark out of 3 (grade D). To achieve full marks, the candidate needs to explain why Katy's parents do not have red hair. The red-hair allele must be recessive. Her parents both have the red-hair allele but their other allele (brown hair) masks it. Katy received two red-hair alleles from her parents during fertilisation so Katy has two red-hair alleles (her genotype) and so therefore has red hair (her phenotype). Also, to show a good understanding of genetics, the candidate should have used the correct terminology. Red hair should be referred to as an allele, not a gene. The red hair allele is a type of hair-colour gene.

Page 71

The height of pea plants is controlled by a single gene. The tall allele is dominant. A tall pea plant was bred with a short pea plant. The offspring formed was in the ratio 1 tall : 1 short. Draw a genetic diagram to show the cross. **AO2 (4 marks)**

Parents	tall pea plant	short pea plant
	TT	tt
Gametes	T and T	t and t

Offspring

	T	T
t	Tt	Tt
t	Tt	Tt

This answer gains 2 marks out of 4 (grade C). The candidate has made a mistake in determining the allele of the tall pea plant. You can see from the cross that all of the offspring are tall, not the ratio of 1 tall : 1 short. The tall plant must have the alleles Tt.

Page 72

Polydactyly is caused by a dominant allele. Jessica has polydactyly. Ryan does not have it. Could their children have polydactyly? Explain your answer. **AO2 (3 marks)**

Yes, because they could inherit the polydactyly gene from Jessica.

This answer gains 2 marks (grade B). To gain full marks, the candidate should go on to explain that the children only need one dominant allele in order to have polydactyly and, as Jessica has the disorder, she must carry at least one dominant allele.

Page 73

Around 300 years ago, the dodo lived on the island of Mauritius. It had no predators. People arrived on the island. They brought predators such as dogs, pigs and rats. Explain why the dodo became extinct. **AO2 (2 marks)**

The dodos got eaten by the dogs, pigs and rats and they all died out.

This answer gains 1 mark out of 2 (grade B). To achieve full marks, the candidate should have added that the dodo was not adapted to escape the new predators and was not able to evolve quickly enough.

C2 Improve your grade

Page 75

Work out the relative atomic mass (A_r) of chlorine. Chlorine has two isotopes, 25 per cent is chlorine-37, 75 per cent is chlorine-35. **AO2 (2 marks)**

25 x 37 = 925, 75 x 35 = 2625 so 3550/100 = 35.5

This answer gets 2 marks (grade B). The mean value is correctly calculated.

Page 76

Use the periodic table to:

a   Work out the electronic structure of these elements: magnesium, chlorine, nitrogen, aluminium, calcium. **AO2 (5 marks)**

b   Work out how many outer electrons each of these elements has: strontium, arsenic, gallium, astatine, germanium. **AO2 (5 marks)**

a   Mg = 2,8,2   Cl = 2,8,7   N = 2,7

Al = 2,8,3   Ca = 2,8,8,2

b   Sr 2, As 5, Ga 1, At 7, Ge 2.

This answer gets 7 marks (grade B). For part a) all except N are correct. The problem is that the first orbit has not been subtracted from the 7, it should be 2,5. For part b) the candidate has correctly found three, but for Ga and Ge has used the periodic table but only counted in from the transition metals instead of using the group number.

Page 77

Draw diagrams to show how calcium and fluorine lose and gain electrons to make the compound calcium fluoride (CaF_2). **AO1 (3 marks)**

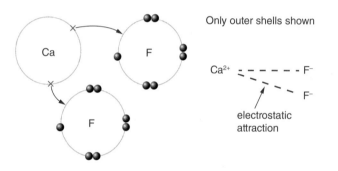

This answer gets 3 marks (grade A). To save time, the candidate has only drawn the outer electron shells. The answer clearly shows what happens to the electrons, states correctly the two ions formed, and gives the ionic bond as an electrostatic attraction.

Page 78

Explain in detail why sodium chloride:

a   has a high melting point

b   conducts electricity when in solution. **AO1 (3 marks)**

a   Sodium chloride has a crystal lattice so the melting point is high.

b   It conducts electricity along the lattice.

This answer gets 0 marks (grade E). The answer recognises the crystal lattice, but should explain that this forms a giant structure where six ions are attracted equally to six others, making a very strong structure that needs lots of energy to break (melt). Part b) fails to mention that the ions are now free to move and can carry the electric charge/current through the solution.

Page 79

Draw diagrams to show the covalent bonds present in these compounds:
HCl, H_2, CO_2, NH_3, and CH_4. **AO1 (5 marks)**

This answer gets 4 marks (grade B). Clear dot-and-cross diagrams are used; but for ammonia, the pair of unbonded electrons has been missed out.

Page 80

Draw diagrams to show the difference between the structure of a thermosetting and a thermosoftening polymer. **AO1 (2 marks)**

thermosetting polymer   thermosoftening polymer

This answer gets 2 marks (grade C). Both chain structures have been clearly drawn and labelled.

Page 81

Explain how the metal lattice allows for metals to be stretched without breaking. **AO1 (2 marks)**

The atoms can be easily pulled slightly apart from each other.

This answer gets 0 marks (grade E). The atoms slide past each other distorting the original shape, but not separating out.

Page 82

Look at Figure 2. It shows the retention–time graph for a gas chromatogram sample.

a How many different components were in the sample?
b List the components in order of the quantity of each present. Start with the most plentiful. **AO3 (2 marks)**

a there are four substances

b A B C D E is the solvent.

This answer gets 0 mark (grade E). Because a) there are five peaks so there are five substances and b) the answer suggests that the candidate thinks E is the solvent. The graph has been wrongly read: the height of the peaks is the important feature. So A C D B E is the correct answer.

Page 83

Calculate the formula of lithium oxide made from 5.6 g of lithium and 6.4 g of oxygen. **AO2 (2 marks)**

Li 5.6/7 = 0.8, O = 6.4/16 = 0.4 ratio is 8:4 or 2:1, so Li_2O

This answer gets 2 marks (grade A). It is a very difficult calculation that only the most able students can get right. The method clearly shows how to get the answer.

Page 84

Calculate the mass of calcium oxide that can be made from 100 tonnes of calcium carbonate.

The balanced equation for the reaction is:

$CaCO_3(s) \rightarrow CaO(s) + CO_2(g)$ **AO2 (3 marks)**

M_r $CaCO_3$ = 40 + 12 + 16 + 16 + 16 = 100

M_r CaO = 40 + 16 = 56, from equation 1 mole $CaCO_3$ produces 1 mole CaO, so 100 tonnes makes 56 tonnes.

This answer gets 3 marks (grade B). A well-presented calculation that shows how to work out M_r, and use the equation to calculate the moles, and then the masses involved.

Page 85

Use Figure 2 to calculate the rate of the reaction at the following times:

a 0 to 0.5 minutes

b 1.0 to 2.0 minutes

c 2.5 to 3.0 minutes. **AO3 (3 marks)**

a 6.4 in 0.5 min, so in 1 min 12.8

b changes from 8.8 to 11.0 in the minute so rate is 2.2

c changes from 12.0 to 12.4 in the minute so rate is 0.4

This answer gets 3 marks (grade B). The answers to a) and b) are correct, but have no units. In this case the units are cm^3 per minute or cm^3/min. For part c) the answer is wrong, the value is only for 0.5 minutes, so must be multiplied by 2 to get the right answer, which needs units.

Page 86

A student was investigating a chemical reaction. She decided to increase the temperature of the reaction, whilst decreasing the concentration of one reactant. Use collision theory to explain why the rate of reaction remained unchanged. **AO3 (2 marks)**

The increased temperature has no effect as she only changed one of the reactant's concentration not both. To change the rate, she needs to change both the reactants' concentration.

This answer gets 0 marks (grade E). Increasing the temperature does increase the rate, but reducing the concentration of one reactant will counteract the change. The increase in reacting collisions caused by the temperature rise is equalled by the reduction in reacting collisions caused by less of the reactant being present.

Page 87

Use Figure 5 to find the rate of the reaction when the concentration is 2 mol/dm^3 after 1 minute. **AO2 (2 marks)**

this is 6

This answer gets 0 marks (grade D). The answer has been simply read off the graph, and has no units. A tangent should have been drawn to enable the gradient of the curve to be found, so that the rate at 1 minute could be found. The x-axis value would be 1.8, y-axis would be 4, so 4/1.8 = 2.2 cm³ per minute.

Page 88

What is activation energy?

Explain how a catalyst lowers the activation energy for a particular reaction. **AO1 (3 marks)**

Activation energy is energy needed for a reaction to start.

A catalyst works by providing an easier path for the reaction.

This answer gets 2 marks (grade C). The first part is just worth a mark. More detail about reacting collisions would have made the answer better. For the second part, the second mark could be gained if there was more detail about it being an energetically lower reaction path.

Page 89

Sketch an energy-level diagram for the exothermic reaction that takes place when nitric acid reacts with sodium hydroxide. Write on the line for 'reactants', the names of the reactants; and on the line for 'products', the products. **AO1 (4 marks)**

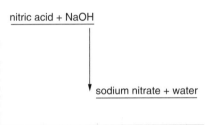

nitric acid + NaOH

sodium nitrate + water

This answer gets 4 marks (grade B). The levels are correctly drawn and the reactants and products all named. NaOH, the formula, is acceptable when asked for a name but if asked for a formula, the name would be marked as wrong.

Page 90

Describe the method to make some solid chromium chloride from chromium metal. **AO3 (4 marks)**

Use hydrochloric acid, add some powdered chromium metal to it, when it stops reacting filter it, then evaporate the solution.

This answer gets 2 marks (grade C). The basic steps are there, but you need to check the solution made is neutral with, for example, pH paper. Warming the mixture would help to ensure complete reaction. More detail is needed about the evaporation technique, including the need to partially evaporate and then to filter it to obtain some crystals to dry.

Page 91

Write half-equations to show the reactions, at both the anode and cathode, for the electrolysis of copper chloride ($CuCl_2$). **AO2 (2 marks)**

$Cu^{2+} + e^- \rightarrow Cu$

$Cl^{2-} + e^- \rightarrow Cl_2$

This answer gets 0 marks (grade D). The answer does not show any understanding. Copper ion is correctly described, but that is all. The need to balance the electrons and charges is attempted, but only as far as wrongly giving chloride ions a 2⁻ charge. Correct equations are $Cu^{2+} + 2e^- \rightarrow Cu$ and $2Cl^- + 2e^- \rightarrow Cl_2$

P2 Improve your grade

Page 93

Sam travels 5 m in 10 s, before stopping for 2 s. He then takes 6 s to return to the start. Draw a distance time graph that shows Sam's journey. **AO2 (3 marks)**

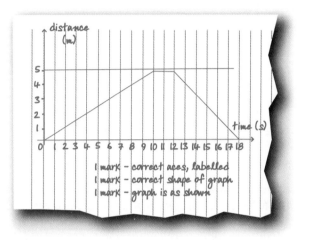

This answer gains 3 out of 3 marks (grade A). The candidate correctly used the information from the question to plot points and then joined the points with a straight line. Distance time graphs like these are one of the few times in physics where a graph "joins the dots".

Page 94

Calculate the total distance travelled by the stone in the first 10 seconds in fig. 3. **AO2 (3 marks)**

distance = area under graph = 80 m

This answer gains 1 out of 3 marks (grade D).

The candidate correctly worked out the area under the graph, but only for the first 4 seconds.

To gain full marks, they should also calculate the distance travelled during 4–10 seconds and add the distances together to get the answer.

Page 95

Describe how drag forces compare with forces from the engine for a car that accelerates, then reaches a steady speed then decelerates. **AO2 (4 marks)**

Drag forces change with speed. When the car accelerates, engine forces are larger than drag forces. At a steady speed there are no drag forces.

This answer gains 2 out of 4 marks (grade C/D).

The candidate forgot to explain what happens when the car decelerates, and was wrong to say there are no drag forces at a steady speed.

To gain full marks, they should say that at a steady speed, forces from engine equal drag forces and as the car decelerates, forces from engine are smaller than drag forces.

Page 96

Sketch the velocity time graph for a marble that is released into a measuring cylinder of water. The marble reaches its terminal velocity. **AO2 (3 marks)**

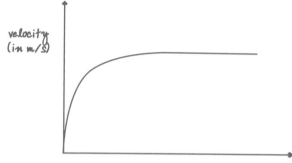

This answer gains 3 out of 3 marks (grade A).

The candidate showed the rapid increase in velocity when the drag forces are small, which gets less as the drag forces increase. Eventually there is no increase in velocity when drag forces equals weight. Remember the marble's speed never decreases.

Page 97

a   Calculate the work done lifting a piano 0.8 m into a lorry. The piano weighs 1850 N. **AO2 (3 marks)**

a   work = force x distance = 1850 x 0.8 = 1480

This answer gains 2 out of 3 marks (grade B).

To gain full marks, always include units.

b   Calculate the force needed to drag the piano into the lorry if a ramp 2 m long is used. **AO2 (2 marks)**

b   force = work/distance = 1480/2 = 740 N

This answer gains 2 out of 2 marks (grade A).

The candidate included the unit and showed working in this question, which allowed them to gain full marks.

Page 98

Calculate the power of a motor that can lift 600 N in 1 minute. **AO2 (3 marks)**

600W

This answer gets 0 marks. Remember 1 minute is 60 seconds. Always show working. Power = 600/60 = 10 W

Page 99

Two trolleys each with a mass of 1 kg roll towards each other. One trolley travels at 2 m/s to the left and one

travels at 3 m/s to the right. They collide and stick together. What is the velocity and direction of the trolleys after the collision? **AO2 (4 marks)**

momentum before = 1 x 2 – 1 x 3 = -1

momentum before = momentum after collision = -1 kg.m/s

speed = 1/1 = 1 m/s

This answer gains 2 out of 4 marks (grade C/D). The candidate did not include a direction for the vehicle after the collision. To gain full marks, a sketch is helpful so you don't miss details – this candidate did not realise the masses stuck together after the collision.

Page 100

Max rubbed a plastic ruler and a metal ruler with a piece of cloth.

a   Which ruler became charged? **AO1 (1 mark)**

b   Max held the charged ruler near a charged balloon. The balloon moved away from the ruler. Explain why **AO2 (2 marks)**

a   the plastic ruler

b   the balloon and ruler are charged

This answer gets 2 marks. The plastic ruler gained a charge. The ruler repelled the balloon because they both have the same charge.

Page 101

When the potential difference across a bulb is 6V, the current through it is 0.1A. What is the resistance of the bulb? **AO2 (3 marks)**

Resistance = voltage/current

This answer gets 1 mark. Voltage and potential difference are the same. Resistance = 6V/0.1A = 60 ohms.

Page 102

A series circuit is set up, using four 1.5 V cells in series **AO2 (5 marks)**

a   What is the total voltage supplied to the circuit?

a   Answer: 6 V (1)

The circuit includes a motor (resistance 10 ohms) and a bulb (resistance 5 ohms).

b   What is the total resistance in the circuit?

b   Answer: 15 ohm (1)

c   Calculate the current in the circuit

c   Answer: 6/15 = 0.4 A (1)

d   What is the potential difference across each component?

d   Answer: 6 V

This answer gains 3 out of 5 marks (grade C).

The candidate has worked through the first points correctly, but forgot that in a series circuit the voltage is shared across components.

To gain full marks, include calculations like these:

Bulb: 5 x 0.4 = 2 V   Motor: 10 x 0.4 = 4 V

Page 103

Explain why a double insulated appliance does not need an earth wire. **AO1 (2 marks)**

It has a fuse instead

This gets 0 marks. The earth wire stops a metal outer casing from becoming live. Double insulated appliances have a plastic outer casing.

Page 104

Fuses rated at 3 A, 5 A and 13 A are available. Which fuse should be used in each case? **AO2 (3 marks)**

a   a lawn mower (power 2000 W)

a   13 A

b   a lamp (power 60 W)

b   3A

c   a toaster (power 800 W)

c   5A

This answer gains 3 out of 3 marks (grade A).

The candidate correctly realised that the best fuse has a rating close to, but larger than the current in the equipment. In some questions, you may be asked to show your calculations.

Page 105

Why are the risks from handling an alpha source safely different from the risks from handling a gamma source? **AO3 (3 marks)**

Alpha sources are not dangerous but gamma radiation is. Gamma radiation cannot be used safely

This answer gets 0 marks. Alpha sources are much more dangerous inside the body because they cannot penetrate skin but are very ionising to cells. Gamma radiation is less ionising but penetrates skin and other materials easily. Both should be handled with care.

Page 106

Explain how the Rutherford scattering experiment provided evidence for the nuclear model of the atom. **AO2 (3 marks)**

Most alpha particles were not scattered so the nucleus is small with a positive charge and most of the mass of the atom is in the nucleus.

This answer gains 1 out of 3 marks (grade D). The candidate included correct statements about the nucleus, but did not explain how the evidence leads to these conclusions. To gain full marks, include extra detail such as most alpha particles were not scattered so the nucleus is small. The nucleus must be massive as it could deflect alpha particles. It has a positive charge as it repelled positively charged alpha particles.

Page 107

The half-life of **substance X** is 20 days, and its original activity is 100 000 counts per minute. **Substance Y** has a half-life of 15 days, and an original count rate of 200 000 counts per minute. What is the count rate of each sample after 60 days? **AO2 (3 marks)**

The count rate of X is 100 000/20 which is 5000 and the count rate of Y is 13 333

This answer gains no marks (grade F).

The candidate does not know what half-life is (the time taken for the original count rate to halve).

For full marks, you should explain that 60 days is 3 half-lives for substance X, so its count rate has halved 3 times and is now an eighth of the original value i.e. 100 000 / 8 = 12 500

60 days is 4 half lives for substance Y so its count rate is a sixteenth of the original value i.e. 200 000 / 16 = 12 500

Page 108

Explain why the presence of iron in the Sun is evidence that the Sun formed from the remains of older stars. **AO2 (3 marks)**

Iron is only formed in the later stages of a star's life cycle. The iron spreads through the Universe during a supernova.

This answer gains 2 out of 3 marks (grade B/C) as the candidate did not explain why iron does not form in the Sun. Iron only forms in the final stages of a massive star's lifecycle.

Page 109

Explain how the stages in a star's life cycle are controlled by its mass. **AO2 (3 marks)**

All stars have a main sequence stage. Small stars become red giants and large stars become red supergiants. Small stars do not have a supernova but change into a black dwarf. Large stars become black holes or neutron stars.

This answer gains 2 marks (grade B) It describes the stages but doesn't explain that more massive stars can undergo different fusion reactions.

Understanding the scientific process

As part of your assessment, you will need to show that you have an understanding of the scientific process – How Science Works.

This involves examining how scientific data is collected and analysed. You will need to evaluate the data by providing evidence to test ideas and develop theories. Some explanations are developed using scientific theories, models and ideas. You should be aware that there are some questions that science cannot answer and some that science cannot address.

Collecting and evaluating data

You should be able to devise a plan that will answer a scientific question or solve a scientific problem. In doing so, you will need to collect data from both primary and secondary sources. Primary data will come from your own findings – often from an experimental procedure or investigation. While working with primary data, you will need to show that you can work safely and accurately, not only on your own but also with others.

Secondary data is found by research, often using ICT – but do not forget books, journals, magazines and newspapers are also sources. The data you collect will need to be evaluated for its validity and reliability as evidence.

Presenting information

You should be able to present your information in an appropriate, scientific manner. This may involve the use of mathematical language as well as using the correct scientific terminology and conventions. You should be able to develop an argument and come to a conclusion based on recall and analysis of scientific information. It is important to use both quantitative and qualitative arguments.

Changing ideas and explanations

Many of today's scientific and technological developments have both benefits and risks. The decisions that scientists make will almost certainly raise ethical, environmental, social or economic questions. Scientific ideas and explanations change as time passes and the standards and values of society change. It is the job of scientists to validate these changing ideas.

How science ideas change

From the information you have learnt, you will know that science is a process of developing, then testing theories and models. Scientists have been carrying out this work for many centuries and it is the results of their ideas and trials that has provided us with the knowledge we have today.

However, in the process of developing this knowledge, many ideas were put forward that seem quite absurd to us today.

During the Middle Ages, *The Miasma Theory* explained how diseases were caused. Miasma was thought to be a poisonous vapour present in the air. This vapour was said to contain particles of decaying matter that created a foul smell. The name of the killer disease malaria is derived from the Italian mala, meaning 'bad' and aria, meaning 'air'.

In the nineteenth century, England was undergoing a rapid expansion of industrialisation and urbanisation. This created many foul-smelling and filthy neighbourhoods, which were focal points for disease. By improving housing, cleanliness and sanitation, levels of disease fell. This fall in the level of disease supported the miasma theory.

In 1854, John Snow confirmed a cholera outbreak in London as originating from a water pump. Within ten years, Louis Pasteur was suggesting the presence of germs in substances such as milk and meat that caused them to go off quickly. He was able to remove the germs by a process we now call pasteurisation. This process is still used to protect perishable foodstuffs today.

Reliability of information

It is important to be able to spot when data or information is presented accurately, and just because you see something online or in a newspaper does not mean that it is accurate or true.

Think about what is wrong in this example, based on a newspaper report. Look at the answer at the bottom of the page to check that your observations are correct.

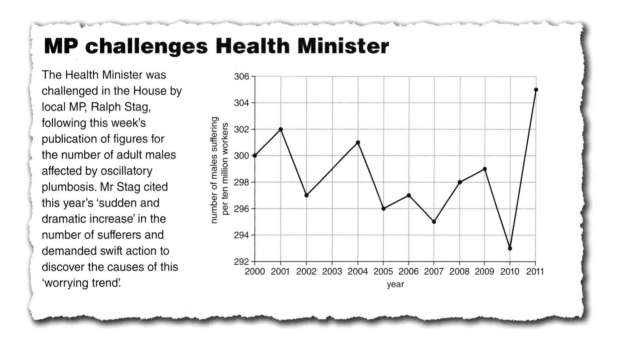

MP challenges Health Minister

The Health Minister was challenged in the House by local MP, Ralph Stag, following this week's publication of figures for the number of adult males affected by oscillatory plumbosis. Mr Stag cited this year's 'sudden and dramatic increase' in the number of sufferers and demanded swift action to discover the causes of this 'worrying trend'.

Answer

The y axis does not start at zero and so a change from 293 to 305 individuals in one year is a difference of only 12 people per ten million workers. This is unlikely to be significant and is not really a 'sudden and dramatic increase'.

How science ideas change

From the information you have learnt, you will know that science is a process of developing, then testing theories and models. Scientists have been carrying out this work for many centuries and it is the results of their ideas and trials that has provided us with the knowledge we have today.

However, in the process of developing this knowledge, many ideas were put forward that seem quite absurd to us today, such as this example from the 17th Century.

> In 1667, Johann Joachim Becher published Physical Education, in which he set out the basis of what was to become the Phlogiston Theory. The theory itself was formally stated in 1703 by Georg Ernst Stahl, a German professor of Medicine and Chemistry. He suggested that in all flammable substances there is something called 'phlogiston', a substance without colour, odour, taste, or mass that is given off during burning. 'Phlogisticated' substances are those that contain phlogiston and, on being burned, are 'dephlogisticated'. The ash of the burned material is thought to be the true material. The theory was widely supported for much of the eighteenth century. Joseph Priestley, who is credited with the discovery of oxygen, defended the theory but it was eventually disproved by Antoine Lavoisier.

Reliability of information

It is important to be able to spot when data or information is presented accurately, and just because you see something online or in a newspaper does not mean that it is accurate or true.

Think about what is wrong in this example, based on a document from a government official to a parliamentary committee. Look at the answer at the bottom of the page to check that your observations are correct.

RECENT DERAILMENTS IN THE STATE

Note that this timeline is for background information only, and is not meant to be a comprehensive analysis of rail safety incidents.

Aug. 5, 2005: 9 cars of a 144-car train derail spilling highly-acidic caustic soda into the Yukon river; a preliminary government investigation states that federal safety regulations were broken by the railroad company…

Answer
Caustic soda is highly alkaline, not highly acidic.

How science ideas change

From the information you have learnt, you will know that science is a process of developing, then testing theories and models. Scientists have been carrying out this work for many centuries and it is the results of their ideas and trials that has provided us with the knowledge we have today.

However, in the process of developing this knowledge, many ideas were put forward that seem quite absurd to us today.

In 1692, the British astronomer Edmund Halley (after whom Halley's Comet was named) suggested that the Earth consisted of four concentric spheres. He was trying to explain the magnetic field that surrounds the Earth and suggested that there was a shell of about 500 miles thick, two inner concentric shells and an inner core. Halley believed that these shells were separated by atmospheres, and each shell had magnetic poles with the spheres rotating at different speeds. The theory was an attempt to explain why unusual compass readings occurred. He also believed that each of these inner spheres, which was constantly lit by a luminous atmosphere, supported life.

Reliability of information

It is important to be able to spot when data or information is presented accurately, and just because you see something online or in a newspaper does not mean that it is accurate or true.

Think about what is wrong in this example from an online shopping catalogue. Look at the answer at the bottom of the page to check that your observations are correct.

FROM BOX TO AIR IN UNDER TWO MINUTES!

Simply unroll the airship and, as the black surface attracts heat, watch it magically inflate.

Seal one end with the cord provided and fly your 8-metre, sausage-shaped kite.

✔ Good for all year round use.

✔ Folds away into box provided.

✔ A unique product – not for the faint hearted.

✔ Educational as well as fun!

Once the airship is filled with air, it is warmed by the heat of the sun.

The warm air inside the airship makes it float, like a full-sized hot-air balloon.

Answer
Black absorbs heat, it does not attract it.

The glossary contains terms useful for your revision. Page numbers are given for items that are covered in this book.

A

absorbed energy 42, 45 an object absorbs energy when the energy from infrared radiation is transferred to the particles of the object, increasing the temperature of the object

acceleration 93–4 rate at which an object speeds up, calculated from change in velocity divided by time

acid 89–90 hydrogen compound that can produce hydrogen ions, H^+

acidic solution aqueous solution with pH less than 7 – the concentration of H^+(aq) is higher than that of OH^-(aq)

activation energy 86, 92 minimum energy needed to break bonds in reactant molecules to allow a reaction to occur

active site 64 a depression in an enzyme molecule into which its substrate fits

actual yield mass of product actually obtained from a reaction

adaptation 15, 23 the way in which an organism evolved to become better able to survive in its environment

addiction 13, 23 when a person becomes dependent on a drug

addition reaction reaction in which a C=C bond opens up and other atoms add on to each carbon atom

adrenaline 10 hormone that helps to prepare your body for action

aerobic respiration 66, 74 a process in which energy is released from glucose, using oxygen

agar 8–9 substance used to make jelly on which bacteria can be grown

aggregate stones, gravel or rock chippings used in the construction industry

air resistance 96 force resisting the movement of an object travelling through air

alcohols 34 a family of organic compounds containing an OH group, for example ethanol (C_2H_5OH)

algae 58 simple, plant-like organisms

alkali 89–90 compound that contains hydroxide ions, OH^-

alkali metal 77 very reactive metal in Group 1 of the periodic table, for example sodium

alkaline solution aqueous solution with pH more than 7 – the concentration of OH^-(aq) is higher than that of H^+(aq)

alkanes 32–3, 41 a family of hydrocarbons: C_nH_{2n+2} with single covalent bonds – found in crude oil

alkenes 33–4, 41 a family of hydrocarbons: C_nH_{2n} with double covalent bonds (C=C), for example ethene (C_2H_4)

allele 69–72 a particular form of a gene

alloy 28, 81 a mixture of two or more metals, with useful properties different from the individual metals

alpha particle 75, 105–7, 110 particle emitted from the nuclei of radioactive atoms and consisting of two protons and two neutrons

alternating current (a.c.) 91, 103, 110 electric current where the direction of the flow of current constantly reverses, as in mains electricity

amino acids 40, 64, 67, 74 small molecules from which proteins are built

ampere (amp) 100, 104 unit used to measure electrical current

amplitude 52, 57, 103 size of wave oscillations – for a mechanical wave, how far the particles vibrate around their central position

amylase 65 an enzyme that breaks down starch molecules to maltose (sugar) molecules

anaerobic respiration 67 process in which energy is released form glucose, without using oxygen

angle of incidence angle between the ray hitting a mirror or lens and the normal

angle of reflection angle between a ray reflecting from a mirror and the normal

anode 78, 91, 233 positive electrode

antibiotic 8, 22 therapeutic drug, acting to kill bacteria, which is taken into the body

antibiotic resistance 8 ability of bacteria to survive in the presence of an antibiotic

antibody 7, 23, 64 protein normally present in the body or produced in response to an antigen which it neutralises, thus producing an immune response

antiseptic 34 substance that kills pathogens

antitoxin 7, 23 substance produced by white blood cells that neutralises the effects of toxins

antiviral 8 therapeutic drug acting to kill viruses

appliance device that transfers the energy supplied by electricity into something useful

aqueous solution 37 substance(s) dissolved in water

argon 39 the most common noble gas – makes up nearly 1% of the air

atmosphere 39–40, 41 thin layer of gas surrounding a planet

atom 24–5, 28–9, 41, 75, 99, 105–6, 110 the basic 'building block' of an element that cannot be chemically broken down

atomic number 24, 41, 105 number of protons in the nucleus of an atom

attraction 32, 78, 110 force that pulls, or holds, objects together

auxin 13, 23 a plant hormone that affects rate of growth

average speed 93 total distance of a journey divided by total time taken

axon 10 long thread of cytoplasm in a neurone, carrying an impulse away from the cell body

B

background radiation 106 low level nuclear radiation that exists everywhere, from rocks and other environmental sources

bacteria 7–9, 21–2, 58 single-celled microorganisms; they do not have a nucleus. They can either be free-living organisms or parasites (they sometimes invade the body and cause disease)

balanced diet 6 eating foods (and drinking drinks) that will provide the body with the correct nutrients in the correct proportions

balanced equation 26 chemical equation where the number of atoms of each element is the same in the reactants as in the products

balanced forces 94–5 system of forces with no resultant force

base 89 substance that neutralises an acid to form a salt and water

bauxite 30 main ore of aluminium – impure aluminium oxide (Al_2O_3)

beta particle 105–7, 110 type of nuclear radiation emitted as an electron by a radioactive nucleus

Big Bang theory 56–7 theory that states that the universe originated from a point at very high temperature, and everything in the universe formed as energy and matter exploded outwards from that point and cooled down

bile 65 a liquid, produced by the liver, which flows into the small intestine where it neutralises the acidic juices from the stomach

biodegradable 35 a biodegradable material can be broken down by microorganisms

biodiesel 31, 36, 41 fuel made from plant oils such as rapeseed

biodiversity range of different living organisms in a habitat

biofuel 31, 34, 36, 41 fuel such as wood, ethanol or biodiesel – obtained from living plants or animals

biological detergent 65 washing powder or liquid that contains enzymes

biomass 17, 48–9 the mass of living material, including waste wood and other natural materials

biosensor sensitive material that detects very low levels of chemical or biological agents in the surroundings

biotic factor 63 something, such as competition, that influences a living organism and is caused by other living organisms

blast furnace 28 furnace for extracting iron from iron ore

blood plasma the liquid part of blood

Bluetooth 54 low energy, short-range radio waves used to connect communications equipment such as mobile phones or laptops to other equipment nearby

boiling 32, 41, 43 change of state from liquid to gas that happens at the boiling point of a substance

braking distance 95, 98, 110 distance a vehicle travels before stopping, once the brakes are applied

brass 28 alloy of copper and zinc

breathing movements that move air into and out of the lungs

brittle easily cracked or broken by hitting or bending

bronze alloy of copper and tin

brownfield site 30 former industrial site that may be redeveloped for other uses

Buckminsterfullerene 81 carbon molecule C_{60} – 60 carbon atoms arranged in the form of a hollow sphere

buckyballs 81 common name for fullerenes, such as C_{60}, buckminsterfullerene

burette 90 item of glassware used to measure the volume of liquid added during a titration

C

calcium carbonate 26, 41, 90 compound with chemical formula $CaCO_3$ – main component of limestone

carbohydrase an enzyme that breaks down carbohydrates; examples include amylase and isomerase

carbon capture technology that filters the carbon and carbon dioxide gas out of waste smoke and gases from power stations and industrial chimneys

carbon cycle 19, 23, 40 the way in which carbon atoms pass between living organisms and their environment

carbon dating 75 method of determining the age of archaeological specimens by measuring the proportion of the carbon-14 isotope present

carbon dioxide 19, 26–7, 30–1, 40 one of the gases emitted from burning fossil fuels that contributes to global warming

carbon-neutral fuel fuel grown from plants so that carbon dioxide is taken in as the plants are growing – this balances out the carbon dioxide released as the fuel is burned

carbon sinks carbon-containing substances, such as limestone and fossil fuels, which formed millions of years ago, removing carbon dioxide from the atmosphere

carnivore an animal that eats other animals

carrier organism that carries a recessive allele for a genetic disorder but does not itself have the disorder

cast iron iron containing 3–4% carbon – used to make objects by casting

casting making an object by pouring molten metal into a mould and allowing it to cool and solidify

catalyst 33, 83, 86, 88, 92 substance added to a chemical reaction to improve the reaction rate, without being used up in the process

catalytic converter 'cat' 31, 88 the section of a vehicle's exhaust system that converts pollutant gases into harmless ones

catalytic cracking 33 cracking hydrocarbons by heating in the presence of a catalyst

cathode 103 negative electrode

cell body the part of a nerve cell that contains the nucleus

cell membrane 58 the outer covering of every cell, which controls the passage of substances into the cell

cell wall 58 a strong covering made of cellulose, found on the outside of plant cells

cellulose 74 a carbohydrate; a polysaccharide used to make plant cell walls

cement 27 substance made by heating limestone with clay – when mixed with water it sets hard like stone

central nervous system (CNS) 10–11, 23 collectively the brain and spinal cord

chain reaction 34, 108 a series of nuclear fission reactions where neutrons released from one reaction cause another nuclear fission reaction and so on

chalcopyrite common ore of copper – formula $CuFeS_2$

charge 24, 78, 101, 104 particles or objects can be positively or negatively electrically charged, or neutral: similar charges repel each other, opposite charges attract

chemical analysis process of performing tests to determine what chemical substances are present in a sample, and/or the amount of each substance present

chemical bond attractive force between atoms that holds them together (may be covalent or ionic)

chemical equation 26 line of chemical formulae showing what reacts and what is produced during a chemical reaction

chemical formula shows the elements present in a compound and the number of atoms of each, such as H_2SO_4

chemical reaction 42, 78, 83 process in which one or more substances are changed into other substances – chemical reactions involve rearranging atoms and energy changes

chlorophyll 61, 74 green pigment inside chloroplasts in some plant cells, which absorbs energy from sunlight

cholesterol 6, 23 chemical needed for the formation of cell membranes, but that increases the risk of heart disease if there is too much in the blood

chromatogram 82 pattern of spots produced by paper or thin-layer chromatography

chromatography 82 analytical technique for separating and identifying the components of a mixture, using paper or a thin layer or column of adsorbent

chromosome 19, 68, 70–1 thread-like structure in the cell nucleus that carries genetic information

climate change changes in seasonal weather patterns that occur because the average temperature of Earth's surface is increasing owing to global warming

clone 20, 23 group of genetically identical organisms

coal 48 solid fossil fuel formed from plant material – composed mainly of carbon

collision frequency 86 number of collisions per second between the particles involved in a chemical reaction

collision theory 86, 92 relates reaction rates to the frequency and energy of collisions between the reacting particles

column chromatography chromatography method using a solvent to carry substances down a column of adsorbent

combustion 31 process where substances react with oxygen, releasing heat

communications satellite 54 artificial satellite that stays above the same point on Earth's surface as it orbits, and used to send communications signals around the world

community all the organisms, of all species, that live together in the same habitat at the same time

compact fluorescent lamp (CFL) light bulb that is efficient at transferring the energy in an electric current as light

competition 15, 23 result of more than one organism needing the same resource, which is in short supply

compost 18 partly rotted organic material, used to improve soil for growing plants

compound 24 substance composed of two or more elements joined together by chemical bonds, for example, H_2O

compression region of a longitudinal wave where the vibrating particles are squashed together more than usual

concentration 83, 87 amount of chemical present in a given volume of a solution – usually measured as g/dm3 or mol/dm3

concrete 27 mixture of cement, sand, aggregate and water

condensation 32, 43 change of state when a substance changes from a gas or vapour to a liquid: the substance condenses

conduction (electrical) 81 flow of electrons through a solid, or ions through a liquid

conduction (thermal) 43–4, 57 heat passing through a material by transmitting vibrations from one particle to another

conductor material that transfers energy easily

consumer an organism that feeds on other organisms

contact process industrial process for making sulfuric acid

continental drift 38 movement of continents relative to each other

continental plate 38 tectonic plate carrying large landmass, though not necessarily a whole continent

convection current 38, 41, 43 when particles in a liquid or gas gain energy from a warmer region and move into a cooler region, being replaced by cooler liquid or gas

convection 43–4, 57 heat transfer in a liquid or gas – when particles in a warmer region gain energy and move into cooler regions carrying this energy with them

conventional current flow direction of flow of electric current around a circuit, from positive to negative – the opposite direction to the flow of electrons

coordination communicating between different parts of the body so that they can act together

core (of Earth) 38, 41 layer in centre of Earth, consisting of a solid inner core and molten outer core

cosmic microwave background radiation (CMBR) 56 microwave radiation coming very faintly from all directions in space

covalent bond 25, 32, 78, 79–80, 92 bond between atoms in which some of their outer electrons are shared

cracking 33–4, 41 oil refinery process that breaks down large hydrocarbon molecules into smaller ones

critical mass 108 the minimum mass of nuclear fuel needed to make a chain reaction happen

crumple zone 97 part of a vehicle designed to absorb energy in an accident, so reducing injuries to passengers

crust 38, 41 surface layer of Earth made of tectonic plates

crystallise form crystals from a liquid – for example, by partly evaporating a solution and leaving to cool

culture 9 a population of microorganisms, grown on a nutrient medium

current 47, 100–1, 104, 110 flow of electricity around a circuit – carried by electrons through solids and by ions through liquids

cuttings 13, 20 small pieces of a plant that can grow into complete new plants

cystic fibrosis 72 a genetic disorder caused by a recessive allele, where lungs become clogged with mucus

cytoplasm 58 the jelly-like material inside a cell, in which metabolic reactions take place

D

decay (biological) 18, 23 the breakdown of organic material by microorganisms

deceleration 93 see negative acceleration

delocalised electrons electrons not attached to any particular atom, so free to move through the structure, allowing electrical conduction – present in metals and graphite

denatured 64 the shape of an enzyme molecule has changed so much that it can longer bind with its substrate

dendrite 10 a short thread of cytoplasm on a neurone, carrying an impulse towards the cell body

dendron long thread of cytoplasm on a neurone, carrying an impulse towards the cell body

diabetes disease in which the body cannot control its blood sugar level

diatomic molecule molecule consisting of just two atoms

differentiation change of a cell from general-purpose to one specialised to carry out a particular function

diffraction 51 change in the direction of a wave caused by passing through a narrow gap or round an obstacle such as a sharp corner

diffusion 59, 74 spreading of particles of a gas, or of any substance in solution, resulting in a net movement from a region where they are in a high concentration to a region where they are in a lower concentration

digestive juices 60 liquids secreted within the digestive system and containing enzymes that help to digest food molecules

digital signal 55 communications signal sent as an electromagnetic wave that is switched on and off very rapidly

diode 101 semiconductor device that allows an electric current to flow through it in only one direction

direct current (d.c.) 91, 103, 110 electric current where the direction of the flow of current stays constant, as in cells and batteries

distance–time graph 93, 110 graph showing how the distance an object travels varies with time: its gradient shows speed

distillation 32 process for separating liquids by boiling them, then condensing the vapours

distribution the transmission of electricity from a power station to homes and businesses

DNA 19, 69, 71 deoxyribonucleic acid – the chemical from which chromosomes are made: its sequence determines genetic characteristics, such as eye colour

dominant (allele) a dominant allele has an effect even when another allele is present

Doppler effect 56 change in wavelength and frequency of a wave that an observer notices if the wave source is moving towards them or away from them

double covalent bond two covalent bonds between the same pair of atoms – each atom shares two of its own electrons plus two from the other atom

drug dependency 13 feeling that you cannot manage without a drug

drug 13–14, 23 a chemical that changes the chemical processes in the body

ductile can be drawn out into thin wires

E

earth wire 110 wire connecting the case of an electrical appliance, through the earth pin on a three-pin plug, to earth

earthed 110 safety feature where part of an appliance is connected to earth to protect users from electrocution if there is a fault

earthquake 39, 51 shaking and vibration at the surface of the Earth resulting from underground movement or from volcanic activity

echo 52 reflection of a sound wave

effector 10, 23 part of the body that responds to a stimulus

efficiency 17, 46, 57 a measure of how effectively an appliance transfers the energy that flows into the appliance into useful effects

egg cell 12, 19 female gamete

elastic 96 material that returns to its original shape when the force deforming it is removed

elastic collision 99 collision where colliding particles or objects bounce apart after collision

elastic potential energy 96, 110 energy stored in an object because it is stretched, compressed or deformed, and released when the object returns to its original shape

electrical power 47–50, 57 a measure of the amount of energy supplied each second

electricity generator device for generating electricity

electrode 29, 91 solid electrical conductor through which the current passes into and out of the liquid during electrolysis – and at which the electrolysis reactions take place

electrolysis 27, 29, 30, 41, 83, 91, 92 decomposing an ionic compound by passing a d.c. electric current through it while molten or in solution

electrolyte 91 solution or molten substance that conducts electricity

electromagnetic (EM) radiation 53–4 energy transferred as electromagnetic waves

electromagnetic spectrum 53 electromagnetic waves ordered according to wavelength and frequency – ranging from radio waves to gamma rays

electromagnetic waves 51, 53–5, 57 a group of waves that transfer energy – they can travel through a vacuum and travel at the speed of light

electron 24–5, 41, 75–7, 99, 105, 110 small particle within an atom that orbits the nucleus (it has a negative charge)

electronic configuration 24 the arrangement of electrons in shells, or energy levels, in an atom

electronic structure (or configuration) 24–5, 41 arrangement of electrons in shells, or energy levels, in an atom

electrostatic induction 92, 100 electric charge induced on an object made of an electrical insulator, by another electrically charged object nearby

element 24 substance made out of only one type of atom

embryo transplant 20, 23, 69 taking an embryo that has been produced from one female's egg and placing it into another female

embryo a very young organism, which began as a zygote and will become a fetus

emit an object emits energy when energy is transferred away from the object as infrared radiation, decreasing the temperature of the object

empirical formula 83 ratio of elements in a compound, as determined by analysis – for example, CH_2O for glucose (molecular formula $C_6H_{12}O_6$)

emulsifier 37, 41 a substance that prevents an emulsion from separating back into oil and water

emulsion 37, 41 a thick, creamy liquid made by thoroughly mixing an oil with water (or an aqueous solution)

endothermic reaction 89, 92 chemical reaction which takes in heat or energy from other sources

energy 6, 17, 42, 89 the ability to 'do work'

energy input 45–7, 57 the energy transferred to a device or appliance from elsewhere

energy levels 24, 89 electrons in shells around the nucleus – the further from the nucleus, the higher the electron's energy level

energy output 46–7, 51 the energy transferred away from a device or appliance – it can be either useful or wasted

environment 16 an organism's surroundings

enzyme 60, 64–6, 74, 88 biological catalyst that increases the speed of a chemical reaction but is not used up in the process

epidemic 7 many people having the same infectious disease

epidermis 61 a tissue covering the outer surface of a plant's leaf, stem or root

epithelial tissue 60 tissue forming a covering over a part of an animal's body

equal and opposite forces balanced forces equal in size but acting in opposite directions

essential oils 36 oils found in flowers, giving them their scent – they vaporise more easily than natural oils from seeds, nuts and fruit

ethanol 31, 34, 41 an alcohol that can be made from sugar and used as a fuel

evaporation 42, 44 change of state where a substance changes from liquid to gas at a temperature below its boiling point

evolution 21–3 a change in a species over time

exothermic reaction 88–9, 92 chemical reaction which gives out heat

extinct 73 no longer existing

extremophile 15 an organism that can live in conditions where a particular factor, such as temperature or pH, is outside the range that most other organisms can tolerate

F

fermentation 34, 41 process in which yeast converts sugar into ethanol (alcohol)

fertilization 20 fusion of the nuclei of a male and a female gamete

fertility drug 12 hormone given to women to cause the ovaries to produce eggs

fibre optic cable 55 glass fibre that is used to transfer communications signals as light or infrared radiation

filament bulb 101 lightbulb giving out light by current flowing through a fine wire and heating the wire until it glows white hot

finite resource material of which there is only a limited amount – once used it cannot be replaced

flammable catches fire and burns easily

flexible (material) can be bent without the material breaking

food chain 17–18, 23 flow diagram showing how energy is passed from one organism to another

force meter device measuring the size of a force, by measuring how much the force stretches a spring

formula (for a chemical compound) 24, 76, 83 group of chemical symbols and numbers, showing which elements, and how many atoms of each, a compound is made up of

forward reaction reaction from left to right in an equation for a reversible reaction

fossil 72, 74–5 preserved remains of a long-dead organism

fossil fuel 19, 31, 40, 49–50 fuel such as coal, oil or natural gas, formed millions of years ago from dead plants and animals

fractional distillation 32, 34, 41 process that separates the hydrocarbons in crude oil according to size of molecules

fractionating column 32 tall tower in which fractional distillation is carried out at an oil refinery

fractions 32 the different substances collected during fractional distillation of crude oil

'free' electrons 43 electrons that move readily from one atom to another to transmit an electric current through a conductor

freezing 43 change of state in which a substance changes from a liquid to a solid

frequency 52, 57, 103 the number of waves passing a set point per second

friction 97 force acting at points of contact between objects moving over each other, to resist the movement

FSH 12 hormone, produced by the pituitary gland, that causes eggs to mature in the ovaries

fuel cell device that generates electricity directly from a fuel, such as hydrogen, without burning it

fuel a material that is burned for the purpose of generating heat

fullerenes 81 cage-like carbon molecules containing many carbon atoms, for example, C_{60}, buckminsterfullerene

fungus (pl. fungi) 18, 58 living organisms whose cells have cell walls, but that cannot photosynthesise

fuse 103–4, 110 a fine wire that melts if too much current flows through it, breaking the circuit and so switching off the current

G

gamete 12, 19–20, 68 sex cell – a cell, such as an egg or sperm, containing the haploid number of chromosomes

gamma rays 54, 105, 107, 110 ionising electromagnetic radiation – radioactive and dangerous to human health

gas chromatography 82 method that uses a gas to carry the substances through a long, thin tube of adsorbent

gasohol mixture of gasoline (petrol) and alcohol (ethanol) used as a vehicle fuel

gene 6, 19, 23, 64, 69, 70–1, 74 section of DNA that codes for a particular characteristic

genetic diagram 71, 74 a format used to describe and explain the probable results of a genetic cross

genetic engineering 21, 23 changing the genes in an organism, for example by inserting genes from another organism

genetically modified 21, 23 organism that has had genes from a different organism inserted into it

genotype 70 the pair of alleles that an organism possesses for a particular gene

geographical isolation 73 the separation of two populations of a species by a geographical barrier, such as a mountain chain

geothermal power station 49 power station generating electricity using the heat in underground rocks to heat water

giant covalent structure 80 solid structure made up of a regular arrangement of covalently bonded atoms – may be an element or a compound

giant ionic structure solid structure made up of a regular arrangement of ions in rows and layers

gland 10 organ that secretes a useful substance

glandular tissue 60 tissue made up of cells that are specialised to secrete a particular substance

global dimming 31 gradual decrease in the average amount of sunlight reaching Earth's surface

Global Positioning System (GPS) navigation system using signals from communications satellites to find an exact position on the surface of Earth

global warming 31, 42, 49 gradual increase in the average temperature of Earth's surface

glucose 61–2, 67, 74 a simple sugar, made by plants in photosynthesis, and broken down in respiration to release energy inside all living cells

glycogen 67   carbohydrate used for energy storage in animal cells

gravitational potential energy 98, 110   energy that an object has because of its position, for example, increasing the height of an object above the ground increases its gravitational potential energy

gravitropism 13, 23   a growth response to gravity

gravity 98   the attractive force acting between all objects with mass – on Earth the attractive force due to gravity pulls objects downwards

green fuel   fuel that does less damage to the environment than fossil fuels

greenhouse gas 40   a gas such as carbon dioxide that reduces the amount of heat escaping from Earth into space, thereby contributing to global warming

H

Haber process   industrial process for making ammonia

haemoglobin 71   chemical in red blood cells which carries oxygen

haemophilia   disease where blood lacks the ability to clot

halide ion   ion of a halogen – halide ions have a 1– charge

halogens 78, 90   reactive non-metals in Group 7 of the periodic table

hard water 90   water supply containing dissolved calcium or magnesium salts – these react with soap, making it hard to form a lather

HDL 6   a type of cholesterol that does not appear to cause heart disease and may help to protect against it

heart disease 6   blockage of blood vessels that bring blood to the heart

herbivore   an animal that eats plants

heterozygous 70   possessing two different alleles of a gene

homozygous 70   possessing two identical alleles of a gene

Hooke's Law 96   for an elastic object, the extension is proportional to the force applied, provided the limit of proportionality is not exceeded

hormones 10   chemicals that act on target organs in the body (hormones are made by the body in special glands)

hot spot   area of Earth's crust heated by rising currents of magma – Hawaii is above a mid-Pacific hot spot

hydrocarbon 32, 41, 83   compound containing only carbon and hydrogen

hydroelectric power station 49–50   power station generating electricity using the energy from water flowing downhill

hydrogen ion H^+ ion – hydrogen ions in solution, $H^+(aq)$, make the solution acidic

hydrophilic 37, 41   water-loving (attracted to water, but not to oil) – used to describe parts of a molecule

hydrophobic 37, 41   water-fearing (attracted to oil, but not to water) – opposite of hydrophilic

hydroxide ion OH^- ion – hydroxide ions in solution, $OH^-(aq)$, make the solution alkaline

hydroxide   ion consisting of an oxygen and a hydrogen atom (written as $OH-$)

hypothesis   an idea that explains a set of facts or observations – a basis for possible experiments

I

image   an image is formed by light rays from an object that travel through a lens or are reflected by a mirror

immiscible 37   liquids that do not mix, but form separate layers, are immiscible

immune system 7, 9   a body system that acts as a defence against pathogens, such as viruses and bacteria

immunity 9   you have immunity if your immune system recognises a pathogen and fights it

incident ray 52   the ray of light hitting a mirror or lens

inelastic collision 99   collision where the colliding particles or objects stick together after collision

infrared radiation 42, 44, 53–4, 57   energy transferred as heat – a type of electromagnetic radiation

inhibitor   substance used to decrease a reaction rate – also called a 'negative catalyst'

initial reaction rate   reaction rate at the start of the reaction

inoculating loop   metal loop that is used to transfer microorganisms

insoluble 90   not soluble in water (forms a precipitate)

insoluble salt 90, 92   salt which is not soluble in water, so forms a precipitate

instantaneous reaction rate 85   reaction rate at a particular instant during the reaction

insulator   material that transfers energy only very slowly – thermal insulators transfer heat slowly, electrical insulators do not allow an electric current to flow through them

insulin 10, 21   hormone made by the pancreas that reduces the level of glucose in the blood

intermolecular forces 79–80   forces between molecules

ion 76, 90, 105, 110   atom (or group of atoms) with a positive or negative charge, caused by losing or gaining electrons

ionic bonding 25, 77   chemical bond formed by attractions between ions of opposite charges

ionic compound 78, 90, 92   compound composed of positive and negative ions held together in a regular lattice by ionic bonding, for example, sodium chloride

ionic equation 89   shows only the ions that actually react – anything that does not change during the reaction is omitted

ionise 105   to cause electrons to split away from their atoms (some forms of EM radiation are harmful to living cells because they cause ionisation)

isomerase 66   enzyme that changes glucose to fructose

isotopes 75, 105, 107   forms of element where their atoms have the same number of protons but different numbers of neutrons

IVF 12   in vitro fertilisation – the fertilisation of an egg by a sperm in a glass container

J

joule 42, 47   unit used to measure energy

K

kilowatt-hour 48   the energy transferred in 1 hour by an appliance with a power rating of 1 kW (sometimes called a 'unit' of electricity)

kinetic energy 97–8, 100, 110   energy an object has because of its movement – it is greater for objects with greater mass or higher speed

kinetic theory 43, 57   model used to explain how energy is transferred by particles in a substance

L

lactic acid 67   a waste product of anaerobic respiration in muscle cells

laterally inverted image 52   left and right are reversed, when seen in a mirror

lattice 80   regular arrangement of ions or atoms in a solid – may be covalent or ionic

lava 39   magma that has erupted onto the surface of Earth

Law of Conservation of Energy 46   energy can be transferred but cannot be created or destroyed

Law of Conservation of Momentum 99   total momentum before a collision is equal to total momentum after the collision, if no outside forces are acting

LDL 6, 101   a type of cholesterol that increases the risk of heart disease

LDR (light dependent resistor)   resistor with a resistance that decreases when light is shone on it

leaching   using a chemical solution to dissolve a substance out of a rock

LED (light emitting diode) 47   diode that gives off light when a current flows through it

LH 12   hormone produced by the pituitary gland, which causes an egg to be released from an ovary

lichen 16   small organism that consists of both a fungus and an alga

limestone 26–8, 41   type of rock consisting mainly of calcium carbonate

limewater 27   calcium hydroxide solution

limit of proportionality (for Hooke's Law)   the point for an elastic material when the extension stops being proportional to the force: materials break or are permanently damaged when stretched beyond this point

limiting factor 63, 74   anything that is in short supply and therefore stops a process from happening faster

lipase 65   enzyme that breaks down fat molecules to fatty acids and glycerol molecules

lithosphere 38   the rocky, outer section of the Earth, consisting of the crust and upper part of the mantle

longitudinal wave 51, 57   a wave in which the direction that the particles are vibrating is the same as the direction in which the energy is being transferred by the wave

low-grade ore   ore containing only a small percentage of metal

lymphocyte 7   type of white blood cell

M

macromolecule   very large molecule made up of hundreds of thousands, or millions, of atoms, for example, a polymer or crystal with a giant covalent structure

magma 38–9   molten rock found below Earth's surface

main sequence star   a star in which nuclear fusion reactions combine small atomic nuclei into elements with larger nuclei

mains supply   domestic electricity supply – in the UK, mains supply is 230 V at 50 Hz

malleable   can be hammered into shape without breaking

malnourished 6   not having a balanced diet

mantle 38, 41   semi-liquid layer of the Earth beneath the crust

mass 6   a measure of the amount of 'stuff' in an object

mass number 24, 75   total number of protons and neutrons in the nucleus of an atom – always a whole number

mass spectrometer 82   instrument for identifying chemicals by measuring their relative formula mass very accurately

mechanical wave   wave in which energy is transferred by particles or objects moving, such as a wave on a string or a water wave

meiosis 68, 74   type of cell division producing four genetically different daughter cells, each with half the normal number of chromosomes

melting 43   change of state of a substance from liquid to solid

menstrual cycle 12, 23   monthly hormonal cycle that starts at puberty in human females

menstruation 12, 23   monthly breakdown of the lining of the uterus leading to bleeding from the vagina

phytoremediation cleaning up contaminated soil by using growing plants to absorb harmful metal compounds

pipette used to measure out an exact volume of liquid

placebo 'dummy' treatment given to some patients, in a drug trial, that does not contain the drug being tested

plane mirror mirror with a flat surface

planet large ball of gas or rock travelling around a star – for example Earth and other planets orbit our Sun

plaque 6, 23 build-up of cholesterol in a blood vessel (which may block it)

plastics compounds produced by polymerisation, capable of being moulded into various shapes or drawn into filaments and used as textile fibres

plate boundaries 39 edges of tectonic plates, where they meet or are moving apart

plum pudding model 106 model that said the atom was like a positively charged 'jelly' with negatively charged electrons dotted through it – later shown to be incorrect

pollution 16, 23 presence of substances that contaminate or damage the environment

poly(ethene) 33–4, 41 plastic polymer made from ethene gas (also called polythene)

polydactyly 72 having more than five fingers or toes on a hand or foot

polymer 34–5, 41, 80 large molecule made up of a chain of monomers

polymerization 34 chemical process that combines monomers to form a polymer: this is how polythene is formed

power 47–8, 50, 98, 110 amount of energy that something transfers each second and is measured in watts (or joules per second)

power rating 47 a measure of how fast an electrical appliance transfers energy supplied as an electrical current

power station 48, 50 place where electricity is generated to feed into the National Grid

precipitate 90 solid product formed by reacting two solutions

precipitation 83 reaction between two solutions to form a solid product (a precipitate)

producer organism that makes its own food from inorganic substances

products chemicals produced at the end of a chemical reaction

progesterone 12 hormone, produced by the ovary, that prepares the uterus for pregnancy

protease 65 enzyme that breaks down protein molecule to amino acid molecules

protein 6, 64 molecule made up of amino acids (found in food of animal origin and also in plants)

proton 24–5, 41, 75, 105, 110 small positive particle found in the nucleus of an atom

protostar 109 dense cloud of dust and gas that can form a new star, if it contains enough matter

pyramid of biomass 17, 23 a diagram in which boxes, drawn to scale, represent the biomass at each step in a food chain

Q

quadrat 63 a square area within which type and numbers of living organisms can be counted or estimated

quarry 26 place where stone is dug out of the ground

R

radio wave 53–4 non-ionising radiation used to transmit radio and TV

radioactive 75, 105–7, 110 materials giving off nuclear radiation

radioisotope 107 a radioactive isotope of an element

rarefaction areas of a longitudinal wave in which the vibrating particles are spread out more than usual

rate of energy transfer a measure of how quickly something moves energy from one place to another

ray diagram 52 diagrams showing how light rays travel

RCCB (residual current circuit breaker) 104 measures the current flowing into and out of an appliance, and switches the current off if they are not equal

reactants 26, 92 chemicals that are reacting together in a chemical reaction

reaction conditions physical conditions under which a reaction is performed, for example, temperature and pressure

reaction rate 85, 87, 92 the speed at which a chemical reaction takes place – measured as the amount of reaction per unit time

reaction rate (average) 85 total amount of reaction ÷ total time

reaction rate (initial) 85 reaction rate at the start of the reaction

reaction time 95 time between when a driver sees a hazard and when they begin to respond to it – increased by tiredness, drugs, or distractions

reactivity series 27, 77 list of metals in order of their reactivity with oxygen, water and acids

receptor 11, 23 nerve cell that detects a stimulus

recessive (allele) a recessive allele only has an effect when a dominant allele is not present

red shift 56–7 when lines in a spectrum are redder than expected – if an object has a red-shift it is moving away from the observer

reduction 27–8, 83, 91 process that reduces the amount of oxygen in a compound, or removes all the oxygen from it – opposite of oxidation

reflected ray 51 ray of light 'bouncing off' from a mirror or reflecting surface

reflex action 11 a fast, automatic response to a stimulus

reflex arc 11 pathway taken by nerve impulse from receptor, through nervous system, to effector

refraction 51 change of direction when a wave hits the boundary between two media at an angle, for example when a light ray passes from air into ƒa glass block

regenerative braking 97 type of braking which transfers some of the kinetic energy wasted by the slowing car into electricity which is used to operate the car brakes

relative atomic mass 75 average mass of all the atoms in an element, taking into account the presence of different isotopes – often rounded to the nearest whole number

relative formula mass 75, 83 total mass of all atoms in a formula = each relative atomic mass × number of atoms present

relative molecular mass 75 same as relative formula mass, but limited to elements or compounds that have separate molecules

renewable resource 49 energy resource that is constantly available or can be replaced as it is used

repeatability consistent results are obtained when a person uses the same procedure a number of times

reproducibility how likely it is that measurements, made again under similar conditions, give the same results

resistance (electrical) 101–2, 110 measure of how hard or how easy it is for an electric current to flow through a component

resistant strain (of bacteria) 8, 22, 23 a population of bacteria that is not killed by an antibiotic

respiration 66, 74 process occurring in all living cells, in which energy is released from glucose

resultant force 94, 110 the single force that would have the same effect on an object as all the forces that are acting on the object

retention factor (Rf) 82 used to help identify individual spots in a chromatogram (Rf = distance moved by the spot ÷ distance moved by the solvent)

retention time (Rt) 82 time taken for a component to travel through the tube of adsorbent during gas chromatography

reverse reaction 85 reaction from right to left in the equation for a reversible reaction

reversible reaction 85, 92 a reaction that can also occur in the opposite direction – that is, the products can react to form the original reactants again

ribosomes 58 tiny structures within a cell, where protein synthesis takes place

rod cell 10 receptor cell in the eye that detects light

rutile an ore of titanium – impure titanium oxide (TiO_2)

S

salt compound composed of metal ions and non-metal ions – formed by acid–base neutralisation

sample take measurements, or make counts, in a small area rather than over the entire area in question

Sankey diagram 46 diagram showing how the energy supplied to something is transferred into 'useful' or 'wasted' energy

saturated fat 6, 33, 37 solid fat, most often of animal origin, containing no C=C double bonds

saturated hydrocarbon hydrocarbon containing only single covalent bonds

scalar quantity quantity that only has size, but not direction, for example, energy is a scalar quantity

secretion 60 production and release of a useful substance

sensory neurone 10, 23 nerve cell carrying information from receptors to the central nervous system

series circuit 100, 102, 110 electrical circuit with only one possible path for the current to flow around

sex chromosomes 70 the X and Y chromosomes

shape memory alloy alloy that 'remembers' its original shape and returns to it when heated

shells 24–5, 41 electrons are arranged in shells (or orbits) around the nucleus of an atom – also known as 'energy levels'

slag 27–8 waste material produced during smelting of a metal – it contains unwanted impurities from the ore

smart material 28, 35, 81 material which changes in response to changes in its surroundings, such as light levels or temperature

smelting 30 extracting metal from an ore by reduction with carbon – heating the ore and carbon in a furnace

soften (water) treat water so as to remove the calcium ions that cause hardness

solar cell device that converts the Sun's energy into electricity

solar panel 57 panel that uses the Sun's energy to heat water

solar power station 49 power station generating electricity using energy transferred by the Sun's radiation

soluble salt 90 salt which dissolves in water

solute substance which dissolves in a liquid to form a solution

solvent liquid in which solutes dissolve to form a solution

speciation 73 formation of a new species

species 73–4 group of organisms that share similar characteristics and that can breed together to produce fertile offspring

specific heat capacity 45, 57 a measure of the amount of energy needed to raise the temperature of 1 kg of a substance by 1 °C

speed of light 57 speed at which electromagnetic radiation travels through a vacuum – 300 000 000 metres per second

speed how quickly an object is moving, usually measured in metres per second (m/s)

sperm cell 19, 59, 67 male gamete

stainless steel 29 steel alloy containing chromium and nickel to resist corrosion

starch 62, 65, 74 a carbohydrate; a polysaccharide that is used for storing energy in plant cells, but not in animal cells

state symbol symbol used in equations to show whether something is solid, liquid, gas or in solution in water

states of matter 43 substances can exist in three states of matter (solid, liquid or gas) – changes from one state to another are called changes of state

static electricity 99 an electric charge on an insulating material, caused by electrons flowing onto or away from the object

statin 14 drug that reduces cholesterol level in the blood

steam cracking cracking hydrocarbons by mixing with steam and heating

steam distillation 36 process of blowing steam through a mixture to vaporise volatile substances – used to extract essential oils from flowers

steel 28–9 alloy of iron and steel, with other metals added depending on its intended use

stem cell 69 a cell that has not yet differentiated – it can divide to form cells that form various kinds of specialised cell

step-down transformer 50 transformer that changes alternating current to a lower voltage

step-up transformer 50 transformer that changes alternating current to a higher voltage

sterile technique handling apparatus and material to prevent microorganisms from entering them

sterile 9, 23 containing no living organisms

stimulus 11 a change in the environment that is detected by a receptor

stopping distance 95, 110 total distance it takes a vehicle to stop – the sum of thinking distance and braking distance

sub-atomic particle 24, 105 particle that make up an atom, such as proton, neutron or electron

subcutaneous just under the skin

subduction zone 39 area of ocean floor in which an oceanic plate is sinking beneath a continental plate

sublime turn directly from solid into a gas without melting

substrate 64 molecule on which an enzyme acts – the enzyme catalyses the reaction that changes the substrate into a product

successful collisions 86 collisions with enough energy to break bonds in the reactant particles, and thus cause a reaction

sugar 66 sweet-tasting compound of carbon, hydrogen and oxygen such as glucose or sucrose

sulfur dioxide 16, 30 poisonous, acidic gas formed when sulfur or a sulfur compound is burned

surface area (of a solid reactant) 87 measure of the area of an object that is in direct contact with its surroundings

sweat 11, 23 liquid secreted onto the skin surface that has a cooling effect as it evaporates

symbol (for an element) 24 one or two letters used to represent a chemical element, for example C for carbon or Na for sodium

synapse 11 gap between two neurones

synthetic artificial or made by people

syrup 66 concentrated solution of sugar

T

target organ the part of the body affected by a hormone

tarnish go dull and discoloured by reacting with oxygen, moisture or other gases in the air

tectonic plate 38, 41 section of Earth's crust that floats on the mantle and slowly moves across the surface

telecommunications 54 communications over long distances using various types of electromagnetic radiation

terminal velocity 96, 110 maximum velocity an object can travel at – at terminal velocity, forward and backward forces are the same

testes 10 organs in a male in which sperms are made

tetrahedral structure structure in which atoms have four covalent bonds to other atoms positioned at the four corners of a tetrahedron, for example, diamond

thalidomide 14 a drug that was originally prescribed to pregnant women but was found to cause deformities in fetuses

theoretical yield   mass of product that a given mass of reactant should produce according to calculations from the equation – the actual yield is always less than this

thermal decomposition 33, 41   chemical reaction in which a substance is broken down into simpler chemicals by heating it

thermistor 101   resistor made from semiconductor material: its resistance decreases as temperature increases

thermochromic 81   thermochromic materials change colour in response to changes in temperature

thermosetting polymer 80, 92   plastic polymer that sets hard when heated and moulded for the first time – it will not soften or melt when heated again

thermosoftening polymer 80, 92   plastic polymer that softens and melts when heated and reheated

thin layer chromatography (TLC) 82   chromatography using a plate coated with a thin layer of powdered adsorbent

thinking distance 95   distance a vehicle travels while a signal travels from the driver's eye to brain and then to foot on the brake pedal: thinking distance increases with vehicle speed

three core cable 103, 110   electrical cable containing three wires, live, neutral and earth

three-pin plug 103, 110   type of plug used for connecting to the mains supply in the UK: it has three pins, live, neutral and earth

tidal power station 49   power station generating electricity using the energy transferred by moving tides

tissue 60, 74   group of cells that work together and carry out a similar task, such as lung tissue

titration   procedure to determine the volume of one solution needed to react with a known volume of another solution

tonne   1 tonne = 1000 kg (1 million grams)

toxin 7   poisonous substance (pathogens make toxins that make us feel ill)

tracer 107   radioactive element used to track the movement of materials, such as water through a pipe or blood through organs of the body

transect 63   line along which organisms are sampled – transects are often used to investigate how the distribution of organisms changes when one type of habitat merges into another

transfer (energy)   energy transfers occur when energy moves from one place to another, or when there is a change in the way in which it is observed

transformer 50   device by which alternating current of one voltage is changed to another voltage

transition metals 88   group of metal elements in the middle block of the periodic table – includes many common metals

transmitter chemical   chemical that transfers a nerve impulse across a synapse

transverse wave 51, 53, 57   a wave in which the vibration of particles is at right angles to the direction in which the wave transfers energy

triple covalent bond   three covalent bonds between the same two atoms – each atom shares three of its own electrons plus three from the other atom

tropism 13   response of a plant to a stimulus, by growing towards or away from it

tsunami   huge waves caused by earthquakes – can be very destructive

turbine 49   device for generating electricity – the turbine has coils of wire that rotate in a magnetic field to generate electricity

U

ultraviolet radiation 53–4   electromagnetic radiation that can damage human skin

unsaturated fats 6, 33, 37, 41   liquid fats, containing $C=C$ double bonds – usually from plants or fish

unsaturated hydrocarbon   hydrocarbon containing one or more $C=C$ double bonds

upright image 52   image that is the same way up as the object

upthrust   upward force on an object in water – for a floating object, the upthrust is equal to the weight of the object

U-value 57   a measure of how easily energy is transferred through a material as heat

V

vaccine 9, 23   killed microorganisms, or living but weakened microorganisms, that are given to produce immunity to a particular disease

vacuole 58   liquid-filled space inside a cell – many plant cells contain vacuoles full of cell sap

vacuum 53   a space in which there are no particles of any kind

validity   how well a measurement really measures what it is supposed to be measuring

Van der Graaff generator   device for investigating static electricity: a large static electricity charge builds up on a metal dome insulated from earth

vaporise   change from liquid to gas (vapour)

variation 19   differences between individuals belonging to the same species

vector quantity   a quantity that has both size and direction: velocity is a vector quantity, having size in a particular direction

velocity 93, 110   measure of how fast an object is moving in a particular direction

velocity–time graph 93, 110   graph showing how the velocity of an object varies with time: its gradient shows acceleration

vent 39   crack or weak spot in the Earth's crust, through which magma reaches the surface

virtual image 52   image that can be seen but cannot be projected onto a screen (a mirror forms a virtual image behind the mirror)

virus 7, 8, 58   very small structure made of a protein coat surrounding DNA (or RNA); viruses can only reproduce inside a living cell

VO_2max   the maximum volume of oxygen the body can use per minute

volcano 39   landform (often a mountain) where molten rock erupts onto the surface of the planet

voltage 101   a measure of the energy carried by an electric current (the old name for potential difference)

W

wasted energy 17, 23, 46   energy that is transferred by a device or appliance in ways that are not wanted, or useful

water of crystallization 85   water molecules present in crystals of some metal salts – shown separately in the formula, for example, hydrated copper sulfate, $CuSO_4.5H_2O$

watt 47   unit of energy transfer – one watt is a rate of energy transfer of one joule per second

wave equation 53   the speed of a wave is always equal to its frequency multiplied by its wavelength

wave power 50, 57   electricity generation using the energy transferred by water waves as the water surface moves up and down

wavelength 52, 56   distance between two wave peaks

weight   the downward force on a mass due to gravity, measured in newtons (N)

wind turbine 49–50, 57   device generating electricity by using the energy in moving air to turn a turbine and a generator

work 97   amount of energy transferred to an object by a force moving the object through a distance: work done = force × distance moved in the direction of the force

X

X-rays 54   ionising electromagnetic radiation – used in X-ray photography to generate pictures of bones

xylem 61   tissue made up of long, empty, dead cells that transports water from the roots to the leaves of a plant

Y

yeast 34, 58   single-celled fungus used in making bread and beer

yield 92   mass of product made from a chemical reaction

Z

zygote 20   a diploid cell formed by the fusion of the nuclei of two gametes

Modern periodic table

Group

1	2											3	4	5	6	7	0
							1 H 1 hydrogen										4 He 2 helium
7 Li 3 lithium	9 Be 4 beryllium											11 B 5 boron	12 C 6 carbon	14 N 7 nitrogen	16 O 8 oxygen	19 F 9 fluorine	20 Ne 10 neon
23 Na 11 sodium	24 Mg 12 magnesium											27 Al 13 aluminium	28 Si 14 silicon	31 P 15 phosphorus	32 S 16 sulfur	35 Cl 17 chlorine	40 Ar 18 argon
39 K 19 potassium	40 Ca 20 calcium	45 Sc 21 scandium	48 Ti 22 titanium	51 V 23 vanadium	52 Cr 24 chromium	55 Mn 25 manganese	56 Fe 26 iron	59 Co 27 cobalt	59 Ni 28 nickel	64 Cu 29 copper	65 Zn 30 zinc	70 Ga 31 gallium	73 Ge 32 germanium	75 As 33 arsenic	79 Se 34 selenium	80 Br 35 bromine	84 Kr 36 krypton
85 Rb 37 rubidium	88 Sr 38 strontium	89 Y 39 yttrium	91 Zr 40 zirconium	93 Nb 41 niobium	96 Mo 42 molybdenum	99 Tc 43 technetium	101 Ru 44 ruthenium	103 Rh 45 rhodium	106 Pd 46 palladium	108 Ag 47 silver	112 Cd 48 cadmium	115 In 49 indium	119 Sn 50 tin	122 Sb 51 antimony	128 Te 52 tellurium	127 I 53 iodine	131 Xe 54 xenon
133 Cs 55 caesium	137 Ba 56 barium	139 La 57 lanthanum	178 Hf 72 hafnium	181 Ta 73 tantalum	184 W 74 tungsten	186 Re 75 rhenium	190 Os 76 osmium	192 Ir 77 iridium	195 Pt 78 platinum	197 Au 79 gold	201 Hg 80 mercury	204 Tl 81 thallium	207 Pb 82 lead	209 Bi 83 bismuth	210 Po 84 polonium	210 At 85 astatine	222 Rn 86 radon
223 Fr 87 francium	226 Ra 88 radium	227 Ac 89 actinium															

Workbook

NEW GCSE SCIENCE

Science and Additional Science

for AQA A Higher

Authors: **Nicky Thomas**
Rob Wensley
Gemma Young

Revision guide +
Exam practice workbook

The key to successful revision is finding the method that suits you best. There is no right or wrong way to do it.

Before you begin, it is important to plan your revision carefully. If you have allocated enough time in advance, you can walk into the exam with confidence, knowing that you are fully prepared.

Start well before the date of the exam, not the day before!

It is worth preparing a revision timetable and trying to stick to it. Use it during the lead up to the exams and between each exam. Make sure you plan some time off too.

Different people revise in different ways and you will soon discover what works best for you.

Some general points to think about when revising

- Find a quiet and comfortable space at home where you won't be disturbed. You will find you achieve more if the room is ventilated and has plenty of light.

- Take regular breaks. Some evidence suggests that revision is most effective when tackled in 30 to 40 minute slots. If you get bogged down at any point, take a break and go back to it later when you are feeling fresh. Try not to revise when you're feeling tired. If you do feel tired, take a break.

- Use your school notes, textbook and this Revision guide.

- Spend some time working through past papers to familiarise yourself with the exam format.

- Produce your own summaries of each module and then look at the summaries in this Revision guide at the end of each module.

- Draw mind maps covering the key information on each topic or module.

- Review the Grade booster checklists on pages 242–247.

- Set up revision cards containing condensed versions of your notes.

- Prioritise your revision of topics. You may want to leave more time to revise the topics you find most difficult.

Workbook

The Workbook allows you to work at your own pace on some typical exam-style questions. You will find that the actual GCSE questions are more likely to test knowledge and understanding across topics. However, the aim of the Revision guide and Workbook is to guide you through each topic so that you can identify your areas of strength and weakness.

The Workbook also contains example questions that require longer answers (**Extended response questions**). You will find one question that is similar to these in each section of your written exam papers. The quality of your written communication will be assessed when you answer these questions in the exam, so practise writing longer answers, using sentences. The **Answers** to all the questions in the Workbook are detachable for flexible practice and can be found on pages 257–280.

At the end of the Workbook there is a series of **Revision checklists** that you can use to tick off the topics when you are confident about them and understand certain key ideas.

Remember

There is a difference between learning and revising.

When you revise, you are looking again at something you have already learned. Revising is a process that helps you to remember this information more clearly.

Learning is about finding out and understanding new information.

Diet and energy

1 Susan and Richard are overweight and want to lose mass. They both go on a diet and eat exactly the same food for a week. Richard loses more mass.

(a) Suggest two possible reasons why.

_____

_____

_____ [2 marks]

(b) Explain why your reasons will lead to Richard losing more mass.

_____

_____

_____ [2 marks]

D–C

2 Explain why a diet high in fat will lead to a bigger weight gain than a diet high in protein.

_____

_____

_____

_____ [2 marks]

B–A*

Diet, exercise and health

1 Outline how high cholesterol levels in the blood can lead to a heart attack.

_____

_____

_____

_____

_____

_____

_____ [5 marks]

D–C

2 Olive oil is an unsaturated fat which can increase the amounts of HDL cholesterol in the blood. Explain why people who eat a diet high in olive oil have a low risk of developing heart disease.

_____

_____

_____

_____ [2 marks]

B–A*

Pathogens and infections

1 Explain how viruses and bacteria make us feel ill.

_____

_____

_____

_____

_____

_____

[3 marks]

D–C

2 It was known by the mid-1800s that microorganisms caused food to go off. In the 1860s, Joseph Lister, a Scottish surgeon, in a attempt to stop so many of his patients dying from infected wounds started spraying carboxylic acid over the wounds.

(a) What was Lister's hypothesis?

_____

_____

_____

[2 marks]

B–A*

(b) What evidence would he have needed to prove his hypothesis?

_____

_____

[1 mark]

Fighting infection

1 State two ways that lymphocytes can destroy pathogens.

_____

_____

_____

[2 marks]

D–C

2 Antibodies are specific.

(a) What does this mean?

_____

_____

[1 mark]

B–A*

(b) Why does the body have to be able to produce millions of different types of lymphocytes?

_____

_____

_____

[2 marks]

Drugs against disease

1 Gita has a lung infection.

(a) Her doctor prescribes her some antibiotics. What is an antibiotic?

_____

_____ [1 mark]

D–C

(b) After 2 weeks her infection has still not cleared up. The doctor takes a swab of the mucus from her lungs and sends it to a pathology lab, so they can find out which antibiotic to give her. Explain how they would do this.

_____

_____

_____

_____

_____ [4 marks]

2 Explain why doctors are less likely to prescribe antibiotics now, compared to 40 years ago.

B–A*

_____

_____

_____

_____ [2 marks]

Antibiotic resistance

1 Explain why numbers of antibiotic-resistant bacteria have increased.

D–C

_____

_____

_____

_____

_____ [4 marks]

2 Describe and explain the trend in deaths from MRSA between 1993 and 2008.

_____

_____

_____

_____

_____

B–A*

_____ [4 marks]

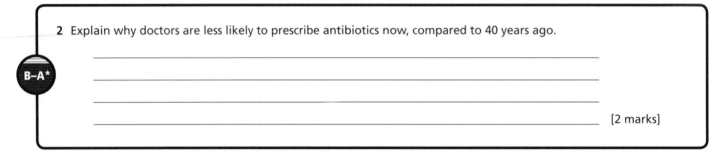

Vaccination

1 The MMR vaccine is given to children in the UK.

(a) What is the function of this vaccine?

_____

_____ [1 mark]

(b) Explain how it works.

_____

_____

_____ [3 marks]

D–C

(c) In the early 21st century, the number of children having the MMR vaccine fell. State the reason for this.

_____

_____ [1 mark]

(d) The number of children having the MMR vaccine has now risen again. Explain why.

_____

_____ [1 mark]

2 Every year, elderly people are offered a flu vaccination. Suggest why they have to have a new one every year.

_____

_____

_____ [2 marks]

B–A*

Growing bacteria

1 Why is it important to grow bacteria at a temperature less than 37 °C?

_____

_____

_____ [2 marks]

D–C

2 In a industrial lab, microorganisms are kept at temperatures of around 40 °C. Suggest why.

_____

_____

_____ [2 marks]

B–A*

Co-ordination, nerves and hormones

D–C

1 What is the central nervous system (CNS)?

_____

_____ [1 mark]

B–A*

2 Michelle is playing in a tennis match. Her body starts producing the hormone adrenaline, which increases her heart rate.

(a) The heart is a target organ of adrenaline. What does this mean?

_____

_____ [1 mark]

(b) Explain why a hormonal response rather than a nervous one is appropriate in this case.

_____

_____

_____

_____ [2 marks]

Receptors

D–C

1 The diagram shows a neurone.

(a) What type of neurone is this?

_____

_____ [1 mark]

(b) Draw an arrow to show the direction of the message sent along this neurone.   [1 mark]

(c) In what form are the messages sent?

_____

_____ [1 mark]

B–A*

2 Paul is playing as a goalkeeper in a football game. The ball is heading towards him.

(a) Which receptors in his eye will be stimulated?

_____

_____ [1 mark]

(b) Describe how a picture of the ball is constructed by his brain.

_____

_____

_____

_____

_____ [3 marks]

Reflex actions

1 Your finger touches a sharp pin. Almost immediately, your finger moves away. Describe the sequence of events that has happened to bring this about.

_____

_____

_____

_____

_____ [6 marks]

D–C

2 You grab onto a hot radiator to stop you from falling over. Ordinarily, you would have pulled away from the radiator in a reflex action but your brain has overridden this.

(a) What kind of control is you brain carrying out?

_____

_____ [1 mark]

B–A*

(b) Explain why your brain has made this decision.

_____

_____ [1 mark]

Controlling the body

1 Why is it important that human body temperature is maintained at around 37 °C?

_____

_____

_____ [2 marks]

D–C

2 If you are stranded in a desert, you should avoid exertion. Explain why.

_____

_____

_____

_____

_____ [4 marks]

B–A*

Reproductive hormones

D–C

1 Explain how the oestrogen in the contraceptive pill stops a woman from getting pregnant.

_____

_____

_____

_____

_____ [3 marks]

B–A*

2 Use the image below to explain how the level of oestrogen affects events in the menstrual cycle.

_____

_____

_____

_____

_____

_____

_____ [3 marks]

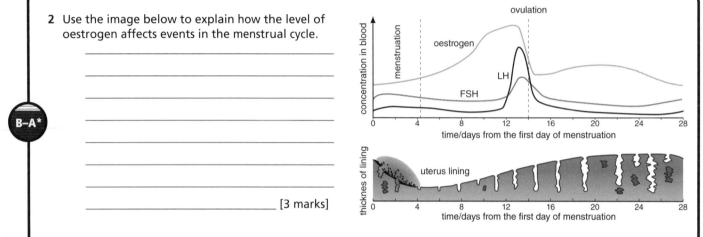

Controlling fertility

D–C

1 Outline the stages in IVF.

_____

_____

_____

_____

_____ [3 marks]

B–A*

2 The number of triplets being born in the UK has risen in the last century. Ruby thinks this is due to more women using fertility drugs.

(a) Explain why her hypothesis could be correct.

_____

_____

_____ [2 marks]

(b) Use what you know about the dangers of multiple births to suggest another hypothesis for this trend.

_____

_____

_____ [2 marks]

Plant responses and hormones

1 Tropisms can be positive or negative. Complete each of these statements with the correct word.

    (a) A shoot growing towards the light is _____ phototropism.   [1 mark]

    (b) A shoot growing away from gravity is _____ gravitropism.   [1 mark]

D–C

2 You are asked to test the hypothesis: The tip of the shoot detects the direction of light.

    (a) Write a plan which will enable you to do this.

    _____

    _____

    _____

    _____

    _____ [3 marks]

    (b) Describe what you will see if the hypothesis is correct.

    _____

    _____ [2 marks]

B–A*

Drugs

1 (a) Name a recreational drug that is:

    (i) legal _____

    (ii) illegal _____ [2 marks]

    (b) Explain why it is difficult to stop smoking.

    _____

    _____

    _____ [2 marks]

D–C

2 State two ways a person could die from alcohol consumption.

    _____

    _____

    _____ [2 marks]

B–A*

Developing new drugs

1 The table of data shows the results from a double-blind trial of a flu drug called zanamivir.

	given zanamivir	given a placebo
number of subjects	293	295
mean age in years	19	19
number of days until their temperature went down to normal	2.00	2.33
number of days until they lost all their symptoms and felt better	3.00	2.83
number of days until they felt just as well as before they had flu	4.5	6.3
average score the volunteers gave to their experience of the major symptoms of flu	23.4	25.3

TABLE 1: Effect of zanamivir and a placebo on soldiers suffering from flu

(a) What is meant by a 'double-blind' trial?

_____

_____ [3 marks]

(b) State one control variable in the trial.

_____ [1 mark]

(c) Use the data to state one reason why the drug should be distributed.

_____ [1 mark]

2 When statins were first trialled they showed no side-effects. However, since then, side-effects have been discovered. Explain why they did not discover the side-effects in the trials.

_____

_____ [4 marks]

Legal and illegal drugs

1 Study the data below and use it to answer the questions that follow.

Year	2000	2001	2002	2003	2004	2005	2006	2007	2008
Men	4483	4938	5069	5443	5431	5566	5768	5732	5999
Women	2401	2561	2632	2721	2790	2820	2990	2992	3032

TABLE 2: The number of deaths from drinking alcohol, in each year from 2000 to 2008, in England and Wales.

(a) Describe two trends as seen in the data.

_____

_____ [2 marks]

(b) Heroin is considered a much more dangerous drug than alcohol. There were around 900 deaths from heroin in 2008. Describe how this compares with the deaths from alcohol and explain the reason for the difference.

_____

_____ [2 marks]

2 State two reasons why professional sportspeople should not take performance-enhancing drugs.

_____

_____ [2 marks]

Competition

1 Suzanne is growing vegetables on her allotment. Explain to her why it is important for her to remove weeds.

_____

_____

_____ [2 marks]

D–C

2 Flamingos are able to feed in lakes that are so alkaline that almost nothing else can survive there – except the shrimps and small aquatic insects that they eat. Explain how this increases their chances of having offspring.

_____

_____

_____

_____

_____ [3 marks]

B–A*

Adaptations for survival

1 Images of a wasp and a hoverfly are shown here.

(a) Wasps can sting. How does this help them to survive?

wasp   hoverfly

_____

_____ [1 mark]

D–C

(b) The hoverfly cannot sting. Explain why it looks like the wasp.

_____

_____

_____ [2 marks]

2 Explain how the discovery of extremophiles on Earth has led scientists to believe that finding life on the planet Mars is possible.

_____

_____

_____

_____

_____ [2 marks]

B–A*

Environmental change

1 Cockatoos are birds that live in central Australia. They feed on seeds and fruit. Global warming has caused rainfall in their habitat to decrease. Explain how this will affect the cockatoo population.

D–C

_____

_____

_____

_____

_____ [3 marks]

2 (a) Predict how the fall in numbers of honeybees will affect food prices in the UK.

_____

_____ [1 mark]

B–A*

(b) Explain your prediction.

_____

_____

_____ [2 marks]

Pollution indicators

1 Cathy works for the Environmental Agency. Her job is to monitor pollution levels. She wants to measure the amount of dissolved oxygen in a stream.

(a) State the instrument she could use to do this.

_____ [1 mark]

(b) She finds high levels of oxygen. What does this tell her about the level of pollution in the stream?

_____

_____ [1 mark]

D–C

(c) Give one example of an invertebrate she may find in the stream and explain how this is a further indicator of the pollution levels in the water.

_____

_____

_____ [2 marks]

2 Bloodworms are red in colour because they contain high levels of haemoglobin, the pigment found in red blood cells. Explain how this adaptation allows them to survive in polluted water.

B–A*

_____

_____

_____ [2 marks]

Food chains and energy flow

1 Only a small amount of the energy from light falling on a plant is used for photosynthesis. State one reason for this.

_____

_____ [1 mark]

D–C

2 **(a)** A plant is 10% efficient. Use the formula:

$$\text{efficiency} = \frac{\text{useful energy transferred}}{\text{original amount of energy}} \times 100\%$$

to calculate how many units of energy are used in photosynthesis if 800 units hit the leaf.

_____

_____ [2 marks]

(b) Explain why plants that grow in shady places usually have bigger leaves.

_____

_____

_____

_____ [2 marks]

B–A*

Biomass

1 Sketch a pyramid of biomass for the following food chain: Grass → rabbit → fox

[2 marks]

D–C

2 Explain why food chains in the ocean are often much longer than food chains on land.

_____

_____

_____

_____

_____

_____ [3 marks]

B–A*

Decay

1 June has a compost heap in her garden.

   (a) In the summer, she waters the compost. Why?

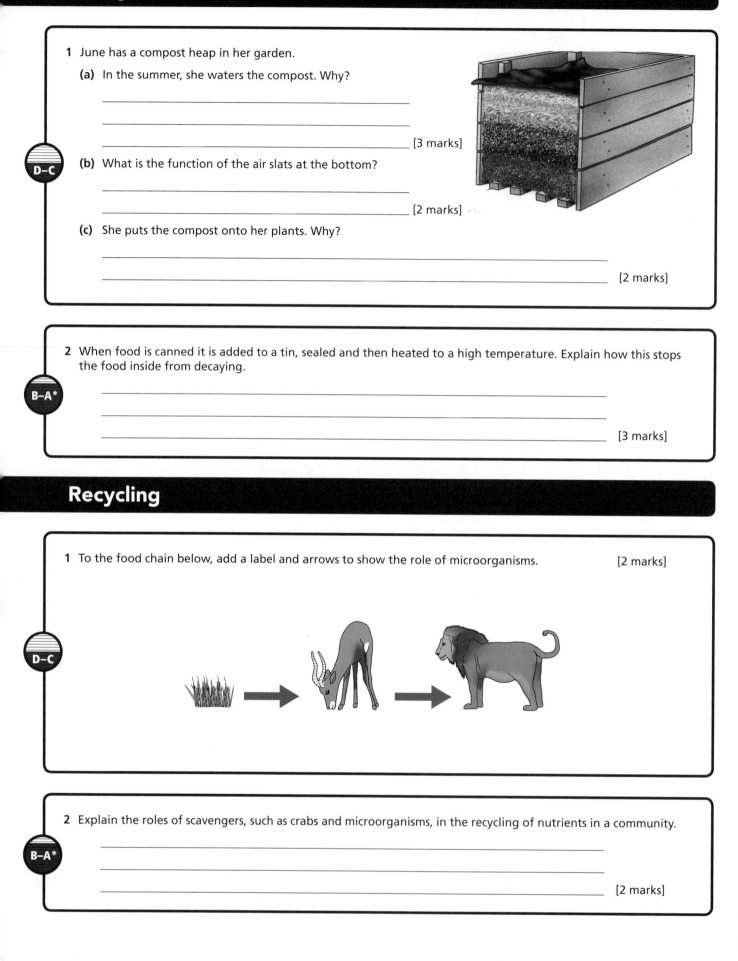

_____

_____

_____ [3 marks]

D–C

   (b) What is the function of the air slats at the bottom?

_____

_____ [2 marks]

   (c) She puts the compost onto her plants. Why?

_____

_____ [2 marks]

2 When food is canned it is added to a tin, sealed and then heated to a high temperature. Explain how this stops the food inside from decaying.

B–A*

_____

_____

_____ [3 marks]

Recycling

1 To the food chain below, add a label and arrows to show the role of microorganisms. [2 marks]

D–C

2 Explain the roles of scavengers, such as crabs and microorganisms, in the recycling of nutrients in a community.

B–A*

_____

_____

_____ [2 marks]

The carbon cycle

1 This diagram of the carbon cycle shows how carbon is recycled around the Earth.

(a) Name the process missing in:

    (i) A _____

    (ii) B _____

                         [2 marks]

(b) Animals need carbon in order to build tissue. How do they get the carbon they need?

_____

_____ **[1 mark]**

(c) A carbon atom that makes up the muscle of an animal will one day return to the atmosphere. Explain how this will happen.

_____

_____

_____ **[4 marks]**

(diagram labels: carbon dioxide in the atmosphere; respiration; respiration in decomposers; carbon in fossil fuels; respiration; organic compounds in animals; death; carbon compounds in dead organic matter; fossilisation; feeding; death; organic compounds in green plants; A; B)

D–C

2 (a) State the energy transfer that takes place in each of these stages of the carbon cycle:

    (i) photosynthesis _____ **[2 marks]**

    (ii) combustion of wood. _____ **[2 marks]**

(b) Explain how some of the energy stored in food ends up as movement energy.

_____ **[2 marks]**

B–A*

Genes and chromosomes

1 Sanjay has genes for blood group A and blood group O. Explain how he got two genes for his blood group.

_____ **[2 marks]**

D–C

2 Leanne picked 10 leaves from two different holly bushes. She measured the length of the leaves. Her results are shown in the table below:

Bush	Length of leaves (mm)
A	64, 59, 76, 56, 72, 68, 73, 77, 58, 64
B	76, 81, 82, 65, 59, 62, 59, 80, 74, 81

(a) Calculate the mean leaf length for each bush. _____ **[2 marks]**

(b) Why did she measure 10 leaves from each bush and not just one?

_____ **[1 mark]**

(c) State reasons for the difference in mean leaf length.

_____ **[2 marks]**

B–A*

Reproduction

1 This single-celled organism is reproducing by dividing into two.

   (a) What type of reproduction is this?

   _____ [1 mark]

   (b) The offspring will be a clone of its parent. Why?

   _____ [1 mark]

   (c) Explain why the offspring of animals are different from their parents.

   _____

   _____ [1 mark]

---

2 (a) State the difference between internal and external fertilisation.

   _____

   _____ [2 marks]

   (b) Most fish use external fertilisation. Explain why female fish produce thousands of eggs at a time.

   _____

   _____ [2 marks]

Cloning plants and animals

1 Margaret has successfully bred an award-winning plant. She wants to clone it.

   (a) Why does she want to clone it?

   _____

   _____ [2 marks]

   (b) State one method she can use and explain how to carry it out.

   _____

   _____

   _____ [3 marks]

---

2 The Pyrenean ibex, a form of wild mountain goat, was officially declared extinct in 2000 but scientists have preserved tissue samples and can use this to create clones.

   (a) Explain why scientists want to clone it.

   _____

   _____ [1 mark]

   (b) Outline the process that could be used to clone the ibex.

   _____

   _____

   _____

   _____ [4 marks]

Genetic engineering

1 Human insulin can be produced by bacteria grown in huge vats.

(a) Why are these bacteria known as a GM organisms?

_____

_____ [1 mark]

(b) Outline the procedure used to make GM bacteria that are able to produce human insulin.

_____

_____

_____

_____ [3 marks]

(c) The culture inside the vat has to be maintained at a temperature of around 37 °C. Why?

_____

_____

_____ [2 marks]

D–C

---

2 A type of GM cotton plant called Bt cotton produces a toxin in its leaves which kills any insect pests that feed on it.

(a) Explain why farmers may want to grow Bt cotton instead of normal cotton.

_____

_____

_____

_____ [3 marks]

(b) State one reason why people may be worried about farms growing Bt cotton.

_____

_____ [1 mark]

B–A*

Evolution

1 Lamarck stated that as a giraffe stretches up to eat leaves from a high tree, its neck gets longer. This characteristic would be passed on to the next generation. Use what you know about how characteristics are inherited to explain why we know this is not true.

_____

_____

_____ [3 marks]

D–C

---

2 'Evolution means that eventually all organisms will become more complex.' Discuss evidence for and against this statement.

_____

_____

_____ [3 marks]

B–A*

Natural selection

1 (a) Peppered moths are a pale colour. They like to rest on the trunks of trees. How does the colour of the moth help it to survive?

_____

_____ [2 marks]

(b) Occasionally there is a mutation and a black moth is born. What is a mutation?

_____ [1 mark]

(c) In the past, air pollution in the UK caused the bark of the trees to turn black. Explain how this meant that the numbers of black moths increased.

_____

_____

_____ [3 marks]

D–C

2 Warfarin can be used as a rat poison. The number of rats that are resistant to warfarin is increasing. Philippa thinks this is because when the rats come into contact with warfarin their body figures out a way of becoming resistant. Explain to her what really happens.

_____

_____

_____ [3 marks]

B–A*

Evidence for evolution

1 The bones in the arms of birds, bats and humans are shown here.

(a) Why are they all slightly different?

_____

_____ [1 mark]

(b) Explain how this is evidence for evolution.

_____

_____

_____

_____ [3 marks]

bat wing   human arm

bird wing

D–C

2 This evolutionary tree includes four of the major classification groups.

(a) Do animals have more features in common with bacteria or archaea? Explain how you know.

_____

_____ [2 marks]

animals

archaea   fungi

bacteria

plants

?

(b) What is represented by the question mark?

_____

_____ [2 marks]

B–A*

Extended response question

As a seedling grows, it is important that its roots become anchored in the ground. Use what you know about how auxins control cell growth to explain how this happens.

The quality of written communication will be assessed in your answer to this question.

_____

_____

_____

_____

_____

_____

_____

_____

_____

_____

_____

_____

_____

_____

_____

_____

_____

_____

_____

_____

_____

_____

_____

_____

_____

_____

_____

_____

_____

_____

_____

_____

_____

_____

_____ [6 marks]

Atoms, elements and compounds

1 Malachite is an ore of copper. It contains both rock, and copper carbonate. When heated with carbon it produces copper, and carbon dioxide.

(a) Name the element obtained from malachite.

_____ [1 mark]

(b) Name the useful compound in malachite.

_____ [1 mark]

D–C

(c) Explain why you know this substance is a compound.

_____ [1 mark]

(d) Explain why malachite is a mixture.

_____

_____ [2 marks]

Inside the atom

1 Phosphorus is an element in the periodic table.

(a) Use information in the box to help you complete these sentences.
An atom of phosphorus contains _____ protons, _____ electrons
and has a mass of _____.
It has _____ neutrons.

[4 marks]

31
P
15

(b) Name or give the symbol of an element that:

(i) has the same number of outer electrons as phosphorus.

_____

D–C

(ii) has two more outer electrons than phosphorus.

_____

(iii) has one less proton than phosphorus.

_____ [3 marks]

(c) Draw a diagram
to show the electronic
structure of phosphorus. [2 marks]

2 Sodium and potassium both react with water to produce hydrogen gas and an alkali.

(a) Draw a diagram
to show the electronic
structure of potassium.

[2 marks]

B–A*

(b) Draw a diagram
to show the electronic
structure of sodium.

[2 marks]

(c) Explain why both sodium and potassium have very similar reactions with water.

_____ [2 marks]

Element patterns

1 (a) How are the elements arranged in the periodic table?

_____ [1 mark]

(b) What are horizontal rows of the periodic table called?

_____ [1 mark]

(c) What are vertical columns of the periodic table called?

_____ [1 mark]

(d) Elements in the same vertical column are sometimes called 'families'. Explain why.

_____ [2 marks]

D–C

2 Elements of Group 0 are known as the noble gases.

(a) Name one noble gas, and give its electronic structure.

_____ [2 marks]

B–A*

(b) Explain why noble gases form very few compounds.

_____ [2 marks]

Combining atoms

1 This is a space filling model of a water molecule (H_2O). The atoms are joined together by covalent bonds.

(a) Describe a covalent bond.

_____

_____ [1 mark]

(b) Draw a dot and cross diagram to show the electron arrangement in the water molecule. [2 marks]

Sodium chloride (NaCl) is a compound where two different elements are present. The atoms transfer electrons to each other.

(c) Give the electronic configuration of a sodium atom.

_____ [1 mark]

(d) Describe how the sodium atom becomes a positively charged ion.

_____ [2 marks]

(e) Explain how this helps the sodium to form a compound with the chlorine.

_____ [3 marks]

D–C

2 Here is a diagram of a carbon dioxide molecule.   O=C=O

(a) Explain what the = symbol means.

_____ [1 mark]

B–A*

(b) Draw a dot and cross diagram to show the covalent bonding in carbon dioxide. [2 marks]

Chemical equations

1 Copper carbonate can be used to make copper. When heated with carbon it produces copper and carbon dioxide.

   (a) Complete this word equation

   copper + carbon → copper + _____   [1 mark]
   carbonate

   (b) This is the balanced symbol equation for the reaction

   $2CuCO_3 + C \rightarrow 2Cu + 3CO_2$

   (i) How many molecules of carbon dioxide are produced in the reaction? _____ [1 mark]

   (ii) How many atoms of copper can be made for each atom of reacting carbon? _____ [1 mark]

2 Copper carbonate has many reactions. Here are unbalanced symbol equations for two of the reactions. Balance each equation by inserting numbers in front of each substance if necessary to balance them.

   (a) $CuCO_3 + HCl \rightarrow CuCl_2 + CO_2 + H_2O$ [1 mark]

   (b) $CuCO_3 + NaOH \rightarrow Cu(OH)_2 + Na_2CO_3$ [1 mark]

Building with limestone

1 A company obtains large quantities of limestone by quarrying.
These quarries are often in beautiful parts of the country such as the Peak District.

The company wants to open a new quarry. Describe the advantages and disadvantages, socially and economically for the local population in opening the new quarry.

_____

_____

_____

_____

_____ [4 marks]

2 Limestone rock areas often have many caves and potholes. These are often caused by the effect of rainwater.

   (a) Describe how the rainwater dissolves the limestone.

   _____

   _____

   _____ [3 marks]

   (b) In limestone caves you often find stalactites and stalagmites, which are made from solid calcium carbonate. Explain how these form.

   _____

   _____ [2 marks]

Heating limestone

1 Heating limestone converts it into calcium hydroxide. Calcium hydroxide dissolves in water to make limewater.

(a) Limewater can be used to test for a gas in the air. Which gas is this? _____ [1 mark]

(b) Describe how to use limewater to test for this gas, and what you would see if the gas were present.

_____ [2 marks]

(c) Cement is made from limestone.

(i) What substance is limestone heated with to make cement? _____ [1 mark]

(ii) Cement is used to make concrete and mortar. What is the difference between mortar and concrete?

_____ [2 marks]

(d) Mortar holds bricks together, concrete can be used to make beams for use in constructing houses. Explain why these two materials made from limestone have these different uses.

_____ [2 marks]

D–C

2 (a) Long beams made from concrete often crack in the middle. Describe how beam manufacturers can prevent beams cracking in use.

_____ [2 marks]

(b) Many people say that concrete dries when it becomes solid. Explain why this statement is wrong.

_____ [2 marks]

force needed to break beam in kN x 1000

percentage of gravel in the concrete

(c) This graph shows how the strength of concrete changes according to the volume of sand and gravel in the concrete. The mass of cement used remained the same throughout the investigation.

(i) Describe the trend of the graph from 0% to 50% gravel in the concrete. _____ [1 mark]

(ii) Suggest why the concrete strength drops rapidly at 75% gravel in the mixture.

_____ [2 marks]

B–A*

Metals from ores

1 Metals such as iron, copper and zinc are obtained from their ores. Here is a table of some common ores.

(a) Name the ore from the table that can be used to obtain

name of ore	formula
haematite	Fe_2O_3
bauxite	Al_2O_3
litharge	PbO
zincite	ZnO

(i) iron _____ [1 mark]

(ii) zinc _____ [1 mark]

(iii) aluminium _____ [1 mark]

(b) Which ore in the table cannot be obtained by reduction of the ore with carbon. Explain your answer. _____ [2 marks]

(c) What is meant by reduction? _____ [1 mark]

D–C

2 Aluminium, sodium and potassium are all more plentiful in the Earth's crust than iron. In the 1850s aluminium was a very rare and expensive metal, and cost more than gold. Today aluminium is still more expensive than iron, but it is cheap enough to be used to make lemonade cans.

(a) How is aluminium extracted from its ore today? _____ [1 mark]

(b) Explain why this method of reduction is more expensive than reduction with carbon.

_____ [2 marks]

(c) Suggest why aluminium was more expensive in the 1850s. _____ [1 mark]

B–A*

Extracting iron

1 Iron ore (Fe_2O_3) is reduced inside a blast furnace to make iron.

(a) Name the two other substances added to the blast furnace.

(i) _____ (ii) _____ [2 marks]

(b) Hot air is blown through the blast furnace. This produces large volumes of carbon dioxide gas. Explain why.

_____ [3 marks]

(c) Slag is formed in the process. Describe how slag is formed.

_____ [1 mark]

2 Iron is produced in a blast furnace by reduction. This chemical equation summarises the reaction.

$$Fe_2O_3(s) + ...CO(g) \rightarrow ...Fe(\ell) + ... CO_2(g)$$

(a) Balance the equation. _____ [1 mark]

(b) What is the name of $CO(g)$? _____ [1 mark]

(c) Explain how the $CO(g)$ has been produced. _____

_____ [2 marks]

(d) Write the formula of the substance that has been reduced. _____ [1 mark]

(e) Write the formula of the substance that has been oxidised. _____ [1 mark]

Metals are useful

1 The diagram represents the structure of a pure metal. Use the diagram to help you answer these questions.

(a) Describe why a metal can be bent without breaking.

_____ [2 marks]

(b) Explain how a metal can conduct electricity.

_____ [2 marks]

(c) Describe how a metal is different from an alloy.

_____ [2 marks]

(d) Explain why an alloy is harder than the original metal.

_____ [2 marks]

2 Nitinol is a nickel titanium memory or smart alloy.

(a) Alloys are often used when each metal they are made from would be unsuitable for that use. Explain why.

_____ [2 marks]

(b) Nitinol is a memory alloy. What does this mean?

_____ [1 mark]

(c) Strips of smart alloy can be cooled, stretched and then used to hold the parts of a broken bone together. Evaluate the benefits of using a smart alloy to hold broken pieces of a single bone together whilst it heals.

_____ [3 marks]

D–C

B–A*

D–C

B–A*

Iron and steel

1 Stainless steel can be made by adding chromium and nickel to iron to make an alloy we call stainless steel.

(a) Give two advantages of using stainless steel instead of iron to make cutlery.

_____

_____ [2 marks]

(b) Vanadium can also be added to iron to make a different stainless steel. Suggest why having different types of stainless steel is useful.

_____ [1 mark]

(c) Which part of the periodic table are all these metals found in?

_____ [1 mark]

D–C

Copper

1 Copper is purified using electrolysis of copper sulfate solution.
This diagram shows the process.

cathode (–ve)　　anode (+ve)

– + electrical supply

copper sulfate solution

(a) What is the anode made from?

_____ [1 mark]

(b) What is the cathode made from?

_____ [1 mark]

(c) Where does the pure copper collect?

_____ [1 mark]

(d) Where do the impurities collect?

_____ [1 mark]

(e) Describe what happens to copper ions at the cathode.

_____ [2 marks]

D–C

2 Copper ores have very low percentages of copper in them. At some old copper mines, the heaps of waste rock contain more copper than the remaining deposits.

(a) Describe how growing plants on the waste rock heaps can be used to obtain copper ores.

_____

_____ [3 marks]

(b) Describe how introducing copper tolerant bacteria can be used to obtain the copper from the rock.

_____

_____ [3 marks]

(c) Although both processes take a long time, suggest why both are now economic to do.

_____ [2 marks]

B–A*

Aluminium and titanium

1 Aluminium is extracted from its ore by electrolysis.

(a) What is meant by electrolysis? _____ [1 mark]

(b) Name the main aluminium ore. _____ [1 mark]

(c) (i) Why is aluminium an expensive metal?

_____ [2 marks]

(ii) Explain how dissolving the aluminium ore in cryolite reduces the cost of aluminium manufacture.

_____ [3 marks]

---

2 This equation represents the reduction of aluminium oxide to aluminium metal. This chemical equation summarises the reaction.

$$...Al_2O_3 \quad \rightarrow \quad ...Al + \quad ... O_2$$

(a) Balance the equation. [1 mark]

(b) Titanium is produced by reacting with magnesium. Suggest why titanium is a very expensive metal to make.

_____ [1 mark]

(c) Both titanium and aluminium are high in the reactivity series, but do not react with acids or alkalis. Explain why.

_____ [2 marks]

(d) Suggest why titanium is a suitable material to use to make artificial joints in the body.

_____ [2 marks]

Metals and the environment

1 Many metals are recycled.

(a) Evaluate the economic benefits of recycling metals against using new metal. Your answer should suggest two advantages and two disadvantages of recycling metals.

_____ [6 marks]

(b) Describe the environmental disadvantages of mining metal ores.

_____ [3 marks]

---

2 Phytomining is a new method of obtaining metal ores from *brownfield* sites. This flow chart shows the process.

land contaminated with toxic metal compounds	→	land used to grow plants	→	plants harvested to become metal ores	→	land fit to build homes on

(a) What is a *brownfield* site? _____ [1 mark]

(b) Explain how the plants remove the metal compounds from the ground.

_____ [2 marks]

(c) Describe how the harvested plants are turned into metal ore.

_____ [2 marks]

(d) What is the social benefit of making the land fit for people to live on?

_____ [2 marks]

A burning problem

1 In 2010, an Icelandic volcano erupted sending large quantities of ash into the atmosphere, and gases such as sulfur dioxide and carbon dioxide. Jet aircraft were prevented from flying over a large part of Europe.

(a) Explain how the ash may cause global dimming.

_____ [2 marks]

(b) Describe how the volcanic eruption may lead to more acid rainfall.

_____ [2 marks]

(c) Suggest how the eruption could lead to global warming.

_____ [1 mark]

D–C

2 Burning dry plant material such as wood in a garden bonfire produces carbon dioxide, lots of ash and a little pure carbon called charcoal.

(a) Explain why some of the carbon in the wood becomes charcoal.

_____ [2 marks]

(b) In a cigarette the same burning process produces another compound called carbon monoxide instead of carbon dioxide.

(i) Write the chemical formula of carbon dioxide. _____ [1 mark]

(ii) Explain how carbon monoxide is different from carbon dioxide.

_____ [1 mark]

(iii) Describe how carbon monoxide is a health risk to smokers.

_____ [2 marks]

B–A*

Reducing air pollution

1 Car exhaust systems are fitted with catalytic converters.

(a) Why are cars fitted with a catalytic converter?

_____ [1 mark]

(b) What environmental benefit is gained by using catalytic converters.

_____ [1 mark]

(c) To reduce vehicle emissions, some cars are designed to use ethanol instead of petrol as a fuel.

Explain why ethanol is a green fuel. _____ [3 marks]

D–C

2 Biofuels are now added to petrol and diesel. Using these fuels made from plants is reducing our dependency on non-renewable resources. Biofuels are considered to be carbon neutral and do not contribute to global warming.

(a) Explain why biofuels are considered to be carbon neutral.

_____ [2 marks]

(b) Describe why biofuels are classed as renewable fuels.

_____ [2 marks]

B–A*

(c) Land that used to grow food crops is now growing crops for biofuels. Describe the problems this may cause now and in the future. _____ [2 marks]

Crude oil

1 Crude oil is a mixture of many different hydrocarbons.

(a) What is meant by *hydrocarbon*?

_____ [1 mark]

(b) Here is a diagram of how the crude oil is separated into more useful substances.

(i) Name the separation process. _____ [1 mark]

(ii) Name the substance with the highest boiling point. _____ [1 mark]

(iii) Name the substance with the lowest boiling point. _____ [1 mark]

(c) Explain how the process separates crude oil into different useful substances.

_____ [4 marks]

D–C

petroleum gas

40 °C petrol
110 °C naphtha
180 °C kerosene
crude oil mixture is added
250 °C diesel
340 °C fuel oil

it is heated and evaporates   bitumen

---

2 The boiling points of several of the substances shown in the diagram are given in this table, with the mean number of carbon atoms in the substance.

(a) Why is the mean number of carbon atoms given for these substances?

_____ [1 mark]

B–A*

(b) Describe the trend shown by the data.

_____ [1 mark]

substance	mean number of C atoms	boiling point in °C
diesel	14	250
kerosene	12	180
naphtha	10	110
petroleum gas	8	20

(c) Explain why the boiling point is related to the mean number of carbon atoms in the mixture.

_____ [2 marks]

Alkanes

1 Here is the general formula for an alkane. $C_n H_{2n+2}$

(a) Explain how you could use this to work out the formula of octane which has 8 carbon atoms.

_____ [2 marks]

D–C

(b) As alkanes have the same basic structure, they have similar properties that change as the molecule gets bigger. Explain how each of these properties changes as the alkane molecule gets bigger:

(i) flammability? _____

(ii) boiling point? _____

(iii) viscosity? _____ [3 marks]

---

2 (a) Describe the type of bonding in alkanes.

_____ [2 marks]

B–A*

(b) Methane (CH_4) and ethane (C_2H_6) are the first two alkanes. Draw their structures.

_____ [2 marks]

(c) Explain why alkanes have similar chemical properties that change as the molecules get larger.

_____ [2 marks]

Cracking

1 Cracking hydrocarbon molecules is a very important part of an oil refinery's work. The longer alkane molecules are broken down into shorter molecules.

(a) Write the formula of decane.

_____ [1 mark]

(b) Name molecule X.

_____ [1 mark]

octane

molecule X

decane

(c) What does the = symbol mean in the structural diagram of molecule X? _____ [1 mark]

(d) Why is molecule X a very useful chemical?

_____ [2 marks]

D–C

2 In cracking, the large alkane molecule is broken down into two smaller molecules by thermal decomposition.

(a) Explain what is meant by thermal decomposition.

_____ [1 mark]

(b) Steam cracking uses steam and a temperature of 850 °C, and catalytic cracking uses a catalyst and a temperature of 600 °C. Describe one advantage and one disadvantage of each type of cracking.

_____ [4 marks]

B–A*

Alkenes

1 Ethene is the simplest alkene

(a) How are alkenes different from alkanes?

_____ [1 mark]

(b) Use the diagram to explain why alkenes are more reactive than alkanes.

_____ [2 marks]

(c) Alkenes are said to be unsaturated hydrocarbons. What do chemists mean by unsaturated?

_____ [1 mark]

D–C

2 A student wanted to check if a hydrocarbon had any double bonds present. The student put 1 cm³ of the hydrocarbon in a test tube.

(a) What should the student add to the hydrocarbon to test whether any double bonds are present?

_____ [1 mark]

(b) What would the student expect to see?

_____ [2 marks]

(c) How would this be different for a hydrocarbon with no double bonds?

_____ [2 marks]

B–A*

Making ethanol

1 Ethanol (C_2H_5OH) can be made by two methods: fermentation from sugars, or from ethene (C_2H_4) obtained from crude oil.

D–C

(a) What is added to ethene to produce ethanol? _____ [1 mark]

(b) Ethanol made from sugars is a renewable biofuel. What is a biofuel?

_____ [1 mark]

(c) Suggest two advantages and two disadvantages of using sugar to produce ethanol for use as a car fuel.

_____ [4 marks]

2 Ballpoint-ink stains on clothing can be difficult to remove. One washing powder manufacturer suggests that the affected area should be rubbed with a solvent other than water before washing.

B–A*

(a) Name a different solvent from water.

_____ [1 mark]

(b) Explain why ballpoint-ink stains may be removed by this solvent but water cannot remove them.

_____ [2 marks]

Polymers from alkenes

1 Alkenes such as ethene can be used to make polymers such as poly(ethene).

ethene → poly(ethene)

D–C

(a) What is meant by a polymer?

_____ [1 mark]

(b) What word do we use to describe the small molecule used to make the polymer?

_____ [1 mark]

(c) Explain how the ethene molecules join together to make poly(ethene).

_____ [2 marks]

2 Thirty years ago thick glass bottles were used to transport dangerous chemicals. This type of bottle is made from poly(ethene) and is now used to transport dangerous chemicals.

(a) Suggest two reasons why the polymer bottle has replaced glass bottles for transporting dangerous chemicals.

_____ [2 marks]

(b) There are many different types of man-made polymers. Guttering is made from PVC, which is made from this small molecule:

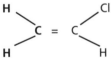

conc. hydrochloric acid

B–A*

(i) Explain why we need different polymers.

_____ [1 mark]

(ii) Compare how PVC guttering's properties need to be different from wrapping polymers such as clingfilm.

_____ [2 marks]

Designer polymers

1 Plastic polymers are easily moulded into shape, and are low density (lightweight), waterproof and resistant to acids and alkalis. Increasingly, car and aircraft bodies are being made from designer plastic polymers such as carbon fibre-reinforced plastic.

   (a) Give three reasons why these expensive plastics are being used in cars and aircraft.

   _____ [3 marks]

   (b) Gore-tex is a designer fabric, it lets water vapour through but not liquid water. Explain why it is a good choice from which to make an outdoor coat.

   _____ [2 marks]

   (c) Suggest why the use of designer polymers is likely to increase in the future.

   _____ [1 marks]

D–C

---

2 Plastic carrier bags from supermarkets are a source of pollution. They do not decompose, and can hurt or kill wildlife that get entangled in them. Recently, cornstarch-polymer carrier bags are being produced that are biodegradable.

   (a) What is meant by biodegradable? _____ [1 mark]

   (b) Suggest why cornstarch bags may be more biodegradable than plastic bags.

   _____ [2 marks]

B–A*

   (c) Surgeons carrying out operations are also starting to use thread made from cornstarch polymers to stitch inside people's bodies. Explain why using biodegradable stitches during an operation is better for the patient.

   _____ [2 marks]

Polymers and waste

1 There are three ways to dispose of polymer waste. It can be sent to landfill sites, it can be burnt in incinerators or it can be recycled.

   (a) There are problems with all three methods of disposal. Suggest a disadvantage of:

      (i) putting the polymer waste in a landfill site. _____

      (ii) incinerating the polymer waste. _____

      (iii) recycling the polymer waste. _____ [3 marks]

D–C

   (b) Evaluate which of the three methods of disposal provides the most effective use of the limited resources available to make polymers. Give reasons for your answer.

   _____ [3 marks]

---

2 Biodegradable polymers are often made from a mixture of ordinary hydrocarbon-based polymers and cornstarch. They are not usually recycled.

   (a) Describe what is likely to happen to a biodegradable polymer when it decomposes.

   _____ [1 mark]

   (b) Suggest why this makes it unsuitable to be used to make compost, even though it contains plant material, such as cornstarch, and is biodegradable.

B–A*

   _____ [2 marks]

   (c) Compare the benefits of disposing of the biodegradable polymer by landfill and by incineration.

   _____ [2 marks]

Oils from plants

1 Vegetable oils are oils that are made from plants such as sunflower, maize, olives, almonds and walnuts. They have many uses, such as in foodstuffs, cosmetics, and as fuels such as biodiesel.

(a) Why are plant oils suitable for use as fuels?

_____ [1 mark]

D–C

(b) From which parts of the plant is the plant oil obtained?

_____ [1 mark]

(c) Describe how the plant oil is obtained from the plant.

_____ [2 marks]

(d) Cooking food in plant oils raises the energy content of the food, and changes the flavour of the food. Explain why cooking a potato in plant oil makes the potato taste different from when it is cooked in water.

_____ [3 marks]

2 Essential oils can be obtained from flowers for perfumes using this equipment.

(a) Describe what is happening at A.

_____

_____ [2 marks]

(b) Describe the substances present in the pipe at B.

_____

_____ [2 marks]

B–A*

(c) Give the scientific name for part C.

_____ [1 mark]

(d) Where has the water at D come from?

_____ [1 mark]

(e) Describe the difference between a vegetable oil and a mineral oil.

_____

_____ [2 marks]

Biofuels

1 Biofuels are fuels made from animal or plant materials. They are considered to be 'greener' than fossil fuels.

(a) Evaluate the economic and environmental benefits of using biofuels. Your answer should suggest two advantages and two disadvantages.

_____

_____

_____

_____

_____ [6 marks]

D–C

(b) Plant oils are far more viscous than diesel. Describe how they can be modified to be used in diesel engines.

_____

_____ [2 marks]

Oils and fats

1 Low-fat spreads and margarines are made from plant oils. Butter is made from milk produced by cows. Some students tested butter, hard margarine, soft margarine and low-fat spread. They tested each spread for saturated fat by dissolving a sample of each in a little ethanol, adding 2 cm³ of orange bromine water and timing how long the reaction took.

(a) What would you expect to see happen? _____ [1 mark]

(b) Explain why the test is not a fair one.

_____ [2 marks]

Here are their results.

(c) Which spread has the most saturated molecules present? Explain your answer.

_____ [2 marks]

(d) Explain the difference between a 'saturated' fat and an 'unsaturated' fat.

_____ [2 marks]

Spread	Mean time for the reaction to complete in seconds
butter	2
hard margarine	4
soft margarine	7
low-fat spread	8

D–C

2 Soft or liquid vegetable fats are often converted into harder vegetable fats by reacting the fat with hydrogen using a catalyst such as nickel.

(a) What is the purpose of using a nickel catalyst?

_____ [1 mark]

B–A*

(b) Draw a diagram to show how this section of a vegetable fat molecule would be hydrogenated.

```
      H H H H H
      | | | | |
  H—C–C–C=C–C—
      | |   |
      H H   H
```

[2 marks]

Emulsions

1 In 2010, there was a large oil spillage off the coast of the USA between New Orleans and Florida. The crude oil did not mix with the seawater but floated on top of it. The crude oil and seawater were mixed together using a dispersant or detergent, to make an emulsion.

(a) What word do we use to describe two liquids that don't mix together? _____ [1 mark]

(b) What type of substance is the dispersant or detergent? _____ [1 mark]

D–C

(c) When making salad cream, egg yolk is added to a mixture of oil, water, vinegar and powdered mustard.

(i) Name the aqueous substances in the mixture. _____ [1 mark]

(ii) What is the purpose of the egg yolk? _____ [1 mark]

2 This is a diagram of a molecule that can be used as a dispersant for crude oil.

hydrophobic end   hydrophilic end

(a) What does the term 'hydrophilic' mean? _____ [1 mark]

(b) What does the term 'hydrophobic' mean? _____ [1 mark]

B–A*

(c) Describe how this molecule allows crude oil to mix with seawater to form an emulsion.

_____ [3 marks]

Earth

1 This is a diagram of the Earth's structure.

Complete the table below.

Part	Name	Its structure
A		solid rock
B		
C		

[5 marks]

2 Look at the diagram in the question above.

(a) Part B has convection currents in it. Explain how these convection currents occur.

_____ [2 marks]

(b) Part A floats on the top of part B, and is broken into small sections called tectonic plates. Explain how these plates and the convection currents form mountains.

_____ [2 marks]

(c) Rivers flow down the mountains. Describe how the rivers form valleys.

_____ [2 marks]

Continents on the move

1 Alfred Wegener suggested in 1915 that originally all the continents were joined together in one continent called Pangea. Most scientists thought he was wrong.

(a) Explain why most scientists disagreed with Wegener's theory.

_____ [2 marks]

(b) Wegener showed that South America could fit together with Africa like jigsaw pieces. Give two other pieces of evidence that were found which persuaded scientists to support his theory.

_____ [2 marks]

(c) New York is slowly moving away from London by a few centimetres each year. Describe how this is happening, using Wegener's theory.

_____ [2 marks]

2 The Mid-Atlantic Ridge is a range of underwater mountains that are forming where the North American plate meets the Eurasian plate. Iceland is an island where this range of mountains has grown so high that it is now above sea level.

Explain how this mountain ridge is formed.

_____

_____ [2 marks]

Earthquakes and volcanoes

1 This diagram shows the main tectonic plates of the Earth's crust.

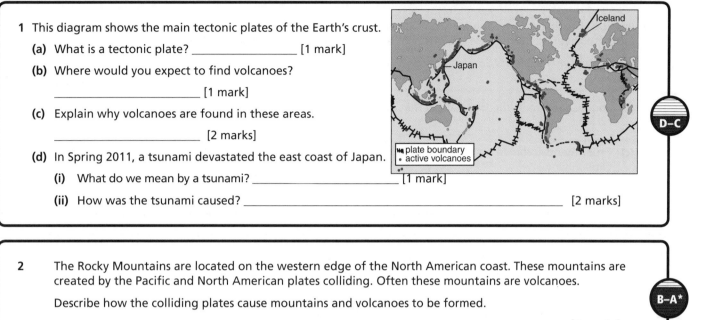

(a) What is a tectonic plate? _____ [1 mark]

(b) Where would you expect to find volcanoes?

_____ [1 mark]

(c) Explain why volcanoes are found in these areas.

_____ [2 marks]

(d) In Spring 2011, a tsunami devastated the east coast of Japan.

   (i) What do we mean by a tsunami? _____ [1 mark]

   (ii) How was the tsunami caused? _____ [2 marks]

D–C

2 The Rocky Mountains are located on the western edge of the North American coast. These mountains are created by the Pacific and North American plates colliding. Often these mountains are volcanoes.

Describe how the colliding plates cause mountains and volcanoes to be formed.

_____ [4 marks]

B–A*

The air we breathe

1 The composition of the air has not always been the same since the Earth formed. It has changed.

(a) Where do we think the early atmosphere came from?

_____ [1 mark]

(b) This graph shows how three gases in the air have changed since the Earth formed. Use information from the graph to help you answer these questions.

   (i) How old is the Earth? _____ [1 mark]

   (ii) Which letter, A to E, shows when plants were first able to photosynthesise? _____ [1 mark]

   (iii) When did life start to live on the land? _____ [1 mark]

(c) Explain why life could exist only in the sea until ozone appeared in the atmosphere.

_____ [2 marks]

D–C

2 Air is a mixture of many gases. We obtain pure gases from the air by fractional distillation.

Here is some information about three important gases in the air.

(a) What is meant by 'fractional distillation'?

_____ [1 mark]

Name of gas	Boiling point in °C
nitrogen	−196
oxygen	−183
argon	−186

(b) Name two other gases in the air that must be removed before the air is cooled. _____ [2 marks]

(c) The air is cooled to −200 °C before distillation.

   (i) Why is it cooled to −200 °C? _____ [1 mark]

   (ii) Which of the three gases distils off first? Explain your answer.

_____ [2 marks]

B–A*

The atmosphere and life

1 Primitive life first evolved about 3.4 billion years ago. No one knows how it evolved, although there are many different theories. Most of the theories suggest that life formed in the seas.

(a) Explain why scientists do not know how life began.

_____ [1 mark]

(b) Suggest why most theories believe that life began in the seas.

_____ [1 mark]

(c) Experiments to prove how life began usually concentrate on devising a method to make amino acids. Explain why.

_____ [2 marks]

D–C

2 Miller and Urey suggested that lightning was a key variable in the formation of life. They used a solution representing seawater from 4 billion years ago, filled their apparatus with the gases present in the air 4 billion years ago and passed electric sparks through the atmosphere. They analysed the seawater after a week.

(a) Name the three gases they used to make the atmosphere of 4 billion years ago.

_____ [1 mark]

(b) Name two types of compound they found in the seawater that had not been there at the start.

_____ [2 marks]

(c) Explain why the experiment did not prove that this was how life was created.

_____ [2 marks]

B–A*

Carbon dioxide levels

1 The diagram below shows the carbon cycle. Some of the processes have been replaced by letters.

(a) **(i)** Name process A. _____

(ii) Name process B. _____

(iii) Name process C. _____

(iv) Name process D. _____ [4 marks]

(b) Explain why process D appears twice. _____ [1 mark]

(c) Explain why burning biofuels is said to be 'carbon neutral'.

_____ [2 marks]

D–C

2 Biofuels and fossil fuels are both carbon-based fuels.

(a) What do we mean by a carbon-based fuel? _____ [1 mark]

(b) Explain why burning fossil fuels raises the levels of carbon dioxide in the atmosphere, but biofuels are considered to be neutral.

_____ [2 marks]

(c) Carbon dioxide is removed from the atmosphere by dissolving in seawater. What problems might this process cause?

_____ [2 marks]

B–A*

Extended response question

Biofuels are fuels made from plant materials. In the UK in 2010, petrol and diesel had to contain at least 5% biofuel.

- Petrol and diesel made from crude oil are non-renewable energy sources.
- Biofuels are renewable sources of energy.
- Burning petrol, diesel and biofuels releases carbon dioxide into the atmosphere.
- Carbon dioxide is thought to cause global warming.
- Much of the carbon dioxide released into the atmosphere dissolves in the sea.

Use the information above, and your knowledge and understanding, to give the positive and negative environmental impacts of increasing the percentage of biofuels in petrol and diesel.

The quality of written communication will be assessed in your answer to this question.

_____

_____

_____

_____

_____

_____

_____

_____

_____

_____

_____

_____

_____

_____

_____

_____

_____

_____

_____

_____

_____

_____

_____

_____

_____

_____

_____

_____ [6 marks]

Energy

1 (a) Describe the energy transfers taking place when a hairdryer is turned on.

_____

_____ [4 marks]

(b) Where is the energy finally transferred to? _____ [1 mark]

(c) A student investigates how quickly hairdryers transfer energy by timing how long it took for each hairdryer to dry a piece of damp cloth. Here are their results:

Model of hairdryer	Time needed to dry damp cloth
Traveller	2 minutes 30 seconds
Olympus1000	1 minute 20 seconds
Hairdry	2 minutes 10 seconds

Explain which hairdryer transferred energy quickest.

_____ [2 marks]

(d) State two variables that the student should control during this investigation.

_____ [2 marks]

2 A company has produced a battery-operated hairdryer which uses rechargeable batteries. Explain whether this hairdryer wastes more energy than a hairdryer operated from the mains supply.

_____

_____ [4 marks]

Infrared radiation

1 Solar panels are fitted on house roofs and are used to heat water.

(a) Explain which is the best colour for the manufacturer to choose for the solar panels

_____

_____ [3 marks)

(b) Why are solar panels normally fitted to south-facing roofs?

_____ [2 marks]

(c) Solar panels should be insulated to avoid heat losses at night or during winter. Explain why.

_____ [3 marks]

2 A student uses a probe to measure the amount of infrared radiation emitted by a beaker of water at room temperature.

(a) Explain whether the probe registers infrared emissions from the beaker.

_____ [2 marks]

(b) She places the beaker of water in a fridge. Explain how this affects the amount of infrared radiation emitted and absorbed.

_____

_____ [4 marks]

Kinetic theory

1 (a) Explain what melting means.

_____ [1 mark]

(b) A piece of ice left on a plate starts to melt. Explain the process of melting by writing about particles.

_____

_____ [4 marks]

(c) Why do the plate and its surroundings cool down when ice melts?

_____ [2 marks]

(d) Explain why the ice melts more quickly in a warmer room.

_____ [1 mark]

D–C

2 Explain what affects the energy of particles by comparing the particles in a piece of copper and particles in the air which are at the same temperature.

_____

_____ [4 marks]

B–A*

Conduction and convection

1 (a) Explain why convection cannot take place in solids.

_____ [2 marks]

(b) Use your ideas about heat transfers to explain how the thermos flask keeps a drink hot.

_____
_____
_____
_____
_____
_____

[6 marks]

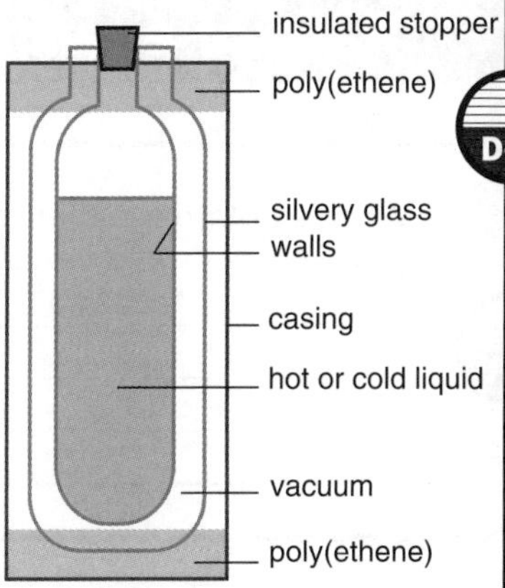

insulated stopper
poly(ethene)
silvery glass walls
casing
hot or cold liquid
vacuum
poly(ethene)

Figure 1: A vacuum flask

D–C

2 Computers can get extremely hot, possibly damaging components inside them. A heat sink is used to cool the computer by transferring heat to the surroundings. Explain why a heat sink in a computer is made from metal rather than plastic.

_____

_____ [3 marks]

B–A*

Evaporation and condensation

1 Suna hangs her washing on a washing line outside.

(a) What is meant by evaporation?

_____ [1 mark]

(b) Write down two differences between evaporation and boiling.

_____ [2 marks]

(c) Write down two weather conditions that make evaporation take place more quickly. Give a reason each weather condition affects the rate of evaporation

(i) Condition 1_____

Why it increases the rate of evaporation

_____ [2 marks]

(ii) Condition 2_____

Why it increases the rate of evaporation

_____ [2 marks]

D–C

2 Many men use aftershave, which contains alcohol. Alcohol evaporates quickly, cooling their skin. Peter is investigating the hypothesis that aftershave will cool a cotton wool pad more quickly than water does Write down a plan for an experiment to test this hypothesis. Include a list of equipment and a method.

_____
_____
_____
_____
_____
_____ [6 marks]

B–A*

Rate of energy transfer

1 When a potato is cut into smaller pieces, it cooks more quickly. Explain why this happens using these terms in your answer: surface area; conduction; radiation.

_____
_____
_____
_____ [4 marks]

D–C

2 The boiling temperature of water is 100 °C, and the boiling temperature of cooking oil is 160 °C Use your ideas about heat transfers to explain why chips cook more quickly when they are fried rather than boiled.

B–A*

_____
_____ [2 marks]

Insulating buildings

1 A homeowner paid for an energy survey and report for their home. The report suggested several ways to reduce energy losses in the home.

(a) Suggest one reason why the homeowner was prepared to pay for a report.

_____ [1 mark]

Here are four suggestions made by the consultant.

Change	Estimated Cost	Estimated annual Saving
Change all light bulbs to energy efficient bulbs	£80	£95
Increase loft insulation to a depth of 15 cm	£200	£120
Set central heating thermostat on to a lower temperature setting	£0	£40
Put the central heating onto an automatic timer switch	£15	£20

All of these measures are put in place in the first year.

(b) Write down the overall savings over five years.

_____ [3 marks]

(c) Loft insulation reduces heat losses because it slows down convection and conduction. Explain whether it has a high or low U-value.

_____ [2 marks]

(d) State two other methods to reduce heat losses in the home. Explain how each method reduces heat losses.

_____
_____ [4 marks]

D–C

2 A local council provides grants for people to install energy saving measures. These measures do not include double-glazing, partly because of the high installation costs and the small average annual energy savings. Explain whether you agree with the council's decision.

_____
_____ [5 marks]

B–A*

Specific heat capacity

1 (a) Explain what is meant by specific heat capacity.

_____ [1 mark]

(b) The specific heat capacity of aluminium is 900 J/kg °C. Calculate the temperature rise when a 250 g block of aluminium is supplied with 1800 J of energy. Write down the equation you use, then clearly show how you calculate your answer.

_____ [3 marks]

D–C

2 Use your ideas of specific heat capacity to explain whether it would be better to use copper or aluminium for a saucepan base.
The specific heat capacity of aluminium is 900 J/kg °C, and the specific heat capacity of copper is 385 J/kg °C.

_____ [3 marks]

B–A*

Energy transfer and waste

1 The picture shows a Sankey diagram for a petrol engine.

   (a) Use the diagram to write down the energy equation for the petrol engine.

   _____

   [2 marks]

   (b) The energy transfer takes place in two stages in the car. State how most of the energy wastage occurs.

   _____

   [1 mark]

   (c) Describe how conservation of energy is shown on a Sankey diagram.

   _____ [3 marks]

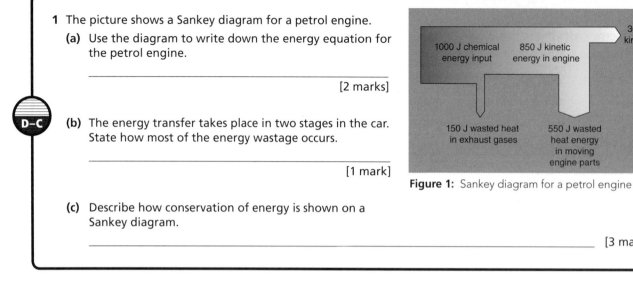

Figure 1: Sankey diagram for a petrol engine

2 An improved engine has been designed which changes 1000 J of chemical energy into 500 J of kinetic energy. The rest is wasted heat. Draw a Sankey diagram to show this transfer

[4 marks]

Efficiency

1 The diagram shows the energy transfer for an electric fan.

   (a) Use the Sankey diagram to calculate the efficiency of the electric fan.

   _____

   [2 marks]

   (b) Explain whether all new electrical appliances should display Sankey diagrams for people to compare their energy efficiency.

   _____
   _____
   _____

   [3 marks]

Figure 1: Sankey diagram for an electric fan

2 A manufacturer claims that a new design of battery-operated motor is 99%. Explain why this claim needs to be checked carefully.

   _____
   _____
   _____ [3 marks]

Electrical appliances

1 A company has designed a wind-up mobile phone battery recharger.

  (a) Describe the useful energy changes that take place when a person winds up the recharger and charges up the battery.

  _____ -> _____ -> _____   [3 marks]

  (b) Explain one advantage of;

     (i) using a wind-up battery recharger

     _____

     _____ [2 marks]

     (ii) using a battery in the mobile phone.

     _____

     _____ [2 marks]

D–C

2 Lighting in gardens is becoming more popular. There are many reasons for this, for example improving safety, security and the appearance of the garden at night. Increasingly, these lighting systems do not use mains electricity.

  (a) Suggest two reasons why manufacturers have been developing an alternative to mains electricity.

  _____

  _____ [2 marks]

B–A*

  (b) Explain why many people chose solar panels instead of batteries for providing energy for garden lights.

  _____

  _____ [2 marks]

Energy and appliances

1 Calculate the energy transferred when an 850 W microwave oven is switched on for 15 minutes.

  _____

  _____

  _____ [3 marks]

D–C

2 The flex for a microwave oven is thinner than the flex fitted to an electric oven. Explain why this is a safety feature.

  _____

  _____

  _____

  _____ [4 marks]

B–A*

The cost of electricity

1

	Filament bulb	Energy efficient bulb
Power	100 W	16 W
Cost when new	30p	£5.00
Lifetime	1 year	5 years

D–C

(a) Each bulb is used for 1500 hours per year. Use this information to calculate the energy used by the filament bulb in a year in kilowatt-hours.

_____ kWh   [2 marks]

(b) If each kilowatt-hour costs 15p, calculate the total cost of using the filament bulb over five years.

_____

_____   [3 marks]

(c) The equivalent cost if energy efficient bulbs are used is less than £19. Explain why the use of energy efficient bulbs is being encouraged for environmental reasons.

_____

_____   [2 marks]

2 Many environmental organisations encourage people to turn equipment off at the plug rather than leaving it on standby. Evaluate the benefits and disadvantages of manufacturing equipment to automatically turn off rather than to remain on standby.

B–A*

_____

_____

_____   [6 marks]

Power stations

1 (a) Describe the useful energy transfers taking place in a gas-fired power station.

_____

_____

_____   [4 marks]

D–C

(b) Explain why a combined heat and power station is more efficient than a power station that just generated electricity.

_____

_____

_____   [4 marks]

2 Many power stations use technology developed over 20 years ago, and are not as efficient as modern power stations. Explain whether we should convert existing older power stations to become more efficient, or whether we should develop new technologies to use in new power stations.

B–A*

_____

_____

_____   [4 marks]

Renewable energy

1 Both Iceland and Norway use renewable sources of energy to generate electricity. Iceland is a volcanic island and Norway is a mountainous country with many valleys and fast flowing rivers. Suggest which is the most suitable renewable energy resource for each country to use to generate electricity, explaining your answers.

_____

_____

_____

_____ [4 marks]

D–C

2 Many wind farms are being built in the UK. These include onshore and offshore wind farms. Critics say that wind farms do not generate electricity when it is calm or stormy, so it is not practical to depend on them. Supporters say that giant batteries under the turbines could store energy generated by wind turbines for future use. Outline one advantage and one disadvantage of using giant batteries in wind farms for the UK.

_____

_____ [2 marks]

B–A*

Electricity and the environment

1 There are plans to build a new power station near a large city. The choice is between a coal-fired power station or wind turbines set on nearby hills in a local beauty spot.
Give one environmental advantage and one environmental disadvantage for each of these schemes.

Type of station	Advantage of the scheme	Disadvantage of the scheme
Coal-fired		
Wind turbines		

[4 marks]

D–C

2 One green campaigner feels that renewable energy does not harm the environment. Explain two reasons that show how using biofuels can damage the local environment.

_____

_____

_____

_____ [4 marks]

B–A*

Making comparisons

D–C

1 (a) Explain two factors that may affect the demand for electricity during the day.

_____

_____

_____ [3 marks]

(b) A reliable energy source can be used to generate electricity whenever it is required.
Put these energy sources in order of reliability, starting with the least reliable first.
gas, wind, waves, tidal

_____ [3 marks]

B–A*

2 Critics of wind turbines say that even with large numbers of wind farms, new fossil-fuel power stations are needed to supply electricity when the weather is unsuitable. Supporters say that it is possible to predict when the weather means wind turbines cannot function. This causes less disruption to the supply of energy than an unexpected failure of a single very large fossil-fuel power station.
Use these ideas to discuss whether large numbers of wind farms are the best solution to the UK's energy problems.

_____

_____

_____

_____

_____ [5 marks]

The National Grid

D–C

1 The National Grid distributes electricity throughout the UK. Overhead high voltage transmission lines transmit electricity across the country at voltages up to 400 000 V. Distribution lines transmit electricity at voltages up to 33 000 V throughout an area.

(a) Why is less energy wasted when electricity is transmitted at higher voltages?

_____

_____ [2 marks]

(b) Calculate the current in a high voltage transmission line that carries 200 000 000 W.

_____

_____

_____ [3 marks]

B–A*

2 Pylons many metres above the ground carry high voltage transmission lines. This reduces risk of damage but causes visual pollution. Many people feel that underground power cables would be a better solution although this is more expensive, and still involves damage to the surroundings.
Explain whether power lines should be carried on pylons or should be buried underground.

_____

_____

_____

_____ [4 marks]

What are waves?

1 Describe two differences and two similarities between transverse waves and longitudinal waves.

Difference 1 _____

Difference 2 _____   D–C

Similarity 1 _____

Similarity 2 _____   [4 marks]

2 Earthquakes cause seismic waves. P-waves are longitudinal waves and s-waves are transverse waves.

   (a) Seismic waves are mechanical waves. What is meant by a mechanical wave?

_____   [1 mark]

   (b) Explain why the same earthquake is detected in different places at different times.

_____

_____   B–A*

_____

_____

_____

_____

_____   [4 marks]

Changing direction

1 Two people had a conversation in a room. The door was open and the conversation could be clearly heard on the other side of the wall. Use your ideas of diffraction to explain why.

_____

_____

_____   D–C

_____

_____   [3 marks]

2 Use your ideas about refraction to explain why a swimming pool looks shallower than it really is.

_____

_____

_____   B–A*

_____

_____   [3 marks]

Sound

1 (a) A musician plays a note with a frequency of 440 Hz. Another musician plays a slightly higher pitched note. Write down a possible frequency of this note.

_____ [1 mark]

(b) How does the wavelength of a sound change when it becomes higher pitched?

_____ [1 mark]

(c) What changes when a sound becomes louder?

_____

_____ [2 marks]

2 The picture shows a trace of a sound from an oscilloscope.

(a) Draw the trace that would be seen if a quieter higher pitched sound was heard. [2 marks]

(b) A sound engineer is making a recording of a band, and will mix the sounds later to make the final version. Explain how oscilloscopes can help the sound engineer.

_____

_____

_____ [3 marks]

Light and mirrors

1 Describe how the image of a child in a plane mirror compares with the real child by choosing the correct words.

The image is *upright/inverted*.
It is *real/virtual*.
It is the *same distance behind/further behind* the mirror than the object is in front. [3 marks]

2 The diagram shows a person looking at some writing in a mirror.
Complete the ray diagram to show why the object is laterally inverted.

abcdef   ʇǝbɔds

[4 marks]

Using waves

1 Match the type of electromagnetic wave with these uses.

Type of electromagnetic radiation	Use
Radio wave	Photography
Microwave	Remote control
Infrared radiation	BBC TV broadcasts
Visible light	Satellite TV

[4 marks]

D–C

2 Mobile phone coverage in many parts of the country can be more unreliable than radio reception. Explain why?

_____

_____

_____

_____

_____ [5 marks]

B–A*

The electromagnetic spectrum

1 Radio waves can be detected from galaxies, such as the Milky Way. If the frequency of radio waves is measured as 2 700 000 kHz, calculate the wavelength of the radiation. Electromagnetic waves travel at 300 000 km/s

_____

_____

_____

_____ [4 marks]

D–C

2 Select the correct answer from the following sentence.
The electromagnetic spectrum is a continuous spectrum of waves whose wavelength varies from more than *1000 / 10 000 / 1 000 000 m* to 10^{-7} /10^{-10} /10^{-15} m

_____

_____ [2 marks]

B–A*

Dangers of electromagnetic radiation

1 Rachel put sun block on her skin before she went out in the Sun.

(a) What type of electromagnetic radiation does sun block protect her from?

_____ [1 mark]

(b) Write down three forms of harm caused by too much exposure to this type of radiation.

_____

_____

_____ [3 marks]

D–C

(c) Write down two other ways that Rachel could reduce her exposure to radiation from the Sun.

_____

_____

_____

_____ [4 marks]

---

2 (a) What effect does shorter wavelength electromagnetic radiation have on cells when absorbed by molecules within the cells?

_____ [1 mark]

B–A*

(b) A patient is undergoing radiotherapy as part of a treatment for cancer. Explain why carefully targeted radiotherapy is a useful treatment for cancer.

_____

_____

_____ [3 marks]

Telecommunications

1 The Olympic Games brings together sportsmen and women from all over the world. Events from the Olympics are transmitted to audiences worldwide. Explain how electromagnetic waves can be used to allow viewers to see live transmissions of the London Olympics in Australia.

D–C

_____

_____

_____ [3 marks]

2 Sky satellite dishes are fixed in one position to receive signals from a communications satellite. The satellite is orbiting the Earth constantly above the equator.
Explain why the signals from the satellite can be received at all times by the dish.

B–A*

_____

_____

_____ [3 marks]

Cable and digital

1 (a) What is the name given to the type of signal shown in figure A?

_____

[1 mark]

A

(b) Explain why the quality of these signals can get worse when they are transmitted long distances through copper cables.

_____

_____

_____

_____

[2 marks]

D–C

Searching space

1 (a) Use the table to explain why we need different telescopes to detect different objects in space.

_____

_____

_____

_____

[2 marks]

Radiation	Objects 'seen' in space
gamma ray	neutron stars
X-ray	neutron stars
ultraviolet	hot stars, quasars
visible	stars
infrared	red giants
far infrared	protostars, planets
radio	pulsars

D–C

(b) Telescopes that detect radiation with long wavelengths need large receiving dishes because the radio signals from objects in space are very weak. Explain whether a radio telescope is larger or smaller than an optical telescope that detects visible light.

_____

_____

_____

_____

[2 marks]

Waves and movement

1 (a) When a car passes a person, its sound appears to change pitch. What is the name of this effect?

_____ [1 mark]

(b) The diagram shows sound waves coming from a car. On the diagram, show how the sound wave would appear if the car was travelling away from the person.

[2 marks]

D–C

2 The Universe is expanding and galaxies in the Universe are moving apart at different speeds. We can see this because of the red shift.

(a) What is meant by the red shift?

_____

_____ [2 marks]

B–A*

(b) Explain how the red shift shows that galaxies are moving apart at different speeds.

_____

_____ [2 marks]

Origins of the Universe

1 Scientists have developed several models of the start of the Universe. One model is the Steady State theory. This theory describes a universe that is constantly expanding. Matter is being constantly created. The Big Bang theory is another theory that is more widely accepted.

(a) Describe the Big Bang theory of the creation of the universe.

_____

_____

_____ [3 marks]

D–C

(b) Describe the evidence that supports the Big Bang theory.

_____

_____

_____ [3 marks]

2 Explain why cosmic microwave background radiation is such strong evidence for the Big Bang.

_____

_____

_____ [3 marks]

B–A*

Extended response question

Use your knowledge of the properties of electromagnetic waves to explain why different members of the electromagnetic spectrum are used for different sorts of communication.

The quality of written communication will be assessed in your answer to this question.

[6 marks]

Animal and plant cells

1 (a) What type of cell is shown in the diagram?

_____ [1 mark]

(b) Name the missing organelles A and B.

_____

_____ [2 marks]

(c) State the function of the ribosomes.

_____

_____ [1 mark]

(d) Explain why leaf cells contain many chloroplasts but root cells contain none.

_____

_____

_____

_____ [3 marks]

D–C

cell membrane

ribosomes

cytoplasm

mitochondrion

A

chloroplast

B

nucleus

---

2 Explain why a scientist studying viruses would use an electron microscope rather than a light microscope.

_____

_____

_____

_____ [3 marks]

B–A*

Microbial cells

1 State one similarity and one difference between:

(a) a plant cell and yeast.

_____

_____

_____ [2 marks]

(b) an animal cell and bacteria.

_____

_____

_____ [2 marks]

D–C

---

2 The smallest virus has a diameter of 0.02 μm. How many of these viruses could fit end to end across the diameter of a typical animal cell (diameter 20 μm)?

_____ [1 mark]

B–A*

Diffusion

1 A cake is cooking in the kitchen. Use ideas about diffusion to explain how the smell travels around the house.

_____

_____

_____

_____

D–C

_____ [4 marks]

2 Glucose diffuses from the blood into cells where it is used for respiration.

(a) State how the concentration of glucose is kept low inside the cells.

_____ [1 mark]

(b) Why is it important that the glucose concentration is kept low in the cells?

_____

_____

_____ [2 marks]

B–A*

(c) This equipment was left for an hour. The inside of the Visking tubing was then tested for the presence of glucose. Predict the results and explain why you think this will happen.

_____

_____

_____

_____

_____ [3 marks]

glucose solution

Visking tubing

distilled (pure) water

Specialised cells

1 State one adaptation of a red blood cell and explain why it has this adaptation.

_____

_____

D–C

_____ [2 marks]

2 Root hair cells are a specialised root cell. State how they are adapted and explain why they have this adaptation.

_____

_____

_____

B–A*

_____

_____ [3 marks]

Tissues

1 (a) What type of tissue is shown in the diagram?

_____ [1 mark]

(b) Explain why this tissue contains many mitochondria.

_____

_____

_____ [2 marks]

D–C

---

2 This diagram shows the human lungs.

(a) Name the process by which oxygen moves across the air sacs and into the bloodstream.

_____ [1 mark]

(b) Suggest how the airs sacs increase the efficiency of this process.

_____

_____ [1 mark]

(c) Explain why getting oxygen to where it is needed is a simpler process in bacteria than it is in humans.

_____

_____

_____ [2 marks]

lung

air sacs

B–A*

Animal tissues and organs

1 Match the digestive system organ to its function.

	Organ		Function
1	stomach	A	contractions of muscle in walls mixes food with digestive juices
2	liver	B	produces digestive juices
3	small intestine	C	is where water is absorbed from the undigested food, producing faeces
4	pancreas	D	produces bile
5	large intestine	E	is where the absorption of soluble food occurs

_____ [5 marks]

D–C

---

2 Explain the function of the epithelial cells which line the inside of the small intestine.

_____

_____

_____

_____

_____ [4 marks]

B–A*

Plant tissues and organs

1 The diagram shows a cross-section of a leaf.

(a) State the function of each of these tissues:

(i) upper epidermis

_____

_____

_____ [2 marks]

(ii) mesophyll.

_____ [1 mark]

(b) Name the tissue that carries water to the leaves.

_____ [1 mark]

D–C

2 (a) The lower epidermis of the leaf contains tiny holes called stomata. What is their function?

_____ [1 mark]

(b) In hot conditions the stomata may close. Suggest why.

_____ [1 mark]

B–A*

Photosynthesis

1 Tyrone kept some geranium plants in a dark cupboard for two days. He then added iodine to their leaves. Iodine turns black if it comes into contact with starch.

(a) Predict what he would see.

_____ [1 mark]

(b) Explain the reasons behind your prediction.

_____

_____

_____ [3 marks]

(c) He also tested the leaves of some variegated (green and white) geranium leaves that had been kept in a sunny place. This is what he saw.
Explain why he got these results.

went black

stayed orange

_____

_____

_____

_____ [4 marks]

D–C

2 Why does chlorophyll look green?

_____

_____

_____ [3 marks]

B–A*

Limiting factors

1 An experiment was set up as shown in the diagram. The lamp was moved towards the plant and the average number of bubbles released by the plant per minute were counted, averaged and recorded at each distance. The results are shown in the table below.

(a) Name the gas in the bubbles.

_____ [1 mark]

(b) Use the results to describe how light intensity affects the rate of photosynthesis.

_____ [1 mark]

(c) Describe what happens to the rate of photosynthesis once the lamp gets closer than 15 cm away.

_____ [1 mark]

(d) Explain why this has happened.

_____

_____

_____ [2 marks]

collected gas
inverted test tube
bubbles of gas
beaker
water
inverted funnel
water-weed

Distance between the lamp and plant (cm)	50	45	40	35	30	25	20	15	10	5
Number of bubbles of oxygen released per minute	1	2	5	10	16	32	54	56	56	56

D–C

2 Greenhouses are used to increase the rate of photosynthesis of the plants growing inside.

(a) Suggest how tomatoes grown inside a greenhouse would compare to those grown outside.

_____ [1 mark]

(b) A greenhouse has a heater and an automatic watering system. Explain how each of these maximises the rate of photosynthesis of the plants growing inside.

_____

_____ [2 marks]

(c) Suggest one more addition to the greenhouse that would increase photosynthesis even more.

_____ [1 mark]

B–A*

The products of photosynthesis

1 Peas are high in protein. Explain how the pea plant makes this protein.

_____

_____

_____ [3 marks]

D–C

2 The seeds of the dandelion plant are dispersed by the wind. The seeds contain oil as a store of energy.

(a) Why do the seeds need a store of energy?

_____ [1 mark]

(b) Explain why oil is a good choice for the energy store in the seed.

_____

_____ [2 marks]

B–A*

Distribution of organisms

1 Explain why the distribution of:

(a) plants in a cave is low

_____

_____ [2 marks]

D–C

(b) egg wrack seaweed at the top of a beach, far from the sea, is low.

_____

_____ [2 marks]

---

2 The grey squirrel was introduced from North America to the UK at the end of the 19th century. In England, red squirrels are now only found in a few places.

(a) Describe how the distribution of red and grey squirrels has changed in the UK since the end of the 19th century.

_____

_____ [1 mark]

B–A*

(b) The grey squirrels are better adapted than the red squirrels. Use this to explain the changes in distribution.

_____

_____

_____ [3 marks]

Using quadrats to sample organisms

1 Rory wanted to estimate the number of snails on his lawn. He threw a quadrat onto it. You can see the result on the right.

(a) Rory's lawn has an area of 2 m². Estimate how many snails there are on the entire lawn.

_____ [2 marks]

(b) Explain how he could increase the validity of his results.

_____

_____ [2 marks]

0.5m

0.5m

D–C

2 A scientist wanted to work out what size quadrat would be best to use to estimate the distribution of different plants growing in a field. She tried different sizes and counted how many different species she could count inside each one. She then drew a graph of her results.
She chose a size of 15 cm. Explain why.

_____

_____

_____

_____ [3 marks]

Number of species
20
15
10
5
0
0 10 25 50 100
Quadrat size (cm)

B–A*

1 The image below represents part of a protein molecule.

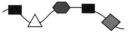

D–C

(a) What are the individual units called?

_____ [1 mark]

(b) Why is it important that these units are arranged in the correct order?

_____

_____ [2 marks]

2 You eat a meal high in protein. Explain how your body uses this protein to form muscle protein in your leg.

B–A*

_____

_____

_____

_____

_____ [5 marks]

Enzymes

1 Emma carried out an investigation where she added amylase to starch and measured how long it took before all the starch disappeared. She repeated the experiment at different temperatures.

(a) In this reaction what is the:

(i) substrate?

D–C

_____ [1 mark]

(ii) product?

_____ [1 mark]

(b) Name her independent variable.

_____ [1 mark]

(c) State one control variable she should use.

_____ [1 mark]

2 The graph shows Emma's results. Explain the shape of the graph.

B–A*

_____

_____

_____

_____

_____

_____ [6 marks]

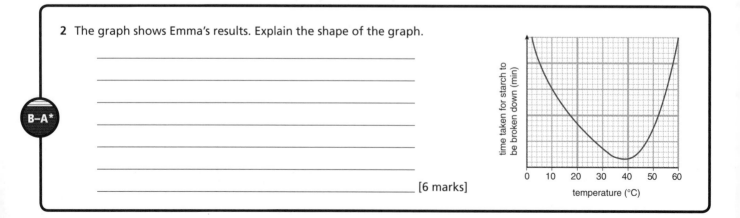

Enzymes and digestion

1 Fill in the gaps in the table. [6 marks]

Enzyme group	Substrate	Product	Where enzyme is produced
amylase	(a)_____	sugars	(b)_____ glands in the mouth and pancreas
(c)_____	protein	amino acids	pancreas, (d)_____ and small intestine
(e)_____	lipids (fats and oils)	fatty acids and (f)_____	pancreas and small intestine

2 (a) Use the graph to state the optimum pH of the enzyme amylase.

_____ [1 mark]

(b) Amylase breaks down starch in the small intestine. The food that enters the small intestine is acidic because it has mixed with the digestive juices in the stomach.

(i) Explain why the digestive juices in the stomach are acidic.

_____ [1 mark]

(ii) Explain how the pH of the small intestine is kept at the optimum pH for amylase. _____

_____

_____ [3 marks]

Enzymes at home

1 (a) Harry was eating a greasy burger and spilt fat down his T-shirt. Explain why washing it in biological washing powder will help remove the fat stain.

_____

_____ [3 marks]

(b) He decided to wash the T-shirt at 60 °C to make sure the stain came out. Explain why this is not a good idea.

_____

_____ [3 marks]

2 Susie wanted to test the hypothesis that biological detergents break down keratin in the skin. She decided to use egg white instead of keratin. Egg white contains the protein albumin.

(a) Write a simple method that she could use to test the hypothesis. State what results she will get if the hypothesis is correct.

_____

_____

_____ [4 marks]

(b) You could argue that her method will lead to results which are not valid. Explain why.

_____ [1 mark]

Enzymes in industry

1 Manufacturers of sports drinks often make sugar syrup for their drinks from starch solution.

(a) State how they turn the starch solution into sugar syrup.

_____ [1 mark]

(b) Why don't they just use sugar syrup?

_____ [1 mark]

(c) Explain why slimming foods may contain fructose instead of glucose.

_____

_____ [3 marks]

(d) Name the enzyme used to turn glucose into fructose.

_____ [1 mark]

D–C

---

2 Bees produce an enzyme called sucrase which they use to produce honey from nectar. Nectar is high in sucrose.

(a) State what two sugars are found in honey.

_____

_____ [2 marks]

(b) State one use of sucrase in the food industry.

_____ [1 mark]

B–A*

Aerobic respiration

1 (a) Why do all your cells need a supply of oxygen?

_____

_____ [2 marks]

(b) Describe how oxygen from the air reaches all your cells.

_____

_____ [2 marks]

D–C

---

2 The amount of oxygen and carbon dioxide absorbed by the leaves of a tree was measured over a day in December.
The results are shown in the diagram.
Explain the shape of the graph.

_____

_____

_____

_____

_____

_____

_____ [6 marks]

B–A*

legend: —— carbon dioxide   ····· oxygen

amount absorbed by the leaves (units): 40, 30, 20, 10

time: 6 a.m, 12 p.m., 6 p.m.

Using energy

1 If you use more energy than you take in you will lose body mass.

   (a) Andrew goes running every day. How does this help him to lose body mass?

   _____

   _____ [2 marks]

   (b) Explain how running on a cold day will help him to lose more mass.

   _____

   _____ [2 marks]

   D–C

2 Explain why muscle cells in the heart contain a lot of mitochondria.

   _____

   _____

   _____

   _____

   _____ [3 marks]

   B–A*

Anaerobic respiration

1 **(a)** State two reasons why it is more beneficial for cells to carry out aerobic respiration than anaerobic.

   _____

   _____ [2 marks]

   (b) Alice has just completed a sprint race. For a few minutes afterwards she continues to breathe deeply and at a fast rate. Explain why.

   _____

   _____ [2 marks]

   D–C

2 Tom is in training for a swimming championship.

   (a) Why does he want to increase his VO_2 max?

   _____ [1 mark]

   (b) This graph shows the VO_2 max for different sports people. Use it to inform him which sport would be best for him to take up in order to improve his VO_2 max.

   _____ [1 marks]

   (c) Explain how taking up this sport will increase his chances of winning the swimming races.

   _____

   _____

   _____

   _____ [3 marks]

   B–A*

Cell division – mitosis

1 Describe why:

(a) mitosis results in two daughter cells.

_____

_____ [1 mark]

(b) the daughter cells are genetically identical to the parent cell.

_____

_____ [1 mark]

(c) skin cells frequently undergo mitosis.

_____

_____ [1 mark]

D–C

2 The diagram shows the stages that happen during mitosis.

(a) Describe what has happened to the chromosomes between stages A and B.

_____

_____ [1 mark]

A   B   C   D

(b) Explain what is happening in stages C and D.

_____

_____ [2 marks]

B–A*

Cell division – meiosis

1 A horse has 64 chromosomes in its body cells. This diagram shows the stages that happen during meiosis in a female horse.

(a) Where in the female horse's body does meiosis take place?

_____ [1 mark]

(b) State the number of chromosomes in each of the numbered cells.

_____ [3 marks]

(c) Explain why meiosis is essential for successful sexual reproduction.

_____

_____

_____ [2 marks]

parent cell — 1

2   3

4 5 6 7 — daughter cells

D–C

2 Outline the differences and similarities between mitosis and meiosis.

_____

_____

_____

_____ [4 marks]

B–A*

Stem cells

1 Parkinson's disease is caused by the death of cells in the brain. Explain how stem cells could be used in the future to treat it.

_____

_____

_____

_____ [3 marks]

D–C

2 You see a website advertising a cure for blindness using stem cell therapy in a clinic in China. Outline reasons why people should be cautious about going there for treatment.

_____

_____

_____

_____ [3 marks]

B–A*

Genes, alleles and DNA

1 DNA fingerprinting can be used to find out who a child's father is. Use the DNA fingerprint below to:

(a) state who the father is.

_____ [1 mark]

(b) explain how the DNA fingerprint is evidence that he is the father.

_____

_____ [2 marks]

D–C

possible father B →
child →
mother →
possible father A →

2 BRCA1 and BRCA2 are genes that belong to a class of genes known as tumour suppressors. Some people carry a mutation in these genes.

(a) Describe two trends as shown on the graph.

_____

_____ [2 marks]

(b) Carly is 35. Both her mother and grandmother died of breast cancer. Explain why she is also at high risk from developing it.

_____

_____ [2 marks]

B–A*

56–87%

33–50%

2%

7%

breast cancer by age 50

breast cancer by age 70

■ BRCA mutation carriers
■ general population

Mendel

1 Explain why scientists did not accept Mendel's ideas until after his death.

_____

_____

_____ [1 mark]

D–C

2 One experiment that Mendel carried out involved breeding together short and tall pea plants. At the time it was widely believed that all characteristics from parents were blended in the offspring.

(a) Use the blending theory to predict the outcome of a cross between a tall pea plant and a short one.

_____ [1 mark]

(b) Mendel carried out this cross and discovered that all the offspring were tall. Explain why.

_____

_____

_____ [3 marks]

B–A*

How genes affect characteristics

1 (a) Natasha has two alleles for hair colour. Why has she got two?

_____ [1 mark]

(b) She has blonde hair. The blonde hair allele (b) is recessive. What pair of alleles has she got?

_____ [1 mark]

(c) Her friend Nisha has brown hair. The brown hair allele is dominant over the blonde hair allele. What possible alleles has Nisha got?

_____

_____ [2 marks]

D–C

2 Some androgens control the development of an embryo so it becomes male. On which chromosome are the genes for these hormones found? Explain your answer.

_____

_____

_____ [2 marks]

B–A*

Inheriting chromosomes and genes

1 A long-haired guinea pig was bred with a short-haired one and produced 12 short-haired offspring.

(a) Which allele is most likely to be dominant?

_____ [1 mark]

(b) What alleles will the offspring have?

_____ [1 mark]

(c) Draw a genetic diagram to show the cross between one of the offspring and a long-haired guinea pig.

D–C

[3 marks]

(d) What is the ratio of homozygous to heterozygous genotypes?

_____ [1 mark]

2 Pea seeds can either be yellow or green. Green is the dominant allele. A grower wanted to find out the genotype of a plant that produces green seeds. Explain how he could use a test cross to find this out.

_____

_____

_____ [3 marks]

B–A*

How genes work

1 Erin has brown eyes. She inherited the brown eye allele from her mother. Describe how this allele produces brown eyes.

_____

_____

_____

_____ [4 marks]

D–C

2 Joshua suffers from type 1 diabetes, which means his body cannot produce insulin. Explain how this could have been caused by a mutation.

_____

_____

_____ [2 marks]

B–A*

Genetic disorders

1 This diagram shows an example of a family tree for cystic fibrosis.

   (a) What genotype does person E have?

   _____ [1 mark]

   (b) Explain why people A and B must be carriers of the cystic fibrosis allele.

   _____

   _____

   _____ [2 marks]

   (c) Calculate the probability of person D also being a carrier of cystic fibrosis.

   _____ [1 mark]

○ normal female   ◐ female with cystic fibrosis

□ normal male   ▨ male with cystic fibrosis

D–C

2 Tay–Sachs disease is a fatal recessive genetic disorder. Children who are born with it usually die by the age of four.
Louise and Stuart's daughter died of the disease. They want to try for another baby. Describe how embryo screening can be used to enable them to have a child free of the disease.

_____

_____

_____

_____ [3 marks]

B–A*

Fossils

1 The bodies of jellyfish are very soft. They have no bones. Explain why scientists have very little evidence of how jellyfish evolved.

_____

_____

_____

_____

_____ [3 marks]

D–C

2 A scientist found what he believed to be the oldest fossil ever found on a beach in western Australia. He dated it at 3.43 billion years old. It seemed to show tubular structures shaped like bacteria, which could be the common ancestor of all life on Earth.

   (a) State one piece of evidence, other than the fossil, that leads scientists to believe that all life on Earth is descended from a common ancestor.

   _____

   _____ [1 mark]

   (b) Scientists have many theories as to how life on Earth evolved. What is needed to prove one of these theories correct?

   _____

   _____ [2 marks]

B–A*

Extinction

1 The Croatian dace is a small fish found only in one small stream in Croatia. People introduced trout into the stream. State two possible reasons why Croatian dace are in danger of becoming extinct.

_____

_____

_____ [2 marks]

D–C

2 Around 65 million years ago a mass extinction occurred. Scientists believe it was due to a massive meteorite hitting the Earth, throwing dust into the air. This blocked the heat from the Sun reaching the surface of the Earth.

(a) What do we mean by a mass extinction?

_____

_____ [2 marks]

(b) What evidence do scientists have that a mass extinction happened at this time?

_____

_____ [1 mark]

B–A*

(c) The meteorite caused the extinction of the dinosaurs but other species such as small mammals survived. Suggest why.

_____

_____ [2 marks]

New species

1 Charles Darwin visited the Galapagos Islands off the coast of South America. He found that on each island there was a different species of finch. We now know that they all descended from one species found on the mainland. Explain the process that led to the evolution of these many different species of finch.

_____

_____

_____

_____

_____

_____ [5 marks]

D–C

2 In one area of a field the soil is contaminated with toxic metals. The type of grass growing in this area starts to evolve to cope with the metals. It also evolves a new flowering time. The grass in the rest of the field remains as it is. Explain how this means that there is a new species of grass in the field.

_____

_____

_____

_____

_____ [3 marks]

B–A*

Extended response question

Control systems are used to keep human temperature at around 37 °C. Explain why it is dangerous for our body temperature to go much higher. Use what you know about enzymes in your answer.

The quality of written communication will be assessed in your answer to this question.

_____

_____

_____

_____

_____

_____

_____

_____

_____

_____

_____

_____

_____

_____

_____

_____

_____

_____

_____

_____

_____

_____

_____

_____

_____

_____

_____

_____

_____

_____

_____

_____

_____

_____

_____

_____

_____

_____ [6 marks]

Investigating atoms

1 (a) Complete this table (**i** to **iv**) about the mass and charge on subatomic particles.

Particle	Relative mass	Relative charge
proton	**(i)**	+1
(iii)	1	**(ii)**
electron	**(iv)**	−1

[4 marks]

28

Si

14

D–C

(b) Use the data in the box to help you answer these questions.

(i) How many protons does this element have? _____

(ii) How many neutrons does this element have? _____

(iii) Write the electronic structure of this element. _____ [3 marks]

2 Ernest Rutherford and his research team suggested, in 1909, that atoms were not solid but had a dense central nucleus. This is a diagram of the apparatus they used.

(a) What is produced at A? _____ [1 mark]

(b) What is B? _____ [1 mark]

(c) What is the purpose of the fluorescent screen?

_____ [1 mark]

(d) Explain how the deflections shown (by labels C to E) led Rutherford to suggest that atoms had a central nucleus.

_____

_____ [3 marks]

fluorescent screen

A

C

B

D

E

B–A*

Mass number and isotopes

1 Many elements have different isotopes. Potassium has two isotopes: potassium-39 and potassium-40.

(a) What is an isotope?

_____ [1 mark]

(b) Describe how potassium-39 is different from potassium-40.

_____ [2 marks]

D–C

(c) In a sample of potassium atoms, 90 were found to be potassium-39, and 10 were potassium-40. Calculate the relative atomic mass of potassium.

_____

_____ [2 marks]

2 Scientists wanted to date some archaeological remains made of wood. Describe how they could use carbon-14 to find the age of the wood.

_____

_____ [2 marks]

B–A*

Compounds and mixtures

1 Barium chloride solution is used to test for the presence of compounds containing sulfate ions. Some barium chloride was reacted with aluminium sulfate solution.

$$3BaCl_2(aq) + Al_2(SO_4)_3(aq) \rightarrow 3BaSO_4(s) + 2AlCl_3(aq)$$

(a) What does (aq) mean?

_____ [1 mark]

(b) Calculate the relative formula mass (M_r) of barium chloride.

(Relative atomic masses: Ba = 137, Cl = 35.5)

_____ [2 marks]

(c) What is the mass of one mole of barium chloride?

_____ [1 mark]

(d) Calculate the relative formula mass (M_r) of aluminium sulfate.

(Relative atomic masses: Al = 27, S = 32, O = 16)

_____ [2 marks]

Electronic structure

1 The diagram shows the electronic structure of a sodium atom.
Draw similar diagrams for the following atoms, writing the electronic structure under each diagram:

(a) calcium

2,8,1

[2 marks]

(b) helium

[2 marks]

(c) aluminium

[2 marks]

(d) fluorine

[2 marks]

(Atomic numbers: Al = 13, Ca = 20, F = 9, He = 2, Na = 11)

2 It is important to know how many outer electrons an element has.

(a) Use a copy of the periodic table to find out the number of outer electrons for the following elements:

(i) bromine (Br) _____ [1 mark]

(ii) strontium (Sr) _____ [1 mark]

(b) Explain why it is important to know the number of outer electrons of an element.

_____ [1 mark]

(c) Noble gases (Group 0) are very unreactive elements. Explain why, in terms of their electronic structure.

_____ [2 marks]

Ionic bonding

1 The compound potassium fluoride (KF) is being investigated by some scientists. They have discovered the following facts about potassium fluoride.

   (a) The scientists think potassium fluoride has ionic bonds. Give two reasons from the table for their conclusion.

property	
melting point	857 °C
boiling point	1502 °C
conduction as a solid	no
conduction as a solution	yes

_____

_____

                                      [2 marks]

   (b) A potassium atom has the electronic structure of 2,8,8,1. A fluorine atom has the electronic structure 2,7.

      (i) Describe the electronic structure of a potassium ion.

      _____ **[1 mark]**

      (ii) Describe the electronic structure of a fluoride ion.

      _____ **[1 mark]**

   (c) Explain how the two ions form the compound potassium fluoride.

     _____ **[2 marks]**

D–C

---

2 A student put some sodium chloride in a beaker and tested it to see if it would conduct electricity, and discovered it did not conduct. The student then added some water to the beaker and found that the solution now conducted electricity.

   (a) Explain why solid sodium chloride does not conduct electricity but when dissolved in water it does.

     _____ **[2 marks]**

   (b) The student took some molten sodium chloride and tested it for electrical conductivity. It was a conductor. Explain why molten sodium chloride can conduct electricity.

     _____ **[1 mark]**

B–A*

Alkali metals

1 Lithium, sodium and potassium are the first three elements of Group 1 or the alkali metals. They all have a single outer electron, and all react easily with water to form hydrogen gas and the metal hydroxide. Potassium has a violent reaction with water, and lithium just floats and fizzes in water.

   (a) Write a word equation for the reaction of sodium with water. _____ **[1 mark]**

   (b) Explain why all the Group 1 metals form positive ions with a 1+ charge.

     _____ **[2 marks]**

   (c) Caesium is another Group 1 metal. It is lower in the group than potassium. Suggest how it would react with water.

     Explain your answer. _____ **[2 marks]**

D–C

---

2 Alkali metals are very common in the Earth's crust. Unlike iron and copper, which were first extracted thousands of years ago, potassium and sodium were only extracted two hundred years ago.

   (a) Why was it possible for iron and copper to be extracted by man much earlier than sodium and potassium?

     _____ **[1 mark]**

   (b) How are potassium and sodium extracted from their metal ores?

     _____ **[1 mark]**

   (c) Why was it only possible to extract them for the first time two hundred years ago?

     _____ **[1 mark]**

B–A*

Halogens

1 The diagram shows the electronic arrangement of a chlorine atom.

(a) Complete the second diagram to show the electron arrangement of a chloride ion. [1 mark]

(b) What is the charge on a chloride ion? _____ [1 mark]

(c) Name the element with the same electronic structure as a chloride ion. _____ [1 mark]

chlorine atom   chloride ion

2 Here is some information about chlorine and its compounds.

(a) Chlorine gas is highly poisonous. Explain why it is safe to use sodium chloride for cooking.

_____

_____ [2 marks]

substance	description	use
chlorine	green gas	killing bacteria in water supplies
sodium chloride	white solid	in cooking
polyvinyl chloride	flexible plastic solid	making plastic window frames

(b) Sodium chloride dissolves in water, but polyvinyl chloride is used to make window frames. Explain why polyvinyl chloride can be used for window frames.

_____ [1 mark]

(c) Chlorine gas is bubbled into drinking water to kill bacteria, but drinking water does not harm people. Explain why this is.

_____ [1 mark]

Ionic lattices

1 The diagram shows part of a sodium chloride crystal lattice

(a) How many chloride ions are attracted to each sodium ion? [1 mark]

(b) Sodium chloride has a melting point of approximately 800 °C, but the molecule hydrogen chloride is a gas at room temperature. Explain why.

_____ [2 marks]

(c) Two electrodes, one positive and one negative, are placed in some molten sodium chloride, and a current passed through. Describe how the sodium and chloride ions react.

● Na⁺   ○ Cl⁻

_____

_____ [3 marks]

2 Sodium sulfate is an ionic compound and has the formula Na_2SO_4.

(a) Write the formulae of the two ions in sodium sulfate.

_____ [2 marks]

(b) Calculate the relative formula mass for sodium sulfate.

(Relative atomic masses (A_r), Na = 23, O = 16, S = 32.)

_____ [2 marks]

Covalent bonding

1 Look at this diagram of two chlorine atoms.

   (a) Draw a diagram to show how the two atoms will make a molecule of chlorine gas (Cl_2)

D–C

[2 marks]

   (b) The atoms are joined together by a covalent bond. What is a covalent bond?

_____ [1 mark]

   (c) Explain why non-metal elements form molecules with covalent bonds.

_____ [2 marks]

---

2 This diagram represents the structure of carbon dioxide. $O = C = O$

   (a) How many covalent bonds are made by each oxygen atom? _____ [1 mark]

   (b) How many covalent bonds are made by the carbon atom? _____ [1 mark]

B–A*

   (c) The bond length of a C=O bond is 0.116 nm. How far apart are the two oxygen atoms' nuclei?

_____ [1 mark]

Covalent molecules

1 The table shows some data about the halogens, Group 7.

   (a) Room temperature is 20 °C. Which of the halogens are gases at room temperature? _____ [1 mark]

   (b) How many outer electrons has a bromine atom? ___ [1 mark]

   (c) Describe the trend in boiling points as the halogen molecule gets larger.

halogen	boiling point in °C	diameter of molecule in nm
fluorine	−220	0.28
chlorine	−34	0.39
bromine	58	0.46
iodine	114	0.53

D–C

_____ [1 mark]

   (d) The boiling point of each halogen is affected by the intermolecular forces between the molecules. Describe, with reasons, how the intermolecular forces between halogen molecules change in the group.

_____

_____

_____ [3 marks]

---

2 Water molecules have a mass of 18. This is the same as fluorine molecules. Fluorine boils at −220 °C, but water has a boiling point of 100 °C.

   (a) Explain why water boils at 100 °C instead of −220 °C.

_____ [2 marks]

B–A*

   (b) Suggest, with reasons, which halogen is likely to have intermolecular forces of about the same strength as water. (The table in question 1 may help.)

_____ [2 marks]

Covalent lattices

1 The diagram shows the structures of diamond and graphite, two covalent lattices of carbon.

graphite   diamond

(a) How many covalent bonds does each carbon atom have in:

(i) diamond? _____ [1 mark]

(ii) graphite? _____ [1 mark]

(b) Diamond is one of the hardest known substances. Describe how its structure enables it to be very hard.

_____

_____ [2 marks]

(c) Graphite is often used as a lubricant. Describe how its structure enables it to be a good lubricant.

_____ [2 marks]

D–C

2 Diamond does not conduct electricity, but graphite does. This is because graphite has de-localised electrons.

(a) What are de-localised electrons?

_____ [1 mark]

(b) Explain how de-localised electrons allow graphite to conduct electricity.

_____

_____ [2 marks]

B–A*

Polymer chains

1 A bottle manufacturer has two different types of plastic bottle. One bottle is made from a thermosetting polymer, the other from a thermosoftening polymer. The thermosetting polymer is cheaper to produce.

(a) Describe the difference between thermosetting and thermosoftening polymers.

_____

_____ [2 marks]

(b) Suggest an environmental benefit from using the thermosoftening polymer bottle.

_____ [1 mark]

(c) Describe how the structure of the thermosetting polymer differs from the thermosoftening polymer.

_____

_____ [2 marks]

D–C

2 Polymer molecules are long chains of atoms held together by covalent bonds. There are weak attractions between the polymer chains. They have no definite melting point, instead they soften over a range of temperatures.

(a) Explain why they soften over a range of temperatures.

_____

_____ [2 marks]

(b) Suggest why polymers with shorter chains have lower melting ranges than polymers with longer chains.

_____ [1 mark]

B–A*

Metallic properties

1 The diagram on the right shows the metal lattice of a pure metal such as titanium.

When making a shape-memory alloy, titanium metal is mixed with nickel.

(a) Draw a diagram in the box on the far right to show the effect of adding nickel atoms to the titanium metal lattice. [2 marks]

(b) What is a shape-memory alloy?

_____ [1 mark]

(c) Describe how a shape-memory alloy is useful in making a dental brace.

_____ [2 marks]

pure metal

D–C

2 The diagram on the right shows the lattice arrangement in metal bonding.

(a) Explain why the nuclei of the metal atoms are spaced out evenly.

_____ [2 marks]

(b) Describe how the 'sea of electrons' allows the metal to conduct electricity.

_____ [2 marks]

(c) Explain how the 'sea of electrons' allows the metal to conduct heat quickly.

_____ [2 marks]

metal atom nuclei in a 'sea of electrons'

B–A*

Modern materials

1 Titanium(IV) oxide is used in sunscreen. The large solid particles used are white in colour and, if applied thickly, coat the skin in a white paste. Chemists are developing nano-sized particles of titanium(IV) oxide to overcome this problem.

(a) How many atoms are there in a nanoparticle?

_____ [1 mark]

(b) How large are nanoparticles?

_____ [1 mark]

(c) Explain why sunscreens' nanoparticles of titanium(IV) oxide will be an improvement on current sunscreens.

_____

_____ [2 marks]

D–C

2 Buckminsterfullerene is a nanoparticle that is like a football. It is possible to trap a drug molecule inside the 'ball' as it is made. The pharmaceutical industry is very interested in this process as it will help in making new cell-killing drugs to treat cancers.

(a) Explain why placing the toxic drug molecule inside buckminsterfullerene will be less damaging to the health of cancer patients.

_____ [1 mark]

(b) Describe what should happen when the new drug reaches a cancerous cell or tumor.

_____

_____ [2 marks]

B–A*

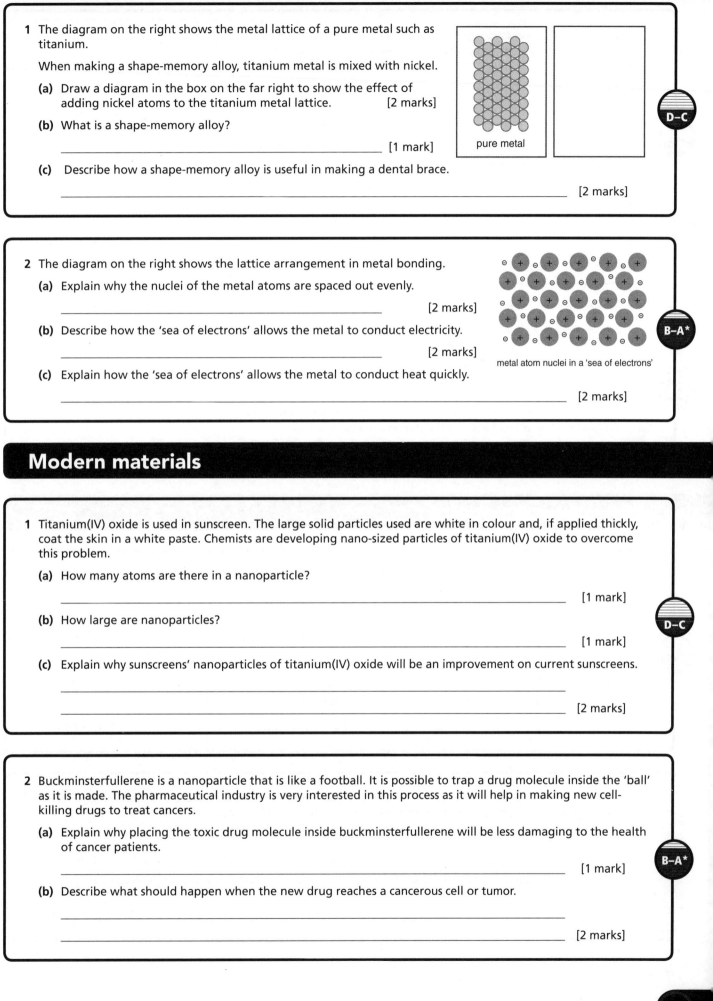

Identifying food additives

1 Some crime scene investigators used chromatography to process some paint found at the scene of a serious hit-and-run accident. They wanted to know the manufacturer of the vehicle. Here is the chromatogram they produced.

(a) How many different chemicals does the sample from the hit-and-run car contain?

_____ [1 mark]

(b) All the spots are red. How many different red pigments are used by all the different manufacturers?

_____ [1 mark]

(c) Which manufacturer made the car?

_____ [1 mark]

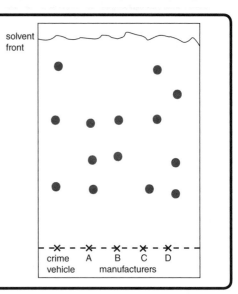

2 To help identify different pigments in a chromatogram the retention factor for a particular solvent and pigment is calculated.

(a) Why is it important to calculate retention factors for different solvents for the same pigment?

_____ [1 mark]

(b) A pigment travels 18 cm up a chromatogram, and the solvent has travelled up 27 cm. Calculate the retention factor for this pigment in this solvent.

_____ [2 marks]

Instrumental methods

1 Gas Chromatography–Mass Spectrometry (GC-MS) is an instrumental method, used to determine the identity of different chemicals.

(a) What is an instrumental method?

_____ [1 mark]

(b) Why are instrumental methods used?

_____ [2 marks]

(c) How does a mass spectrometer help analyse the results of the gas chromatogram?

_____ [2 marks]

2 Mass spectrometers work by removing an electron from the molecule to be identified, and then finding the mass of the molecular ion. Sometimes in removing the electron the molecule is broken up into fragments.

(a) What charge does a molecular ion always have?

_____ [1 mark]

(b) What is the advantage of the molecule breaking into fragments when analysing two molecules with the same mass?

_____ [2 marks]

Making chemicals

1 Chemicals can be made by a variety of different reactions such as neutralisation, oxidation, reduction, and precipitation.

(a) What is a neutralisation reaction? _____ [1 mark]

(b) What is a precipitation reaction? _____ [1 mark]

(c) What is an oxidation reaction? _____ [1 mark]

(d) Why are oxidation and reduction reactions sometimes described as redox reactions?

_____ [2 marks]

D–C

2 In a chemical equation, the number of moles of each substance needed in the reaction is shown as a figure in front of the substance's formula in the balanced equation. Here is a balanced equation.

$$Na_2CO_3(s) + 2HCl(aq) \rightarrow 2NaCl + CO_2(g) + H_2O(l)$$

If one mole of sodium carbonate is used how many moles of:

(a) hydrochloric acid are needed? _____ [1 mark]

(b) sodium chloride are produced? _____ [1 mark]

(c) carbon dioxide are produced? _____ [1 mark]

B–A*

Chemical composition

1 A student heated some copper oxide using this apparatus to reduce some copper oxide to copper metal.
The student weighed the 'boat', then the copper oxide and 'boat' before heating the copper oxide strongly. When cooled, the student then weighed the copper produced in the 'boat'.

burning hydrogen gas

hydrogen gas

porcelain 'boat'   heat   copper oxide

(a) What was the mass of copper oxide used?

_____ [1 mark]

(b) What was the mass of copper produced?

_____ [1 mark]

Mass of boat = 15.44 g,
boat and copper oxide = 19.42 g,
boat and copper produced = 18.62 g

D–C

(c) What was the mass of oxygen removed from the copper oxide?

_____ [1 mark]

(d) Use the masses to calculate the formula of copper oxide. Show your working.
(Relative atomic masses, (A_r): Cu = 63.5, O = 16)

_____ [3 marks]

2 A group of students wanted to find out if a hydrocarbon was an alkene. They burnt 3.36 g of the hydrocarbon and found it produced 10.56 g of carbon dioxide (CO_2), and 4.32 g of water (H_2O).
(Relative atomic masses, (A_r): C = 12, H = 1, O = 16)

(a) What is the relative molecular mass of carbon dioxide?. _____ [1 mark]

(b) What is the relative molecular mass of water? _____ [1 mark]

(c) Calculate the empirical formula of the hydrocarbon. _____ [3 marks]

(d) Is the hydrocarbon an alkene? Explain your answer.

_____ [1 mark]

B–A*

Quantities

1 The reaction for the thermal decomposition of calcium carbonate to calcium oxide is shown by this equation.

$$CaCO_3(s) \rightarrow CaO(s) + CO_2(g)$$

(Relative atomic masses, (A_r): Ca = 40, C = 12, O = 16)

(a) Calculate the relative formula mass of calcium carbonate.

_____ [1 mark]

(b) Calculate the relative formula mass of calcium oxide.

_____ [1 mark]

(c) If 125 tonnes of calcium carbonate are used, what is the maximum theoretical yield of calcium oxide?

_____ [2 marks]

D–C

---

2 Lead iodide is obtained by precipitation of lead nitrate and sodium iodide solutions. The reaction shown by this equation:

$$Pb(NO_3)_2(aq) + 2NaI(aq) \rightarrow PbI_2(s) + 2NaNO_3(aq)$$

(Relative atomic masses, (A_r): Pb = 207, Na = 23, O = 16, I = 127, N = 14)

(a) Calculate the mass of sodium nitrate that can be made from 16.5 g of lead nitrate.

_____ [4 marks]

(b) Explain why this yield is unlikely.

_____ [1 mark]

B–A*

How much product?

1 A student planned to make some copper sulfate by reacting copper oxide with sulfuric acid.

$$CuO(s) + H_2SO_4(aq) \rightarrow CuSO_4(aq) + H_2O(\ell)$$

(Relative atomic masses, (A_r): Cu = 63.5, H = 1, S = 32, O = 16)

The student used 1.59 g of copper oxide with an excess of sulfuric acid. The student weighed the copper sulfate made, and found he had made 2.2 g of copper sulfate.

(a) What is meant by an excess? _____ [1 mark]

(b) Calculate the theoretical yield of copper sulfate. _____ [3 marks]

(c) Calculate the percentage yield obtained. _____ [1 mark]

D–C

---

2 A plant making sulfuric acid reacts sulfur with oxygen to make sulfur(VI) oxide according to this equation.

$$2S(s) + 3O_2(g) \rightarrow 2SO_3(g)$$

(Relative atomic masses, (A_r): S = 32, O = 16)

(a) Calculate the theoretical yield that can be obtained from burning 100 tonnes of sulfur.

_____ [3 marks]

(b) Calculate the mass of oxygen required to burn the sulfur.

_____ [1 mark]

(c) If only 187.6 tonnes of sulfur(VI) oxide are made, what is the percentage yield?

_____ [2 marks]

B–A*

Reactions that go both ways

1 Cobalt chloride paper can be used to test for the presence of water. It uses this reversible reaction:

$$CoCl_2 \cdot 6H_2O(s) \rightleftharpoons CoCl_2(s) + 6H_2O(l)$$

pink   blue

(a) What does this symbol $\rightleftharpoons$ mean?

_____ [1 mark]

D–C

(b) Describe the colour change when water is added to cobalt chloride paper.

_____ [1 mark]

(c) Explain why cobalt chloride paper may be dried and used again.

_____

_____ [2 marks]

Rates of reaction

1 A group of students wanted to measure the rate of reaction when calcium carbonate dissolves in hydrochloric acid.

$$CaCO_3(s) + HCl(aq) \rightarrow CaCl_2(aq) + H_2O(l) + CO_2(g)$$

They used the apparatus shown in the diagram. They placed 20 g of calcium carbonate in a conical flask then added 25 cm³ of dilute hydrochloric acid.

(a) Describe the measurements they need to take to follow the rate of the reaction.

_____ [2 marks]

Another group of students measured the volume of gas produced, and plotted a graph of their results.

(b) Explain why the graph is a curve.

_____

_____ [2 marks]

D–C

(c) What is the maximum volume of carbon dioxide obtained from the reaction?

_____ [1 mark]

2 (a) Use the graph above to calculate the rate of the reaction after 1.5 minutes.

_____

_____ [2 marks]

B–A*

(b) At the end of the experiment there is still calcium carbonate left in the flask. Explain why this has happened.

_____

_____ [2 marks]

Collision theory

1 Magnesium ribbon reacts with dilute hydrochloric acid like this:

$$Mg(s) + 2HCl(aq) \rightarrow MgCl_2(aq) + H_2(g)$$

(a) If a 2 g piece of magnesium ribbon was reacted with an excess of hydrochloric acid, what would happen to the rate of reaction if the temperature was increased? Explain your answer.

_____ [2 marks]

D–C

(b) If a 2 g piece of magnesium ribbon was cut into six pieces then reacted with an excess of hydrochloric acid, what would happen to the rate of reaction? Explain your answer.

_____ [2 marks]

(c) If a 2 g piece of magnesium ribbon was reacted with a more dilute solution of hydrochloric acid, what would happen to the rate of reaction? Explain your answer.

_____ [2 marks]

2 Hydrogen gas can be made in the laboratory by reacting zinc metal with hydrochloric acid. Copper sulfate is often used as a catalyst.

(a) What is a catalyst? _____ [1 mark]

B–A*

(b) A teacher uses a 2 cm³ cube of zinc metal for a first reaction, then repeats the reaction using the same-sized cube cut into eight smaller 1 cm³ cubes. Calculate the change in surface area, and so the change in the rate of the reaction. _____ [2 marks]

(c) The teacher wants to check if the results of the second experiment are repeatable, so repeats the experiment using manganese oxide as the catalyst. Give two reasons why the results obtained cannot confirm the repeatability of the experiment. _____ [2 marks]

Adding energy

1 A student planned to make some iron sulfate by reacting 2 g of iron with an excess of sulfuric acid.

$$Fe(s) + H_2SO_4(aq) \rightarrow FeSO_4(aq) + H_2(g)$$

The student carried out the reaction at two different temperatures. The results are shown on the graph.

(a) What is the maximum volume of hydrogen gas the student can obtain? _____ [1 mark]

D–C

(b) The student decided to repeat the experiment at 40 °C.

    (i) Sketch on the graph what the results should look like. [2 marks]

    (ii) Explain your curve, using collision theory.

_____ [2 marks]

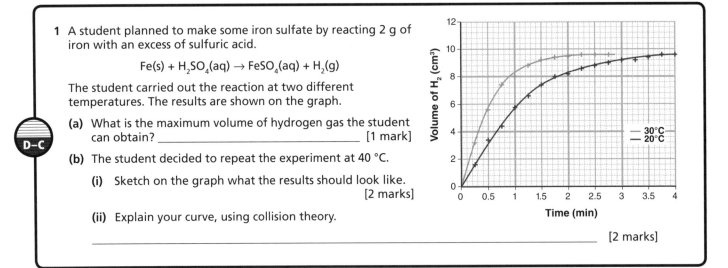

2 For a chemical reaction to take place, the reactant particles must have sufficient energy to react with each other when they collide. This energy is called the activation energy.

(a) Magnesium ribbon burns vigorously in air if briefly heated in a Bunsen flame. It does not burn without heating. Explain in terms of activation energy why only a little heat energy makes magnesium burn.

_____ [2 marks]

B–A*

(b) Suggest why increasing the temperature will increase the number of particles that have sufficient activation energy to react. _____ [2 marks]

Concentration

1 The graph shows the volume of gas produced when 3 g of aluminium reacted with hydrochloric acid at three different concentrations.

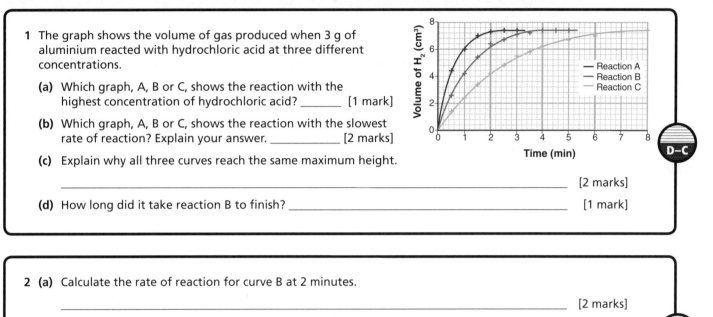

(a) Which graph, A, B or C, shows the reaction with the highest concentration of hydrochloric acid? _____ [1 mark]

(b) Which graph, A, B or C, shows the reaction with the slowest rate of reaction? Explain your answer. _____ [2 marks]

(c) Explain why all three curves reach the same maximum height.

_____ [2 marks]

(d) How long did it take reaction B to finish? _____ [1 mark]

2 (a) Calculate the rate of reaction for curve B at 2 minutes.

_____ [2 marks]

(b) If the concentration of hydrochloric acid in Reaction B was doubled, how long would it take the reaction to complete? Explain your answer.

_____ [2 marks]

Size matters

1 A student looked at burning iron in a Bunsen flame. The student tried burning a small nail, some iron wool, and some iron filings in a Bunsen flame. Here are the observations of the experiment.

type of iron	observations
nail	glowed red, did not burn, did not change size
iron wool held by tongs	bright red glow that spread through the wool, as it spread the wool disappeared, brown bits appeared on the heat mat
iron filings sprinkled into flame	bright sparks as filings sprinkled onto flame, brown bits on heat mat

(a) Which of the three types of iron reacted fastest? _____ [1 mark]

(b) Explain your answer using collision theory.

_____ [2 marks]

(c) Which of the types of iron did not react? What evidence is in the table for this?

_____ [2 marks]

2 Ships use fuel oil to heat their boilers, at normal speeds they spray fuel oil into the boilers. To rapidly increase their speed, they use additional sprayers that produce a much finer spray.

(a) Suggest why ships use sprayed fuel oil rather than liquid fuel oil.

_____ [2 marks]

(b) Explain why ships use a finer spray when they want to heat the boilers up faster and increase speed.

_____ [2 marks]

Clever catalysis

1 Hydrogen peroxide decomposes slowly to form water and oxygen. It can be speeded up by using manganese(IV) oxide as a catalyst.

A student wanted to prove that manganese(IV) oxide was a catalyst, so did an experiment. Here are the results.

volume of hydrogen peroxide in cm³	mass of manganese(IV) oxide at the start in grams	mass of manganese(IV) oxide at the end in grams	time for 50 cm³ of oxygen gas to be produced in seconds.
100	0.0	0.0	250
100	2.3	2.3	50

D–C

(a) By how much did the manganese(IV) oxide affect the rate of reaction? Explain your answer.

_____ [2 marks]

(b) What evidence is there that the manganese(IV) oxide is a catalyst? Explain your answer.

_____ [2 marks]

(c) Suggest how the manganese(IV) oxide may have affected the reaction rate.

_____ [2 marks]

2 Many motor vehicles now have catalytic converters fitted. Vehicles fitted with them cannot use petrol that contains lead.

(a) Explain why a car fitted with a catalytic converter cannot use leaded petrol.

_____ [2 marks]

B–A*

(b) Catalytic converters are expected to last for at least 100 000 kilometres of use with unleaded petrol. Suggest why after this a catalytic converter may no longer work.

_____ [2 marks]

Controlling important reactions

1 Ammonia is a very useful gas. It is used to make fertilisers. Controlling the rate of the reaction is very important to maximise the ammonia made, whilst keeping the costs economic. Ammonia is made by a reversible reaction between hydrogen and nitrogen.

(a) Write a word equation for the manufacture of ammonia.

_____ [1 mark]

D–C

(b) The reaction uses an iron catalyst. What does this suggest about the rate of the reaction between nitrogen and hydrogen? _____ [1 mark]

(c) At 450 °C only 20% of the nitrogen and hydrogen become ammonia. Explain why a temperature of 450 °C is used for the reaction.

_____ [2 marks]

(d) Explain why a pressure of 200 atmospheres is used for the reaction.

_____ [2 marks]

2 In industry ammonia is made at 450 °C and at a pressure of 200 atmospheres. The speed of reaction is more important than the yield.

(a) Use your knowledge of the Haber Process to suggest why yield is not an important economic consideration.

B–A*

_____ [2 marks]

(b) Explain why this combination of temperature and pressure is chosen so that the process is economic.

_____ [3 marks]

The ins and outs of energy

1 A student dissolved 5 g of ammonium nitrate in 50 cm³ of water. The temperature of the solution fell by 6 °C.

(a) What type of reaction is this? _____ [1 mark]

(b) Explain why the temperature dropped.

_____ [2 marks]

(c) The diagram shows a self-heating can of coffee. Which two chemicals must react for the coffee to start to heat up?

_____ [1 mark]

(d) Where does the heat energy come from to heat the coffee?

_____ [2 marks]

coffee product

insert

quicklime

foil separator

water

plastic button

D–C

2 The reaction of anhydrous copper sulfate with water is a reversible reaction.

anhydrous copper sulfate + water ⇌ hydrated copper sulfate

(a) What do we mean by a reversible reaction?

_____ [1 mark]

B–A*

(b) What is meant by the forward reaction?

_____ [1 mark]

(c) If the forward reaction is exothermic, explain why the backward reaction is endothermic.

_____ [2 marks]

Acid–base chemistry

1 Alkalis and bases can both neutralise an acid.

(a) Describe the difference between an alkali and a base.

_____ [1 mark]

(b) When sodium hydroxide solution reacts with nitric acid it forms two new compounds. Complete the equation to show the two products.

sodium hydroxide + nitric acid → _____ + _____ [1 mark]

D–C

(c) Copper carbonate can act as a base and react with hydrochloric acid. Complete the equation to show the products formed.

copper carbonate + hydrochloric acid → _____ + _____ + _____ [2 marks]

2 A teacher demonstrated the dissolving of ammonia gas (NH_3) in water to some students. The water had some universal indicator in it. The water turned from green at the start to blue at the end.

(a) What did the experiment show about ammonia gas?

_____ [1 mark]

(b) The ammonia dissolved in the water to form a positive and a negative ion. Write the formulae of the two ions formed.

_____ [2 marks]

B–A*

(c) Explain why the solution made is able to neutralise acids.

_____ [1 mark]

Making soluble salts

1 A student wanted to make some cobalt sulfate. The student planned to use cobalt metal, but a friend said this would be too hazardous and suggested using cobalt oxide instead.

(a) Name the acid needed to make cobalt sulfate. _____ [1 mark]

(b) The student added some cobalt oxide to the acid.

(i) What should the student do to help the cobalt oxide and acid to react?

_____ [1 mark]

(ii) Suggest two ways that the student could use to show all the acid had been used up.

_____ [2 marks]

(c) How could the student make sure that the cobalt sulfate made was pure, and had no cobalt oxide left in it?

_____ [1 mark]

(d) How could the student obtain a solid sample of cobalt sulfate crystals?

_____ [2 marks]

2 Calcium sulfate is used to make Plaster of Paris. A teacher decided to make some Plaster of Paris by reacting some pieces of calcium carbonate with sulfuric acid.

(a) What did the teacher expect to see when the calcium carbonate reacted with the sulfuric acid?

_____ [1 mark]

(b) This happened for a short while and then stopped. Explain why the reaction stopped.

_____ [2 marks]

Insoluble salts

1 Calcium carbonate obtained from limestone or chalk is too impure to be used in kitchen cream cleaning liquids. Instead it is obtained by using this reaction.

$$CaCl_2(aq) + Na_2CO_3(aq) \rightarrow CaCO_3(s) + 2NaCl(aq)$$

(a) Name the products in the reaction. _____ [1 mark]

(b) After the reaction is complete, how could you separate the calcium carbonate from the solution?

_____ [1 mark]

(c) Explain how you could ensure that the calcium carbonate would be pure when dried.

_____ [1 mark]

2 Silver nitrate ($AgNO_3$) solution is used to test solutions thought to contain halogen compounds or halides such as chlorides, bromides and iodides.

(a) If a solution of calcium iodide (CaI_2) is tested with silver nitrate solution:

(i) What would you see happen? _____ [1 mark]

(ii) What precipitate would be formed? _____ [1 mark]

(b) If a solution containing magnesium chloride ($MgCl_2$) is tested with silver nitrate solution, what colour would you expect the precipitate to be? _____ [1 mark]

(c) Write a balanced symbol equation for the reaction of silver nitrate solution with magnesium chloride solution.

_____ [2 marks]

Ionic liquids

1 Potassium dichromate ($K_2Cr_2O_7$) is a yellow ionic solid that dissolves in water to form potassium ions (K^+) and dichromate ions ($Cr_2O_7^{2-}$). The yellow colour is caused by the dichromate ions. If an electric current is passed through a filter paper soaked in water with a pipette of potassium dichromate added as shown on the right, this is what happens.

(a) What is the name given to the process of splitting a compound using electricity?

_____ [1 mark]

(b) Suggest why the yellow colour travels towards the positive electrode or anode.

_____ [2 marks]

(c) Explain what will happen to a potassium ion when it reaches the negative electrode.

_____ [2 marks]

2 (a) Explain why the movement of the dichromate and potassium ions allows electricity to flow round the circuit.

_____ [2 marks]

(b) A student tried to do this experiment but connected the electrodes to a power pack that only provided alternating current. The circuit conducted electricity, but the colour stayed in the middle of the filter paper. Explain why?

_____ [2 marks]

Electrolysis

1 The diagram shows how copper is purified using electrolysis.

(a) Which electrode has the impure copper on? _____ [1 mark]

(b) What happens to the impure copper at this electrode?

_____ [2 marks]

(c) Explain why the electrolyte solution has to contain copper ions.

_____ [1 mark]

(d) What are substances X? _____ [1 mark]

2 When molten sodium chloride is electrolysed, the sodium ions [$Na^+(\ell)$] are attracted to the cathode where they gain electrons. This can be represented by the half-equation:

$$Na^+(\ell) + e^- \rightarrow Na(\ell)$$

(a) Why is this called a half-equation?

_____ [1 mark]

(b) Write a similar half-equation for the reaction happening at the anode with chloride ions.

_____ [1 mark]

(c) Explain why the reaction at the anode is called an oxidation reaction?

_____ [1 mark]

(d) Write a balanced equation showing the overall reaction.

_____ [1 mark]

Extended response question

Sodium hydroxide is made by the electrolysis of brine (sodium chloride) at room temperature. Sodium chloride is an ionic compound containing sodium ions (Na^+) and chloride ions (Cl^-).

Sodium hydroxide can also be made by the electrolysis of molten sodium chloride and calcium chloride at 600 °C, and then reacting the sodium produced with water.

The figure shows the apparatus used to electrolyse the brine.

Use information in the box, and your knowledge and understanding of this process, to answer this question.

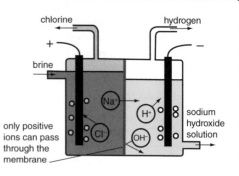

Explain, as fully as you can, why sodium hydroxide is made by the electrolysis of brine rather than from molten sodium chloride.

The quality of written communication will be assessed in your answer to this question.

_____

_____

_____

_____

_____

_____

_____

_____

_____

_____

_____

_____

_____

_____

_____

_____

_____

_____

_____

_____

_____

_____

_____

_____

_____

_____

_____

_____ [6 marks]

See how it moves

1 The distance time graph shows the motion of a bus.

(a) Describe the motion of the bus at;

(i) section A _____

(ii) section B _____

[2 marks]

(b) In which section did the bus travel fastest?

_____

[1 mark]

(c) Describe the appearance of the graph if the bus is accelerating.

_____ [2 marks]

D–C

2 (a) Explain the difference between average speed and instantaneous speed.

_____ [2 marks]

B–A*

(b) A speed camera takes two photographs at half-second intervals of a car driving along a road. Explain how these could be used to calculate the speed of a driver.

_____
_____
_____ [3 marks]

Speed is not everything

1 (a) Explain what is meant by velocity.

_____
_____ [2 marks)

(b) Explain why the velocity of a car changes as it drives round a corner at a constant speed.

_____
_____ [2 marks]

D–C

(c) The car initially travels at 40 m/s. After braking, the car slows down and stops 5 seconds later. What is its acceleration?

_____
_____ [3 marks]

2 The velocity time graph shows the motion of Alice on her bicycle.

(a) Describe Alice's motion in each section.

_____
_____ [4 marks]

B–A*

(b) Explain how you could use the graph to calculate the distance Alice travels.

_____ [2 marks]

Forcing it

1 A van is trying to pull another car out of a ditch. Both vehicles cannot move.

(a) Describe how the force from the van compares with the force from the car.

_____

_____ [2 marks]

D–C

(b) A tractor is used to pull the car out of the ditch.
How does the force from the tractor compare with the force from the van?

_____ [1 mark]

2 (a) A car manual includes these tips for drivers to help them drive more economically:

- *do not drive at high speeds with your windows open – air conditioning is more efficient*
- *remove the luggage rack from the roof of your vehicle if you do not need it*
- *try to drive at a steady speed and avoid braking and accelerating.*

Use your ideas about forces to explain why these tips help to reduce fuel consumption.

B–A*

_____

_____

_____

_____ [5 marks]

Forces and acceleration

1 The table shows results from an experiment to investigate the link between force and acceleration.

(a) Describe the pattern shown in the results.

Force	Acceleration
1 N	2 m/s^2
2 N	4 m/s^2
3 N	6 m/s^2
4 N	7 m/s^2
5 N	10 m/s^2

_____

_____ [2 marks]

D–C

(b) Explain whether the results have been taken over a large enough range.

_____ [2 marks]

(c) Why should results from an experiment be repeated?

_____ [2 marks]

2 The mass of a rock is 4 kg.

(a) Calculate the force needed to accelerate the rock by 5 m/s^2.

_____

_____ [3 marks]

B–A*

(b) Explain how the mass of another rock could be found using a forcemeter and an accelerometer.

_____

_____ [3 marks]

Balanced forces

1 The diagram shows a tug of war between two teams. Neither team is winning so nobody is moving.

  (a) Write down two pairs of balanced forces that act on the person labelled X.

    (i) Pair 1 _____ and _____

    (ii) Pair 2 _____ and _____   **[4 marks]**

  (b) Describe how forces change when the team on the right hand side starts to win by pulling the other team towards them.

    _____

                            [2 marks]

X

D–C

2 A person enters a lift and travels to the floor above. Describe and explain how the forces acting on this person change during this journey.

_____

_____   **[5 marks]**

B–A*

Stop!

1 (a) Stopping distance can be split into two parts. What are these called?

    _____   **[2 marks]**

  (b) Nicole is taking her first driving lesson. Why will her thinking distance be longer than her driving instructor's?

    _____   **[1 mark]**

  (c) The table shows how thinking distance changes with speed

Thinking distance (m)	0	6	9	12
Speed (km/h)	0	32	48	64

    (i) Describe how thinking distance changes with speed.

    _____   **[2 marks]**

D–C

    (ii) A driver drinks a glass of wine and sometime later their thinking time is calculated as 1.0 seconds. Calculate their thinking distance when travelling at 9 m/s.

    _____   **[1 mark]**

    (iii) A speed of 9 m/s is the same as 32 km/h. Use information from the table to describe the effect of the drink on the thinking distance of a driver.

    _____   **[1 mark]**

2 Explain how the condition of a car can affect its stopping distance, giving examples. Explain which factor you think is most important.

_____

_____   **[6 marks]**

B–A*

Terminal velocity

1 (a) What is meant by terminal velocity?

_____ [1 mark]

(b) A ball bearing is dropped into a cylinder of oil.
Describe how each of the forces (including the resultant force) changes:

(i) when the ball bearing is first released in the oil.

_____ [2 marks]

(ii) when the ball bearing reaches terminal velocity.

_____ [2 marks]

D–C

2 (a) Explain in terms of forces why a skydiver slows down when they open their parachute.

_____
_____
_____ [5 marks]

B–A*

(b) A designer claims to have invented a parachute that allows a skydiver to hover in calm weather conditions. Explain if this is possible in terms of forces.

_____
_____
_____ [4 marks]

Forces and elasticity

1 Alex set up the experiment shown in the diagram to test the hypothesis that the extension of a spring depends on the force applied to it. The extension of the spring is its change in length.

(a) What is the independent variable _____ [1 mark]
(b) What is the dependent variable _____ [1 mark]
(c) Explain why it is important that Alex writes down all the data he collects, and not just the result of calculations.

_____
[1 mark]

(d) Explain two steps Alex should take to make sure his results are valid.

_____
[2 marks]

D–C

spring

mass

pointer

ruler

clamp and stand

2 A catapult manufacturer is testing different types of rubber to find the most suitable for their product. State two qualities of the rubber that would make it suitable for a catapult. Describe a test that the manufacturer could use to evaluate the most suitable choice.

B–A*

_____
_____
_____ [5 marks]

Energy to move

1 Complete these sentences:

   (a) A moving car has kinetic energy transferred from _____ energy in its fuel. This energy is transferred to the surroundings in the form of _____ energy and _____ energy.

   [3 marks]

   D–C

   (b) Explain why the energy transferred to the surroundings every second is less for a vehicle travelling on a smooth road than if it is travelling on a rough road at the same speed.

   _____

   _____ [2 marks]

2 A flywheel is a heavy wheel that stores kinetic energy when it spins. Evaluate factors that affect whether a flywheel could be used in a wind turbine to store energy when the wind is not blowing.

   _____

   _____

   _____

   _____

   _____ [5 marks]

   B–A*

Working hard

1 Muhammad did 12 J of work when he lifted an apple.

   (a) How much energy was transferred to the apple?

   _____ [1 mark]

   (b) What is the name of the force he was working against when he lifted the apple up?

   _____ [1 mark]

   D–C

   (c) Muhammad dragged a box of apples 2 m along the ground. He measured the force needed as 25 N. How much work did he do?

   _____

   _____

   _____ [3 marks]

2 A car is travelling at 30 m/s along a level road. At this speed, the frictional force is 600 m

   (a) How far does the car travel in 5 s?

   _____ [1 mark]

   (b) How much work is done by the engine in this time?

   _____

   _____ [2 marks]

   B–A*

   (c) Explain why the engine must work harder when the car starts to drive up a slope at the same speed.

   _____

   _____ [2 marks]

Energy in quantity

1 A box is stored on a shelf in a warehouse.

(a) Calculate the gravitational potential energy stored by a 6 kg box placed on a shelf 0.8 m above the ground.

_____

_____

_____ [3 marks]

D–C

(b) How much energy does the box have when it is lifted to a shelf that is 0.6 m higher?

_____

_____

_____ [3 marks]

2 Explain which of these two objects has most kinetic energy: a person (mass 80 kg) jogging at 6 m/s or child on a skateboard (total mass 40 kg) moving at 8 m/s.

_____

_____

_____

B–A*

_____

_____

_____ [6 marks]

Energy, work and power

1 A man uses a motor to lift some machinery. The motor lifts 400 kg of machinery by 3 m in 60 seconds.

(a) What is the power of the motor?

_____

_____

_____

_____ [4 marks]

D–C

(b) Explain what could change if the man uses a more powerful motor.

_____

_____

_____

_____ [4 marks]

2 Use your idea about work to explain why it is easier to pull a piano up a ramp compared with lifting it directly to the same final height.

_____

_____

B–A*

_____

_____

_____ [6 marks]

Momentum

1 The diagram shows two trolleys travelling towards each other and colliding. The two trolleys stick together after the collision.

Inelastic collision

(a) Calculate the total momentum before the collision.

_____

_____ [3 marks]

D–C

(b) Write down the momentum after the collision.

_____ [1 mark]

(c) Calculate the speed that the joined up trolleys moves off with.

_____

_____ [3 marks]

2 Explain how conservation of momentum allows the position of a satellite in space to be altered by small amounts using jet packs attached to the satellite.

_____

_____

_____

_____ [5 marks]

B–A*

Static electricity

1 Owen rubs a balloon on some cloth and the balloon becomes charged.

(a) Explain how rubbing the balloon makes it negatively charged.

_____

_____

_____ [3 marks]

(b) What charge does the cloth have after it is rubbed on the balloon?

_____ [1 mark]

D–C

(c) How can the balloon be used to find the charge of another object?

_____

_____ [3 marks]

2 (a) What is a gold leaf electroscope used to detect?

_____ [1 mark]

(b) Describe how a gold leaf electroscope can be used to show if a plastic comb has a positive or negative charge.

_____

_____ [3 marks]

B–A*

Moving charges

1 A negatively charged balloon can stick on an uncharged painted wall.

   (a) What is the name of this effect?

   _____ [1 mark]

   (b) Explain why the balloon can stick on an uncharged painted wall.

   _____

   _____ [2 marks]

D–C

2 (a) Explain how electrical charge behaves differently in electrical conductors and in electrical insulators.

   _____

   _____ [2 marks]

B–A*

   (b) An electrical sub-station contains equipment at very high voltages. A person inside the substation could receive an electric shock large enough to kill them without touching the equipment. Explain why.

   _____

   _____

   _____ [3 marks]

Circuit diagrams

1 The diagram shows a series circuit and a parallel circuit, which use identical bulbs and cells. The ammeter reading in the series circuit is 0.4 A when the switch is closed.

   (a) What is the ammeter reading in the parallel circuit when the switch is closed?

   _____

   [1 mark]

Series circuit

Parallel circuit

D–C

   (b) The two bulbs are identical in the circuit, and each cell supplies 1.5 V. What is the voltage supplied to each circuit?

   _____ [1 mark]

   (c) Suggest a voltage reading for the voltmeter in the series circuit, giving a reason.

   _____

   _____ [2 marks]

2 Describe the energy transfers taking place in the series circuit.

B–A*

   _____

   _____

   _____

   _____ [4 marks]

Ohm's law

1 Alex measured the resistance of a piece of wire.

(a) Explain what is meant by resistance.

_____ [1 mark]

(b) The resistance of the piece of wire was 10 ohms.
Suggest a value for the resistance of a longer piece of the same wire.

_____ [1 mark]

(c) Write down two other factors that affect the resistance of a piece of wire.

_____
_____ [2 marks]

2 The graph shows the results of an experiment to find out how the current changes when the voltage changes in a wire.

(a) What pattern does the graph show?

_____
_____ [1 mark]

(b) Use the graph to write down the current when the voltage is 6 V.

_____ [1 mark]

(c) Use your answer to part b to calculate the resistance of the wire at 6 V.

_____
_____ [3 marks]

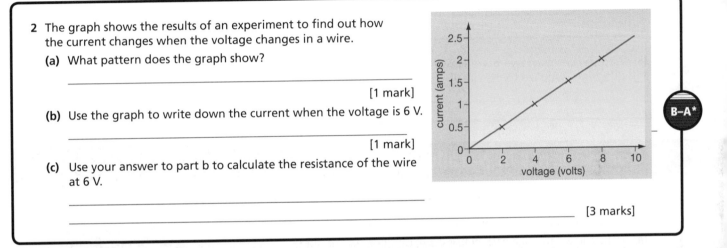

Non-ohmic devices

1 (a) How could you change the resistance of component 1?

_____ [1 mark]

(b) Write down one use for a circuit that contains component 1.

_____ [1 mark]

(c) A diode only allows the current to flow in one direction. Which symbol shows a diode?

_____ [1 mark]

2 Semiconductor materials are used in different components. Explain two ways that circuits including these components respond to their surroundings.

_____
_____ [4 marks]

Components in series

1 The circuit shows three bulbs wired in series.

(a) Fred measures the current in four places in the circuit. What can you say about the readings?

_____ [1 mark]

(b) There are two cells each supplying 1.5 V. What is the value of the potential difference supplied to the circuit? ____V   [1 mark]

(c) The bulbs are all identical. Fred measures the potential difference across two of the bulbs. What is the reading on the voltmeter? _____ V   [2 marks]

(d) An extra bulb, identical to the others, is added to the circuit in series. Explain how the resistance of the circuit changes in as much detail as possible.

_____ [2 marks]

D–C

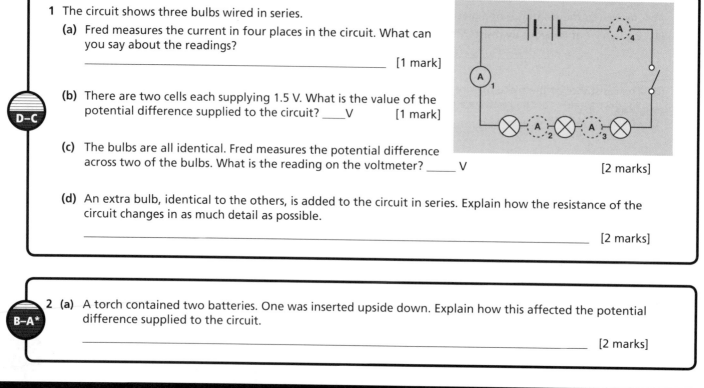

2 (a) A torch contained two batteries. One was inserted upside down. Explain how this affected the potential difference supplied to the circuit.

_____ [2 marks]

B–A*

Components in parallel

1 The circuit shows three bulbs wired in parallel. The circuit is switched on.

(a) If the current through each bulb is 0.5 A, calculate the current through the battery.

_____ [1 mark]

(b) The voltage supplied by the battery is 6 V. Write down the voltage across each of the three bulbs.

_____ [1 mark]

(c) Write down two advantages of wiring bulbs in parallel rather than in series.

_____ [2 marks]

D–C

2 The diagram shows a circuit that includes three identical bulbs, which each have a resistance of 2 ohms. The battery supplies 6 V.

(a) Calculate the resistance of each path:

   (i) upper path _____ ohms

   (ii) lower path_____ ohms

(b) Calculate the combined resistance of both branches.

_____ [2 marks]

(c) Explain which branch will carry the largest current.

_____ [2 marks]

B–A*

Household electricity

1 The diagram shows two traces from an oscilloscope.

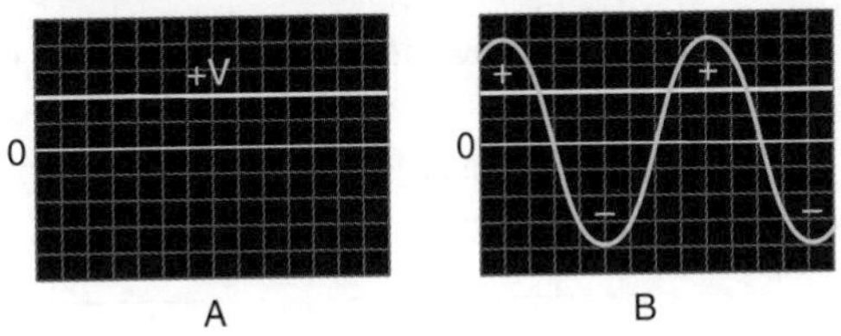

(a) Which diagram shows the trace you would get from a battery?

_____

[1 mark]

A

B

D–C

(b) Explain what is meant by "the frequency of the current is 50 Hz".

_____

_____

[2 marks]

---

2 Each square on the x-axis represents a time period of 0.01seconds. Calculate the frequency of the wave shown on the trace.

_____

_____

_____

[3 marks]

B–A*

Plugs and cables

1 Alf has finished wiring an electric plug. Describe three checks that he should make before using the equipment.

_____

_____

_____

_____

[4 marks]

D–C

---

2 Explain why electric sockets are not installed in bathrooms.

_____

_____

_____

[4 marks]

B–A*

Electrical safety

1 When Alex was mowing the lawn using an electric lawn mower, he accidentally mowed over the cable, cutting it.

(a) A fuse box is installed in the house. Explain why Alex must not pick up the cable.

_____

_____ [3 marks]

(b) It is safer to use a residual current circuit breaker (RCCB) than to rely on a fuse box. Explain why.

_____

_____ [3 marks]

D–C

2 The power of an iron is 800 W. Nathan has a choice of fuses rated at 3 A, 5 A, 13 A.

(a) One fuse has 13 A printed on it. What does that tell you about the fuse.

_____

_____ [2 marks]

(b) Which fuse should Nathan use in the plug? Explain your choice.

_____

_____

_____ [3 marks]

(c) Why are fuses used?

_____

_____ [2 marks]

B–A*

Current, charge and power

1 The power of an electric fan is 1000 W.

(a) How much energy is transferred by the fan in 5 minutes?

_____

_____

_____ [3 marks]

(b) The current through the fan is 4.3 A. Calculate the charge transferred by the fan in 5 minutes.

_____

_____

_____ [4 marks]

D–C

2 Explain what affects the amount of energy transferred by electrical charge in a certain time.

_____

_____

_____ [4 marks]

B–A*

Structure of atoms

1 (a) Complete the table to show the relative charges and masses of particles in the nucleus.

	Mass	Charge
electron	(i)	−1
proton	1	(ii)
(iii)	1	0

[3 marks]

D–C

(b) Carbon has several isotopes.

(i) What is different for different isotopes of carbon?

_____ [1 mark]

(ii) What is the same for all atoms of carbon?

_____
_____ [2 marks]

---

2 Describe how forces interact inside the atom.

_____
_____
_____
_____ [4 marks]

B–A*

Radioactivity

1 (a) What is meant by the term "radioactive"?

_____ [1 mark]

(b) Which form of radiation is most ionising: alpha, beta or gamma?

_____ [1 mark]

D–C

(c) Describe the changes that take place in the nucleus when a beta particle is emitted.

_____ [1 mark]

---

2 (a) Americium-241 decays by losing an alpha particle. Complete the values of X and Y in this equation.

$$_{X}^{241}\text{Am} \rightarrow _{93}^{Y}\text{Np} + _{2}^{4}\alpha$$

_____
_____ [2 marks]

B–A*

(b) Explain why smoke detectors containing Americium-241 are not a danger to people.

_____
_____ [2 marks]

More about nuclear radiation

1 The diagram shows one of the first models of the atom.

(a) Write down the three main features of the nuclear model of the atom.

_____

_____

_____

[3 marks]

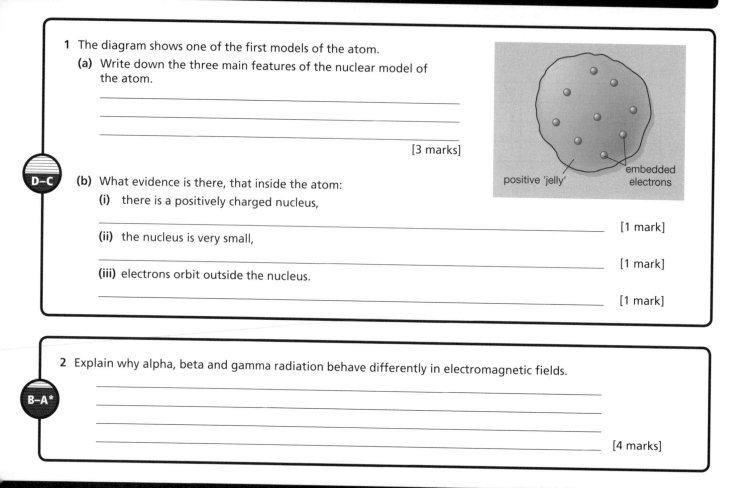

positive 'jelly'   embedded electrons

D–C

(b) What evidence is there, that inside the atom:

(i) there is a positively charged nucleus,

_____ [1 mark]

(ii) the nucleus is very small,

_____ [1 mark]

(iii) electrons orbit outside the nucleus.

_____ [1 mark]

---

2 Explain why alpha, beta and gamma radiation behave differently in electromagnetic fields.

B–A*

_____

_____

_____

_____ [4 marks]

Background radiation

1 The diagram shows the sources of background radiation in the UK.

(a) What proportion of background radiation comes from radon and thoron?

_____ [1 mark]

D–C

(b) Explain why people may be exposed to different levels of background radiation.

_____

_____

_____ [3 marks]

0.5% fallout from nuclear weapons testing

0.1% total discharge from nuclear power industry

11.5% medical, mainly X-rays

0.4% work related

0.5% other - mainly air travel

14% cosmic rays from outer space

17% radiation from the bodies of all living things and their food

19% gamma rays from rocks and soil

radon and thoron gas released from soil, rocks and building materials

---

2 Explain whether background radiation is a significant health risk.

B–A*

_____

_____

_____

_____

_____ [5 marks]

Half-life

1 (a) Explain what is meant by half-life.

_____

_____

_____ [3 marks]

D–C

(b) The proportion of radioactive carbon in a wooden arrow has been measured. Explain whether the proportion of radioactive carbon in an arrow 10 000 years old will be more or less than that found in an arrow made from modern wood.

_____

_____ [2 marks]

2 A sample of cobalt has a half-life of 5 years. Its count rate is measured as 1200 counts per second.

(a) 5 years later, what will the count rate be?

_____ [1 mark]

(b) How long will it take for the count rate to fall from 1200 counts per second to 150 counts per second?

B–A*

_____

_____

_____ [3 marks]

Using nuclear radiation

1 The diagram shows one method of controlling thickness of cardboard in a factory.

(a) Write down one reason why it is important to control the thickness of cardboard.

_____ [1 mark]

(b) Why can't alpha radiation be used to monitor the thickness of cardboard?

thickness detector

source of Beta radiation

cardboard sheet

D–C

_____

_____ [1 mark]

(c) Explain what adjustments the machine should make to the separation of the rollers if the amount of radiation reaching the detector decreases.

_____ [2 marks]

2 Explain whether there are any ethical issues with the medical use of radioactive tracers.

B–A*

_____

_____

_____

_____ [5 marks]

Nuclear fission

1 (a) What is meant by nuclear fission?

_____

_____ [2 marks]

(b) Write down one use for nuclear fission in the UK.

_____ [1 mark]

D–C

(c) The diagram shows a chain reaction.
How many neutrons are produced from the first stage in this chain reaction?

_____

[1 mark]

(d) Explain why a chain reaction must be controlled.

_____

_____

_____

[3 marks]

2 Pace makers are inserted in a patient with irregular heartbeats. They control the heartbeat so it beats in a regular pattern. Suggest advantages and disadvantages of using a small nuclear power source for pace makers.

B–A*

_____

_____

_____

_____ [4 marks]

Nuclear fusion

1 (a) Describe two differences between nuclear fission and nuclear fusion.

_____

_____ [2 marks]

D–C

(b) Write down one place in our solar system where nuclear fission takes place.

_____ [1 mark]

2 Explain how nuclear fusion can explain the presence of different elements in the Universe.

B–A*

_____

_____

_____

_____

_____ [5 marks]

Life cycle of stars

1 Describe how the forces acting in a star control how it moves between different stages in its life cycle.

_____

_____

_____

_____

[5 marks]

D–C

2 Explain why the size of a star affects its life cycle.

_____

_____

[3 marks]

B–A*

Extended response question

Describe two safety features, found in modern cars, which are designed to protect the driver and passengers in the event of a crash. Use your knowledge of energy and momentum to explain how each feature protects people in the car from serious injury or death.

The quality of written communication will be assessed in your answer to this question.

_____

_____

_____

_____

_____

_____

_____

_____

_____

_____

_____

_____

_____

_____

_____

_____

_____

_____

_____

_____

_____

_____

_____

_____

_____

_____

_____

_____

_____

_____

_____

_____

[6 marks]

1 I can describe how lifestyle can affect health ☐

2 I can explain how white blood cells defend us against pathogens ☐

3 I can describe how Semmelweiss reduced the number of deaths in hospitals and explain the key evidence that led him to his conclusions ☐

4 I can explain how vaccination produces immunity against a disease ☐

5 I can explain the limitations of antibiotics and the implications of their overuse ☐

6 I can describe how to safely grow cultures of microorganisms in the laboratory ☐

7 I can describe the pathway taken by a nerve impulse during a reflex action, including how the impulse crosses a junction between two neurones ☐

8 I can explain why it is important to keep temperature and blood sugar level constant ☐

9 I can describe how FSH, LH and oestrogen control the menstrual cycle ☐

10 I understand plant growth responses (tropisms) ☐

11 I can explain how drugs are tested in the laboratory and in clinical trials ☐

12 I can use the terms legal, illegal, dependency and addiction when discussing drugs ☐

13 I can describe adaptations of plants and animals and explain why they have these ☐

14 I can describe how energy is moved along a food chain ☐

15 I can describe what a pyramid of biomass shows us ☐

16 I can explain how carbon is constantly cycled ☐

17 I understand that different characteristics depend on different genes and the environment ☐

18 I can describe different cloning techniques ☐

19 I can describe the advantages and disadvantages of GM crops ☐

20 I can explain how natural selection works ☐

B2 Biology Checklist

1 I can describe the main features of animal, plant, bacterial and yeast cells ☐

2 I can describe the function of the different parts (organelles) of a cell ☐

3 I can describe how oxygen needed for respiration passes through cell membranes by diffusion ☐

4 I can explain that, in multicellular organisms, cells may differentiate and become specialised ☐

5 I can describe the function of a number of animal and plant tissues, organs and systems ☐

6 I can write down the word equation for photosynthesis ☐

7 I can describe the three factors that limit the rate of photosynthesis and how they affect the process ☐

8 I can list the range of substances, together with their uses, that plants can make from glucose ☐

9 I can describe how and why you would use quadrats and transects when investigating the distribution of organisms ☐

10 I can list some examples of the different types of protein found in the human body ☐

11 I can explain that enzymes act as catalysts, increasing the rate of chemical reactions ☐

12 I can describe at least one domestic and one industrial use of enzymes, including its advantages and disadvantages ☐

13 I can explain why anaerobic respiration releases less energy than aerobic respiration and results in an 'oxygen debt' ☐

14 I can explain what happens to chromosomes during mitosis and meiosis ☐

15 I can explain the inheritance of single characteristics by constructing genetic diagrams and using the terms homozygous, heterozygous, phenotype and genotype ☐

16 I can explain how different genes result in the production of different proteins ☐

17 I can describe the chromosome combinations that determine sex ☐

18 I can describe some of the ways in which fossils are formed ☐

19 I can give some of the reasons that organisms become extinct ☐

20 Explain the four key requirements in the formation of a new species ☐

C1 Chemistry Checklist

1 I can name and position the three subatomic particles in an atom of any of the first 20 elements using the periodic table, including writing the electronic structures ☐

2 I can describe how non-metal compounds share electrons to form molecules, and how the atoms of metal and non-metal compounds transfer electrons to form ions ☐

3 I can understand and use a symbol equation to determine the number of atoms in a reaction, and calculate the mass of a reactant or product from the masses of the other reactants and products ☐

4 I can explain how limestone can be used to produce quicklime, limewater, cement, mortar and concrete ☐

5 I can evaluate the social, economic and environmental impacts of quarrying limestone and metal ores ☐

6 I can describe how metals – such as iron, copper and aluminium – are extracted from their ores by reduction ☐

7 I can use the reactivity series to identify whether electrolysis or heating with carbon is the best method of extraction for a metal ☐

8 I can explain why electrolysis is an expensive method of extraction, and list the benefits of recycling metals ☐

9 I can describe how copper can be obtained from low-grade ores, by phytomining or bioleaching, and explain the environmental and economic benefits of these processes ☐

10 I can describe the differences between iron, low- and high-carbon steel, and how making alloys produces metals that are more useful than the pure elements ☐

11 I can describe that crude oil is a mixture of a large number of compounds that can be separated into fractions by fractional distillation, using the differences in boiling points between molecules ☐

12 I can state the trends in boiling points, viscosity and flammability, as the mean molecule size in each fraction gets larger ☐

13 I can recognise that a molecule is an alkane (C_nH_{2n+2}) from its structural formula; and draw the structural formulae of methane (CH_4), ethane (C_2H_6), propane (C_3H_8) and butane (C_4H_{10}) ☐

14 I can describe the process of combustion, and describe the polluting effects of carbon, nitrogen and sulfur compounds ☐

15 I can evaluate the economic, ethical and environmental issues surrounding the use of both crude-oil-based fuels and biofuels ☐

16 I can describe how hydrocarbons can be broken down (cracked) to produce alkanes and unsaturated hydrocarbons called alkenes (C_nH_{2n}) that contain a double carbon–carbon bond ☐

17 I can describe how alkenes can be used to make polymers such as poly(ethene) and poly(propene) ☐

18 I can describe some useful applications of polymers; and new uses that are being developed, including biodegradable plastic bags ☐

19 I can describe two ways of making ethanol ☐

20 I can explain how plant material can be processed to produce plant oils ☐

21 I can describe how mixtures of oil and water can be emulsified, and explain how an emulsifier works ☐

22 I can describe how to test if an alkene or a plant oil contains carbon=carbon bonds using bromine water ☐

23 I can describe the structure of the Earth, and explain how tectonic plates cause earthquakes and volcanic eruptions ☐

24 I can describe the chemical content of the Earth's first atmosphere, how it formed, and how it has changed to the atmosphere today ☐

25 I can describe one theory of how the building blocks of life could have been made, and explain why we cannot be sure that this theory is correct ☐

26 I can explain how carbon is stored in rocks, fossil fuels and the sea, and the consequences of increasing the concentration of carbon dioxide in the air ☐

C2 Chemistry Checklist

1 I can describe how atoms lose or gain electrons to gain noble gas structures ☐

2 I can describe how atoms join using ionic and covalent bonds to make compounds ☐

3 I can describe the following structures: sodium chloride crystal lattice; metallic lattices; and macromolecules, such as diamond, graphite and silicon dioxide ☐

4 I can explain the different properties of compounds, including electrical conductivity, in terms of their structure ☐

5 I can describe the differences between thermosetting and thermosoftening polymers; how shape memory alloys work; and the benefits and uses of nano science ☐

6 I can describe the atomic structure of the first 20 elements using the periodic table, and use the group number to find the number of outer electrons of the other elements ☐

7 I can calculate the relative atomic mass from data on isotopes, the relative formula mass of a compound, and know that each of these masses in grams is known as a mole of substance and has identical numbers of particles ☐

8 I can calculate the percentage composition of an element in a compound; and calculate the empirical formula of a compound using the masses or percentage of each element present in a sample ☐

9 I can calculate the mass of reactants or products when given data about a reaction from the balanced equation, and can calculate theoretical yields and actual yields ☐

10 I can explain why the theoretical yield in a reaction is not always achieved, and explain that many chemical reactions are reversible ☐

11 I can explain why using instrumental methods, such as gas chromatography and mass spectrometry, provide a quick and reliable analysis of substances ☐

12 I can calculate the rate of a chemical reaction, and know that concentration (and pressure in gases), temperature, particle size, and catalysts all affect the rate of a reaction ☐

13 I can use collision theory to explain how and why the rate of a chemical reaction will change when one variable is altered ☐

14 I can plot, use and interpret rate-of-reaction graphs ☐

15 I can identify exothermic and endothermic reactions and describe their use in hand warmers, sports injury packs and self-heating food applications ☐

16 I can choose the correct acid for making a salt, and describe how to make soluble and insoluble salts ☐

17 I can describe the difference between an alkali and a base ☐

18 I can describe the process of electrolysis of both solutions and molten liquids in terms of electron transfers; and the uses of the products, including electroplating and obtaining sodium and chlorine from molten sodium chloride, and sodium hydroxide, chlorine and hydrogen from a solution of sodium chloride ☐

P1 Physics Checklist

1 I can compare how heat is transferred by conduction, convection and radiation ☐

2 I know what factors have an effect on the way we heat and insulate buildings ☐

3 I can use kinetic theory to explain the properties of different states of matter ☐

4 I know what is meant by efficient use of energy, and how to compare the efficiency of energy transfers ☐

5 I can calculate the energy transferred by electrical appliances ☐

6 I can describe different methods of generating electricity including renewable and non-renewable energy sources ☐

7 I can evaluate the most suitable ways of generating electricity in different circumstances ☐

8 I can explain why electricity is transmitted at high voltages in the National Grid ☐

9 I can describe properties of transverse waves, including reflection and refraction ☐

10 I can describe the uses and hazards of electromagnetic waves ☐

11 I can calculate the speed of waves ☐

12 I can describe the properties of longitudinal waves such as sound waves ☐

13 I know what is meant by red-shift ☐

14 I can describe the Big Bang theory and explain why scientists believe the Universe is still expanding ☐

P2 Physics Checklist

1 I can calculate a resultant force, and describe its effect on objects including elastic objects ☐

2 I can calculate the acceleration of an object ☐

3 I can draw and interpret distance time graphs and velocity time graphs ☐

4 I can describe different factors that affect the stopping distance of a vehicle ☐

5 I can explain why a moving object reaches a terminal velocity ☐

6 I can carry out calculations involving power and energy ☐

7 I can explain momentum changes during collisions and explosions using calculations ☐

8 I can describe how a static electric charge builds up and its effects ☐

9 I know standard circuit symbols and can draw and interpret voltage-current graphs for different components ☐

10 I can use Ohm's law ☐

11 I can calculate current, voltage and resistance in series and parallel circuits ☐

12 I can describe safe practices in the use of electricity ☐

13 I can describe the structure of atoms ☐

14 I can compare different properties of ionising radiation ☐

15 I know the origins of background radioactivity ☐

16 I can carry out half-life calculations ☐

17 I know what nuclear fission is and what is meant by a chain reaction ☐

18 I know what nuclear fusion is and how it is involved in the formation of different elements ☐

19 I can describe the life cycle of stars ☐

B1 Answers

Pages 148–149

Diet and energy

1a Richard is exercising as well (more) [1]. Richard has a higher metabolic rate [1].

b If he exercises then he will use up stored fat to release energy [1]. If his metabolic rate is higher than Susan's, then chemical reactions in his cells will use up more energy than hers [1].

2 Fat releases double the amount of energy per gram compared to protein [1]. This energy gets stored as fat if it is not used up [1]. Proteins are used for growth and repair rather than energy [1].

Diet, exercise and health

1 A high level of cholesterol in the blood increases the risk of developing plaques [1] in the walls of the arteries that supply the heart with blood [1]. Plaques make it more likely that a blood clot will form in the artery. The clot can block the artery [1]. This will stop blood carrying oxygen getting to the heart muscle so the heart cells die [1]. This area of the heart no longer functions [1]. This is a heart attack.

2 HDL cholesterol helps to remove cholesterol from the walls of blood vessels [1]. So blood vessels are less likely to get blocked with blood clots and heart disease is less likely to develop. [1]

Pathogens and infections

1 Bacteria produce toxins [1]. Viruses reproduce inside our body cells [1]. When the cell gets filled up with viruses it bursts open, killing the cell [1].

2a Microorganisms are causing the infections [1] and the acid will kill them [1].

b The death rate to decrease. [1]

Fighting infection

1 By producing antibodies [1] and by producing antitoxins [1].

2a They can only attach themselves to one type of pathogen [1].

b Each lymphocyte produces one type of antibody [1]. Millions of different antibodies are needed so that most pathogens can be attacked and destroyed [1].

Pages 150–151

Drugs against disease

1a A drug that kills bacteria. [1]

b They spread the mucus onto a jelly [1] and add paper discs soaked in different antibiotics [1]. After a while they inspect the dish. The paper disc with the biggest clear area around it is the most effective antibiotic [1] because it stopped the growth of the bacteria in the mucus [1].

2 Overuse of antibiotics is making bacteria more resistant [1]. We need to reduce the amount of resistant bacteria, otherwise we will not be able to treat bacterial infections [1].

Antibiotic resistance

1 One bacteria in a person undergoes a random mutation which means it is resistant to an antibiotic [1]. The person takes antibiotics to kill the bacteria. It works, except for the resistant one [1]. The bacterium now has no competitors and grows rapidly [1]. It divides and makes lots of identical copies of itself [1]. There is now a population of antibiotic-resistant bacteria.

2 Deaths rose between 1993 and 2006 [1], and decreased between 2006 and 2008 [1]. They rose because MRSA was multiplying rapidly and nothing was put in place to stop this [1]. After 2006, people became more aware of the threat of MRSA and started putting in preventative measures, especially in hospitals, e.g. staff washing hands more thoroughly, hospital wards kept cleaner, visitors cleaning hands on entrance to hospital [1].

Vaccination

1a It is to stop children getting measles, mumps and rubella. [1]

b A small amount of the dead or inactive viruses that cause the diseases is injected into the blood [1]. The white blood cells attack them, just as they would attack living pathogens [1]. They remember how to make the antibody [1], so the child is now immune to the diseases.

c There was a study published that linked the vaccine to autism. [1]

d It was discovered that there was never any evidence to prove the link. [1]

2 The flu is a new type that has formed from the virus mutating [1]. A new vaccine has been developed, as the old vaccine would not be effective against the new virus [1].

Growing bacteria

1 If it was 37 °C or more, it would encourage the growth of pathogens that live in the body [1]. If the dish is then opened, the pathogens could escape and cause illness [1].

2 This temperature will promote rapid growth [1], which will make the product that you are culturing available quickly and so maximise profits [1].

Pages 152–153

Co-ordination, nerves and hormones

1 The brain and spinal cord. [1]

2a The heart responds to adrenalin. [1]

b It is longer lasting [1] so the response [increased heart rate] is maintained throughout the tennis match [1].

Receptors

1a sensory [1]

b arrow drawn to the right [1]

c (electrical) impulses [1]

2a cones [1]

b Impulses [1] from several cones [1] travel along the optic nerve [1] to the brain.

Reflex actions

1 Receptors in the skin detect the pin/pain [1], impulse sent along sensory neurone to spinal cord [1], impulse travels across synapse to relay neurone [1], impulse travels across synapse to motor neurone [1], impulse travels along motor neurone to arm muscle [1], muscle contracts pulling hand away [1].

2a conscious [1]

b Falling over is more dangerous than touching the hot radiator. [1]

Controlling the body

1 This is the temperature that enzymes work best at [1]. Without enzymes, important chemical reactions in the body would not take place [1].

2 Exertion will cause your body temperature to rise [1], and you will sweat to bring it back down to normal [1]. This will cause a loss of water [1]. A low water concentration in the body is dangerous because cells will stop working [1].

Pages 154–155

Reproductive hormones

1 High levels of oestrogen stop the production of FSH [1]. Without FSH, eggs cannot mature [1] so no eggs are released to be fertilised [1].

2 Low oestrogen levels between days 0–4 cause the uterus lining to break down and menstruation to occur [1]. A rise between days 4–14 makes the lining build up again [1]. A decrease around day 13 triggers ovulation [1].

Controlling fertility

1 Fertility drugs are given to stimulate the maturation of several eggs [1]. Eggs are collected and fertilised by sperm from the father [1]. The embryos formed are inserted into the mother's uterus (womb) [1].

2a Fertility drugs increase the amount of eggs released at a time [1], so it's more likely that three get fertilised [1].

b Triplets are more likely to have problems developing in the uterus than a single fetus [1] but medical treatment has improved in the last century so more survive [1].

Plant responses and hormones

1a positive [1]

b negative [1]

2a Cover or remove the tip of the seedling [1]. Use a control that has had no treatment [1]. Grow both in the presence of light coming from one direction [1].

b The plant with no tip does not grow towards the light (straight up) [1]. The shoot of the control plant does grow towards the light [1].

Drugs

1 i One mark for any legal drug: e.g. alcohol or nicotine.

ii One mark for any illegal drug: e.g. cannabis or heroin.

b Nicotine is addictive [1], so a person will suffer withdrawal symptoms if they stop smoking [1].

2 Any two reasons from, for example: liver failure [1], car accident [1], alcohol poisoning [1].

Pages 156–157

Developing new drugs

1a Some of the volunteers are given a placebo [1], which does not contain the drug [1], but neither they nor their doctor knows whether they have been given it [1].

b the age of the soldiers [1]

c One reason, for example: the recovery time for the soldiers who took the drug was less than for those who took the placebo. [1]

2 They only tested the drug on one group of people [1], so these people may not have been predisposed to develop the side-effects that were seen when the general population started taking the drug [1]. The trial did not last very long [1], and the side-effects are only seen when the person has been taking the drug for several months or years [1].

Legal and illegal drugs

1a The number of deaths from drinking alcohol has risen (between the years 2000 and 2008) [1]. More men die from drinking alcohol then women [1].

b The number of people who died from alcohol is much higher than those who died from heroin (9031, which is around 10 times higher) [1]. This is because many more people drink alcohol than take heroin [1].

2 They give them an unfair advantage over their competitors (they could be banned from the sport) [1]. They could cause long-lasting damage [1].

Competition

1 The weeds will compete with the vegetable plants (for light, water, space and nutrients) [1], so the vegetable plants will not be able to grow as well [1].

2 They do not have to compete for resources [1] so are more likely to get what they need to survive [1] and breed [1].

Adaptations for survival

1a It deters predators from eating it. [1]

b The yellow and black are warning colours [1] so predators of the hoverfly think it is dangerous and leave it alone [1].

2 Extremophiles can survive in very difficult environments [1] so microorganisms could survive on Mars where conditions are difficult [1].

Pages 158–159

Environmental change

1 Numbers will decrease [1] because plants will die through lack of water [1] so there will be less food for them [1].

2a Food prices will increase.

b With fewer honeybees, fewer plants will be pollinated [1] so less food crops will develop [1].

Pollution indicators

1a oxygen meter [1]

b It is low. [1]

c Mayfly larvae / freshwater shrimp / stonefly larva [1]. They can only live in unpolluted water [1].

2 They can maximise the uptake of oxygen [1] from polluted water where oxygen is low [1].

Food chains and energy flow

1 One reason from: some of the light misses the leaves altogether / hits the leaf and reflects / hits the leaf but goes all the way through without hitting any chlorophyll / hits the chlorophyll but is not absorbed because it is of the wrong wavelength (colour). [1]

2a $800 \times (10/100) = 80$ units

b To increase the chance of capturing light [1] so they can photosynthesise more efficiently [1].

Biomass

1 The sketch should show: a pyramid with the levels labelled with the name of the organism, starting with

grass at the bottom [1], each level should then get progressively smaller [1].

2   Most animals in the ocean are cold-blooded so they do not use energy to regulate their temperature [1]. This means that less energy is lost from the food chain at each stage [1], so energy can be passed along to more organisms [1].

Pages 160–161

Decay

1a To increase the moisture content [1] so the numbers of microorganisms in the compost increase [1], which speeds up decay and the production of compost [1].

b   To let in oxygen [1] so the microorganisms responsible for decay can live (respire) [1].

c   It contains nutrients [1] that increase their growth [1].

2 The high temperatures kill any microorganisms in the food [1], the sealing prevents any more from entering [1] so decay cannot happen [1].

Recycling

1   Added to food chain diagram: a label saying 'microorganisms' [1], with arrows from all grass, antelope and lion pointing towards it [1].

2   Scavengers eat the bodies of dead animals [1]. Microorganisms break down waste and dead bodies and release the nutrients (decay) [1].

The carbon cycle

1a i combustion/burning [1]
     ii photosynthesis [1]

b   from eating plants or other animals [1]

c   The animal will die [1] and the tissue will be decayed by microorganisms [1] which will carry out respiration [1] and release carbon dioxide [1].

2a i light → chemical [2]
     ii chemical → heat and light (and sound) [2]

b   The chemical energy in food is released during respiration [1] and transferred to movement energy in the muscles of an animal [1].

Genes and chromosomes

1   He inherited one from his mother in her egg [1] and one from his father in his sperm [1].

2a Bush A: 66.7 cm [1]; Bush B: 71.9 cm [1].

b   To improve the reliability of her results. [1]

c   Bushes have each inherited a different gene for leaf length [1]. Environment will also be a factor as the bushes will each receive different amounts of water/light [1].

Pages 162–163

Reproduction

1a asexual [1]

b   It has the same genes. [1]

c   They reproduce sexually [1] so have a different mix of genes [1].

2a In internal fertilisation, the egg and the sperm meet inside the female's body [1]. In external fertilisation, they meet outside of the female's body [1].

b   To increase the chance of fertilisation [1] so at least a few offspring develop [1].

Cloning plants and animals

1a So she has many identical copies of it. [1]

b   Either: use tissue culture [1], take a piece of the tissue from the plant [1] and grow it in a sterile nutrient liquid or gel [1].
Or: take a cutting [1], cut off a small piece of the stem [1] dip it in hormone rooting powder and place it in soil [1].

2a To produce lots of animals so it is no longer extinct. [1]

b   Take eggs from a goat of a similar species and remove the nucleus [1], then take a preserved cell from the ibex and remove the nucleus [1]. Insert the nucleus from the ibex cell into the egg [1], then put the formed embryo into a donor mother goat [1].

Genetic engineering

1a They have genes from another organism.

b   The gene for insulin is cut from a human chromosome; and a chromosome from bacterium is cut open using enzymes [1]. The insulin gene is inserted into the bacterium's chromosome [1]. The chromosome is put back into the bacterium [1].

c   The bacteria grow best at this temperature [1] so insulin is produced more quickly [1].

2a They will not have to use pesticides [1], will get higher yields of cotton as less are destroyed by pests [1] and make more profit [1].

b   Any sensible answer, for example: The toxins may affect other wildlife / disrupt food chains / the Bt gene may be transferred to other plants.

Evolution

1   Giraffes will have a gene for long necks [1] which would be passed on to their offspring [1]; and reaching for leaves cannot change the gene [1].

2   For: We have evidence that all life on Earth, including complex organisms such as humans, evolved from simple single-celled organisms [1].
Against: There are millions of species of bacteria on Earth [1] that are adapted to their environment and so will not evolve [1].

Page 164

Natural selection

1a It's the same colour as the tree trunk (camouflaged) [1]. Makes it difficult to be spotted by predators [1].

b   a change in a gene [1]

c   The black moths are now camouflaged [1], so they do not get eaten [1] and they live to reproduce and pass on the gene to their offspring [1].

2   The resistance is caused by a mutation [1] which happens by chance [1]. A rat that has a mutation survives, and passes it on to its offspring [1].

Evidence for evolution

1a Because each animal uses its limbs differently/they are adapted [1].

b   They all have the bones in the same place [1] which shows that birds, bats and humans all have a common ancestor [1] from which they evolved [1].

2a Archaea [1] because they are closer on the tree [1].

b   A common ancestor [1] from which all life on Earth evolved [1].

Page 165

Extended response question

5–6 marks
A detailed description of how a difference in

concentration of auxins in the developing root will cause the roots to bend downwards. The answer will include information such as that in the roots a high concentration of auxins inhibits cell elongation. Auxin tends to accumulate on the lower side of a root. The lower side of the shoot grows more slowly than the upper surface. This causes the root to bend downwards. This is a tropism called gravitropism (or geotropism).

All information in answer is relevant, clear, organised and presented in a structured and coherent format. Specialist terms are used appropriately. Few, if any, errors in grammar, punctuation and spelling.

3–4 marks

Limited description of why a root bends downwards. The answer will state that auxins inhibit cell growth in roots. The upper side of the root will grow more, causing it to bend and this causes the root to bend downwards.

For the most part the information is relevant and presented in a structured and coherent format. Specialist terms are used for the most part appropriately. There are occasional errors in grammar, punctuation and spelling.

1–2 marks

An incomplete description, stating that the upper side of the root will grow more, causing it to bend downwards.

Answer may be simplistic. There may be limited use of specialist terms. Errors of grammar, punctuation and spelling prevent communication of the science.

C1 Answers
Pages 166–167
Atoms, elements, and compounds
1a copper [1]

b copper carbonate [1]

c It is made of more than one element. [1]

d Malachite contains both rock and copper carbonate [1], so it is a mixture [1].

Inside the atom
1a An atom of phosphorus contains 15 protons [1], 15 electrons [1] and has a mass of 31. [1] It has 16 neutrons. [1]

b **i** N, As, Sb, or Bi [1]

ii Cl [1]

iii Si [1]

c 3 concentric circles with two crosses on inner most circle, eight crosses on middle circle [1] and five crosses on outer circle. [1]

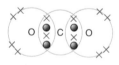

2a

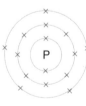

b

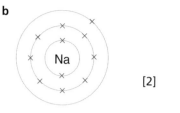

[2]

c They both have one outer electron [1], this means that their reactions will be similar [1].

Element patterns
1a by proton number [1]

b periods [1]

c groups [1]

d They have similar properties/characteristics, [1] just like people in the same family. [1]

2a helium (2), neon (2,8), argon (2,8,8), krypton (2,8,18,8), xenon (2,8,18,18,8), or radon (2,8,18,32,18,8), [1] for choosing a noble gas and with electronic structure [1]

b They have a filled outer electron shell [1], that makes them unreactive [1].

Combining atoms
1a a shared pair of electrons [1]

b

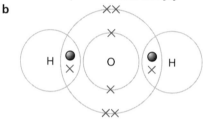

One mark for showing a pair of electrons between O and H atom, [1] for eight electrons on the outer shell of O atom.

c 2,8,1 [1]

d It loses its outer electron [1], and loses a negative charge, so it becomes a positively charged ion.[1]

e The chlorine gains the electron lost by the sodium[1], and becomes a negative ion [1], which is then attracted to the sodium ion.[1]

2a a double covalent bond. [1]

b One mark for showing two pairs of electrons joining each atom together; one mark for showing eight electrons on each outer shell.

Pages 168–169
Chemical equations
1a Carbon dioxide. [1]

b **i** 3 [1]

ii 2 [1]

2a $CuCO_3 + 2HCl \rightarrow CuCl_2 + CO_2 + H_2O$ [1]

b $CuCO_3 + 2NaOH \rightarrow Cu(OH)_2 + Na_2CO_3$ [1]

Building with limestone
1 Social advantages: more employment, more money to pay for community facilities. [1]

Social disadvantages: pollution effects on local

population (not on wildlife). [1]

Economic advantages: more money in the local economy, better roads and railways. [1]

Economic disadvantages: reduction in tourism. [1]

To gain full marks you must mention both advantages and disadvantages.

2a Carbon dioxide [1] dissolves in the rain to make carbonic acid [1] that dissolves the limestone [1].

b The water with the dissolved limestone evaporates [1] and the carbon dioxide from the air makes calcium carbonate [1] that then becomes stalactites and stalagmites.

Heating limestone

1a Carbon dioxide. [1]

b Bubble the gas through limewater [1] a white precipitate appears in the solution, or it goes milky/cloudy white. [1]

c i Clay [1].

ii Mortar is sand and cement [1], concrete is sand, cement and gravel/stones/aggregate. [1]

d Mortar is good at sticking things together [1], concrete has greater strength than mortar when made into a beam. [1]

2a Add iron or steel rods to the inside of the beam [1] to reinforce them [1].

b The water does not evaporate, but becomes part of the cement crystals [1] that sticks the sand and gravel together. [1]

c i The increase in strength of the concrete is (directly) proportional to the percentage of gravel. [1]

ii The cement can no longer stick together the gravel and sand [1] so the concrete crumbles easily. [1]

Metals from ores

1a i Haematite [1]

ii Zincite [1]

iii Bauxite. [1]

b Bauxite because aluminium is higher in the reactivity series than carbon [1], so it cannot be reduced by carbon. [1]

c Removal of oxygen [1] from a compound.

2a It is reduced by electrolysis/electricity. [1]

b It requires electricity [1], and this is more expensive than carbon. [1]

c No-one knew how to use electricity to obtain aluminium. [1]

Pages 170–171

Extracting Iron

1a i Limestone [1], ii coke [1] accept carbon.

b The oxygen in the air reacts with the carbon in the coke [1] to make carbon dioxide. More carbon dioxide is produced by the iron oxide (haematite reacting with carbon monoxide [1] made from carbon dioxide reacting with more coke. [1]

c Slag is formed from the impurities in the ore used. [1]

2a $Fe_2O_3(s) + 3CO(g) \rightarrow 2Fe(\ell) + 3CO_2(g)$ [1]

b Carbon monoxide. [1]

c Either carbon combusts incompletely [1] with insufficient oxygen to make carbon monoxide [1], or some carbon dioxide reacts with more carbon/coke [1] to make two molecules of carbon monoxide [1].

d Fe_2O_3

e CO

Metals are useful

1a The atoms are in a regular arrangement/layers, [1] that can slide past each other without braking apart. [1]

b The outer electrons are only loosely held to each atom [1], so they are free to move within the layers [1], allowing electricity to flow.

c Metals are made from a single type of atom [1], alloys are mixtures of different metals and have different types of atoms in them [1].

d The different atoms are often different sizes, distorting the structure [1], and making it harder for the atoms to slide past each other. [1]

2a Alloys have different properties [1] that might be more suitable for that use. [1]

b It can remember a shape, and return to it. [1]

c Smart alloys will shrink inside the body as they warm up, pulling the bone pieces together very strongly [1]. Ordinary metal plates do not do this. This means the gap between the bones is smaller [1], and less bone will need to grow to fill the gap, so the break will heal sooner. [1]

Iron and steel

1a It does not rust [1], the metal looks attractive in use. [1]

b They can be used for different uses. [1]

c They are in the transition metals.[1]

Copper

1a Impure copper. [1]

b Copper. [1]

c At the cathode. [1]

d Under the anode. [1]

e They gain an electron, become atoms [1] and stick to the cathode [1]

2a The plant absorbs compounds of the metal [1] by its roots. The plants when harvested have large quantities of the metal compounds in them [1] when burnt, the ash can be used as an ore for the metal. [1]

b The bacteria produce enzymes that dissolve the insoluble metal compounds into the rain water that falls on the rocks [1]. The collected rainwater can then be concentrated to produce a solution containing the metal ions [1] that can be electrolysed to obtain the metal. [1]

c Once started both methods require little or no labour costs [1]. The plants need to be harvested [1], (or) but the bacteria continue until there is no more metal ore to extract [1].

Pages 172–173

Aluminium and titanium

1a Using electricity to break down a compound into simpler substances. [1]

b bauxite [1]

c i It needs electricity [1], which is expensive to produce [1].

ii Aluminium ore melts at a high temperature. [1] Adding cryolite reduces this temperature [1], saving energy. [1]

2a $2Al_2O_3 \rightarrow 4Al + 3O_2$ [1]

b The magnesium is very expensive. [1]

c Both metals form a layer of oxide on their surfaces [1], which prevent the acids and alkalis reacting with the metal [1].

d It is unreactive [1], so it does not corrode in body fluids [1].

Metals and the environment

1a Advantages, two from: recycled metal uses less energy to purify, saves on world resources, there's no need to buy fresh ore, waste material is relatively cheap. [2]

Disadvantages, two from: an efficient collection system is needed that costs money, it's often mixed with other rubbish so needs to be separated, it needs to be transported for processing. [2]

Comparison of merits of recycling and using new metals [1] is given in answer, that justifies a clear preference [1].

b Any three from: causes dust, destroys land for later use, increase of traffic in area, causes noise pollution, may produce environmentally damaging by-products/ waste. [3]

2a A previously used piece of industrial or housing land. [1]

b Absorb dissolved compounds into plant [1] through the roots [1].

c Plants are dried and burnt [1], leaving an ash that is rich in the metal compound for use as ore [1].

d Any two from: returns land to productive use, removes wasteland, makes it safe for people to use. [2]

A burning problem

1a It remains in the air for some time [1], reducing the amount of sunlight that reaches the ground [1].

b Volcanic gases contain sulfur dioxide, which dissolves in rainwater [1] and makes the rain acidic [1].

c It may increase the quantity of carbon dioxide gas in the atmosphere. [1]

2a Carbon in the plant burns with oxygen to make CO_2 [1]. If not enough oxygen is present then the carbon remains as carbon due to incomplete combustion. [1]

b **i** CO_2 [1]

ii It has only one atom of oxygen, not two as in carbon dioxide. [1]

iii It combines with haemoglobin in preference to oxygen. [1]

Reducing air pollution

1a To convert nitrogen oxides to nitrogen before the exhaust gases are released, or carbon monoxide to carbon dioxide. [1]

b It reduces acid rain. [1]

c Ethanol is renewable as it is grown [1]; and when burnt only produces CO_2 that was just removed from the air [1] by photosynthesis, so doesn't increase the amount of CO_2 in the air [1].

2a They absorb only as much CO_2 from the air as they produce when burnt [1], so they do not add or reduce the amount of CO_2 in the air [1].

b They are made from plant material [1]. When used, more plants can be grown to replace them [1].

c Any two from: less food will be grown/more famines, food will cost more, allow us to continue to expand transport use with more motorways etc., more land will be needed to grow crops. [2]

Crude oil

1a a compound containing only carbon and hydrogen [1]

b **i** fractional distillation [1]

ii bitumen [1]

ii petroleum gas [1]

c The crude oil is heated up [1] so the liquids become gases/evaporate [1], as they rise up the column they cool and condense/become liquids [1], so they are separated by their boiling points [1].

2a These substances are a mixture of molecules with different numbers of carbon atoms. [1]

b As the number of carbon atoms increases, the boiling point increases. [1]

c The more carbon atoms, the larger the molecule [1], so more energy is needed to separate the individual molecules from each other [1].

Alkanes

1a If there are 8 C atoms, then the H atoms will be $2 \times 8 = 16$ [1] plus another 2, so 18 in total [1].

b **i** harder to burn/less flammable [1]

ii increases [1]

iii increases [1]

2a All the bonds are single [1] covalent bonds [1]

b

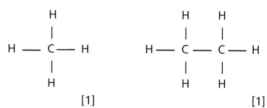

[1] [1]

c They all have the same structure [1], but as the molecule gets larger the size affects the properties [1].

Cracking

1a $C_{10}H_{22}$

b ethene [1]

c a double bond [1]

d It is very reactive [1] and can be used to make polymers [1].

2a Broken down into simpler substances by heat. [1]

b Steam cracking:

advantage – steam is relatively cheap [1]

disadvantage – higher temperature means that more energy is needed [1].

Catalytic cracking:

advantage – lower temperature needed so saves energy or may take place quicker [1]

disadvantage – cost of catalyst needs to be taken into account [1].

Alkenes

1a Alkenes have a double bond; alkanes do not. [1]

b The double bond can react with other substances [1], it is easy to break the double bond [1].

c There are double bonds that can be broken to allow more atoms to join into the molecule. [1]

2a bromine water [1]

b The orange colour of the bromine water [1] would rapidly go colourless [1]. Or accept: two layers would form (1 mark only).

c Two layers would form [1] and the bottom layer would remain orange [1]. Or accept: there would be no change (1 mark only).

Pages 176–177
Making ethanol

1a water or steam [1]

b a fuel made from biological material [1]

c Advantages: the fuel is renewable/can be replaced, does not contribute to global warming/CO_2 in the air. [2]
Disadvantages: uses land that could grow crops for people/forces food prices up, hinders development of non-carbon based transport fuels. [2]

2a ethanol [1], or another named solvent [1]

b Ballpoint ink dissolves in ethanol/other named solvent [1], but not in water. [1]

Polymers from alkenes

1a a chain of repeating molecules bonded together [1]

b monomer [1]

c The double bond breaks apart [1], each end of the broken double bond joins with a different monomer molecule [1] to make a chain.

2a Any two from: stronger/more robust, won't break, cheaper, flexible. [2]

b i Guttering needs to be stronger than poly(ethene), or we need materials of different properties for different functions/jobs/purposes. [1]
ii Guttering needs to be (fairly) rigid [1], clingfilm needs to be flexible [1].

Designer polymers

1a Lightweight means cars and plane weigh less [1], use less fuel [1] and are easier to mould into the shapes required [1].

b It will stop you getting wet [1], but allow evaporated sweat to pass out of the coat [1].

c To make materials better suited to their use, or different applications. [1]

2a broken down by natural decay (by fungi, small animals and plants and microbes) [1]

b The cornstarch will provide food/material for the organisms to eat and so break down the bags [1]. The hydrocarbon-based bags cannot be eaten by organisms [1].

c The stitches will quickly breakdown in the body after healing [1]. This saves the expense and trauma of a second operation to remove the stitches [1].

Polymers and waste

1a i won't degrade quickly/will fill up landfill sites [1]
ii use up hydrocarbon resources/produce poisonous or obnoxious gases [1]
iii lots of different polymer types so sorting would be costly [1]

b Using landfill doesn't allow the resources to be easily used again [1], burning prevents the hydrocarbons being used again as they become CO_2, H_2O, and other gases [1], recycling allows the plastic to be re-used, saving crude oil resources [1].

2a Microbes digest the cornstarch polymer leaving small pieces of hydrocarbon polymer not degraded. [1]

b The hydrocarbon-based polymers that remain after the cornstarch has degraded [1] may not be good to grow plants in [1].

c Landfill does not pollute the air [1], incineration provides heat energy that can be used for electrical generation or heating homes [1].

Pages 178–179
Oils from plants

1a They are rich in energy. [1]

b seeds, nuts and fruits [1]

c The plant parts are crushed [1], and the oil filtered from the pulp [1].

d The oil is added to the potato [1], the potato is cooked at a higher temperature [1] causing different changes to the molecules than if cooked at a lower temperature [1].

2a The oils in the flowers dissolve into the steam [1] and are removed from the flowers [1].

b steam [1] and vaporised flower oil [1]

c condenser [1]

d From the steam that has condensed. [1]

e Vegetable oils come from plants [1], mineral oils are from crude oil or fossil fuels [1].

Biofuels

1a Advantages, any two from: cheaper than fossil fuels, renewable resource, carbon neutral, could use agricultural waste products such as manure/waste from food crops. [2]
Disadvantages, any two from: reduction in land for food use, rainforest turned into land to grow fuel crops, increase in costs of food crops as reduced land available to grow food. [2]
Comparison of merits of biofuels against fossil fuels [1] is given in an answer, that justifies a clear stated preference [1].

b Plant oils can be reacted with alcohols such as methanol [1], to make less viscous esters [1] that can be used in diesel engines.

Oils and fats

1a The orange colour disappears. [1]

b There's no mention of the mass of the spread used [1], and no mention of the volume of ethanol used [1].

c Butter [1]. The reaction is quickest [1].

d An unsaturated fat has some double bonds [1], a saturated fat has very few double bonds [1].

2a It speeds up the reaction. [1]

b
$$H-\overset{\overset{\displaystyle H}{|}}{\underset{\underset{\displaystyle H}{|}}{C}}-\overset{\overset{\displaystyle H}{|}}{\underset{\underset{\displaystyle H}{|}}{C}}-\overset{\overset{\displaystyle H}{|}}{\underset{\underset{\displaystyle H}{|}}{C}}-\overset{\overset{\displaystyle H}{|}}{\underset{\underset{\displaystyle H}{|}}{C}}-\overset{\overset{\displaystyle H}{|}}{\underset{\underset{\displaystyle H}{|}}{C}}-$$
[2]

Emulsions

1a immiscible [1]

b It's an emulsifying agent or emulsifier. [1]

c i water and vinegar [1]
ii to act as an emulsifying agent [1]

2a water loving [1]

b water fearing [1]

c The hydrophilic end attaches/is attracted to water/aqueous molecules [1]. The hydrophobic molecules are attracted to oily molecules [1], so this molecule allows different molecules to be held in a mixture rather than separating. [1]

Pages 180–181

Earth

1

Part	Name	Its structure
A	crust [1]	solid rock
B	mantle [1]	plastic/jelly-like [1] (with convection currents)
C	core [1]	made from hot iron and nickel [1], (solid inner, liquid outer)

2a They are caused by heat from radioactive processes [1] in the core [1].

b The plates are moved by the convection currents [1]. Where they collide they move/slide over each other causing mountains [1].

c Water wears the rock away/causes erosion [1], leaving a channel that becomes a valley over many hundreds/thousands of years [1].

Continents on the move

1a There was no evidence for the theory [1], no one could explain how the plates would move [1].

b The rock types and structures were similar in Africa and South America [1]. Rocks containing similar fossils were found in both Africa and South America [1].

c The North American plate is moving west and the Eurasian plate is moving east [1], carried by convection currents [1].

2 As the North American plate moves west and the Eurasian plate moves east, a small gap is created [1]. This gap is filled by magma [1].

Earthquakes and volcanoes

1a It's a piece of the Earth's crust. [1]

b at the edges of tectonic plates [1]

c The meeting of the plates creates pressure in the mantle [1], this is relieved by magma being forced to the surface [1].

d i a tidal wave [1]

ii An earthquake took place in the Earth's crust beneath the sea [1]. The movement of the crust caused a shockwave in the water [1].

2 The ocean plate is thinner/lighter [1] than the continental plate, so it goes underneath the continental plate [1]. This forces the continental plate higher on the mantle, raising up mountains, the thin plates allow magma to come through causing a volcano [1].

The air we breathe

1a volcanoes [1]

b i 4.7 billion years [1]

ii B [1]

iii E [1]

c Ultraviolet light from sunlight would kill it [1]. The ozone protected the life from the ultraviolet light [1].

2a It means turning the gas mixture into a liquid and then collecting/separating the different gases by their boiling points. [1]

b carbon dioxide [1] and water vapour [1]

c i to liquefy all the gases [1]

ii Nitrogen [1], as it has the lowest boiling point [1].

Page 182

The atmosphere and life

1a There are no fossil records of early life/It was too small to leave traces [1].

b Seawater allows all the compounds to be able to meet each other/Ultraviolet light could not damage or kill the emerging life. [1]

c All living things/proteins are made from amino acids [1]. If we know how to make amino acids then it is a possible route to making life [1].

2a ammonia (NH_3), methane (CH_4) and hydrogen (H_2) (all three for the mark) [1]

b sugars/ribose sugar [1] and amino acids [1]

c It didn't make a living thing [1], it only made building blocks/compounds that are present in living things [1].

Carbon dioxide levels

1a i A = photosynthesis [1]

ii B = burning [1]

iii C = death and decay [1]

iv D = respiration [1]

b Both plants and animals carry out respiration. [1]

c The carbon dioxide produced by burning, and given off to the air [1], has recently been removed from the air by photosynthesis [1].

2a It's a fuel made from compounds containing a lot of carbon atoms. [1]

b Fossil-fuel carbon has been locked up for millions of years and released quickly [1]; whereas biofuel carbon has only recently been removed from the air and is returned as quickly as photosynthesis takes place [1].

c Carbon dioxide is an acidic gas, so it will affect the pH of the water [1]. This change in pH may dissolve sea shells/make it hard for plants to photosynthesise/kill some living organisms [1].

Page 183

Extended response question

5 or 6 marks:

There is a clear, balanced and detailed description of the positive and negative environmental impacts involved with increasing the percentage of biofuels, with 5–6 points from the examples given. The answer shows almost faultless spelling, punctuation and grammar. It is coherent and in an organised, logical sequence. It contains a range of appropriate or relevant specialist terms used accurately.

3 or 4 marks:

There is some description of the positive and negative environmental impacts involved with increasing the percentage of biofuels, with 3–4 points from the **examples** given. There are some errors in spelling, punctuation and grammar. The answer has some structure and organisation. The use of specialist terms has been attempted, but not always accurately.

1 or 2 marks:

There is a brief description of the environmental impacts involved with increasing the percentage of biofuels, with 1–2 points from the **examples** given. The spelling, punctuation and grammar are very weak. The answer is poorly organised with almost no specialist terms and/or their use demonstrates a general lack of understanding of their meaning.

Possible points to make:

positive

- carbon neutral
- growing plants will remove carbon dioxide from atmosphere
- less crude oil will be needed
- less risk of oil spills if less oil needed
- biofuels are natural, and spillages will be easier to deal with.

negative

- will not reduce carbon emissions as still burning carbon dioxide
- more land will be needed to grow crops
- rainforest may be destroyed
- animal habitats may be disrupted or destroyed.

P1 Answers

Pages 184–185

Energy

1a Electrical energy [1]; changes to heat [1]; kinetic energy [1]; and sound [1].

b It is transferred to the surroundings. [1]

c Olympus 1000 [1], it took least time to dry the cloth [1].

d Any two valid points, for example: distance from hairdryer to cloth [1]; dampness of cloth initially [1]; dryness of cloth finally [1]; size of cloth [1].

2 More energy transfers are involved [1], as electricity is used to recharge the batteries [1]. Energy is wasted at each transfer [1], so more energy is wasted [1].

Infrared radiation

1a Black [1]. Solar panels absorb infrared radiation from the Sun [1]. Black is the best colour to absorb infrared radiation [1].

b Any two from: radiation from the sun is more intense from the south [1], more infrared radiation will be absorbed [1], the water will heat up quicker [1].

c The house loses heat by radiation when it is warmer than the surroundings [1]. At night/during winter, it is cooler outside the house compared to inside [1]. Black surfaces emit infrared radiation well [1].

2a Yes [1], the beaker absorbs and emits infrared at the same rate [1].

b More infrared is emitted than absorbed [1] so the beaker cools down [1], until it reaches the same temperature as the fridge [1] when infrared is absorbed and emitted at the same rate [1].

Kinetic theory

1a Melting is when a solid changes to a liquid. [1]

b Particles are in fixed positions in a solid [1] but change places in a liquid [1]. In a solid, ice bonds hold the particles together [1]. When ice melts, bonds between particles start to break and reform as the ice melts to liquid [1].

c Energy is needed to break bonds [1]. Energy is absorbed from the surroundings [1].

d Energy is absorbed more quickly. [1]

2 Any four from: particles in solids are held by strong bonds [1]; but particles in air have absorbed enough energy to break bonds [1]; particles in solids vibrate slightly in fixed positions [1]; but particles in air move rapidly [1]; copper particles are heavier than particles in air (mainly nitrogen and oxygen) [1].

Conduction and convection

1a Convection is when particles move through a substance, transferring energy [1]; particles in solids cannot change places and move through a solid [1].

b The flask reduces conduction [1] with the poly(ethene) base/vacuum/insulated stopper, which is an insulator [1]. Reduces convection [1]: through the top using the stopper [1] OR through the sides using the vacuum [1]. Reduces radiation [1] with the silvered walls [1].

2 Heat sinks need to conduct heat rapidly away from components [1]. Metals contain free electrons [1] and conduct heat faster than plastics [1].

Pages 186–187

Evaporation and condensation

1a Evaporation is when a liquid changes to a gas. [1]

b Boiling takes place throughout a liquid; evaporation takes place at the surface. [1] Boiling takes place at the boiling temperature, evaporation takes place at any temperature. [1]

c Any two from: warmer weather – molecules have more energy and can break free of bonds more easily [2], windy or drier weather – surrounding air does not become saturated [2].

2 Credit sensible methods.

Equipment: cotton wool, thermometer, stopwatch [1]; water, aftershave [1].

Method: soak cotton wool pads in water and aftershave [1]; monitor temperature of each over a period of time [1].

Control/change variables: Use same sized pads and soak up same mass of liquid [1]; keep pads under same conditions [1].

Rate of energy transfer

1 If the potato is cut, the surface area is larger [1] and the distance to the centre is smaller [1]. More heat can be transferred by radiation at the surface [1]. Heat is not transferred as far by conduction through the potato [1].

2 Heat is transferred quicker if there is a larger temperature difference [1]. Boiling oil is hotter than boiling water [1].

Insulating buildings

1a The report can suggest savings that are greater than the cost of the report. [1]

b Total savings are £275 per year [1] or £1375 in 5 years [1]. Overall savings are £1080 [1].

c Loft insulation has a low U-value [1] as it is a good insulator/reduces heat losses [1].

d Maximum of 2 marks for two suggestions, and maximum 2 marks for two explanations, for example: cavity wall insulation or double glazing [1] reduces heat losses by conduction [1]; installing draught excluders [1] reduces losses by convection [1]; putting foil behind radiators [1] reduces heat losses by radiation [1].

2 Credit any valid response with explanations, for example:

Yes [1], the council has limited money [1]; it is better to have some energy-saving measures in as many homes as possible [1]; installing double glazing may use more energy in production and installation [1] than it saves [1].

No [1], badly fitting windows waste a large amount of energy [1]; in some cases the savings will be greater than the average shown in the table [1]; double glazing saves some energy every year [1]; it will last for a long time [1].

Specific heat capacity

1a Specific heat capacity is the energy absorbed by 1 kg of a material when its temperature increases by 1 °C. [1]

b Temperature rise = energy/(mass x specific heat capacity) [1] = 1800/(0.25 x 900) [1] = 8 °C [1]

2 A saucepan needs to heat up quickly and evenly [1]. Copper has a lower specific heat capacity so it heats up quicker than aluminium [1], so copper would be a better choice [1].

Pages 188–189

Energy transfer and waste

1a chemical energy -> kinetic energy + heat energy (2 marks if all forms of energy are correct; 1 mark if any form of energy is incorrect or missing)

b As wasted heat energy. [1]

c Conservation of energy states that energy is not lost or created during an energy transfer. [1] The width of the arrows shows the relative proportion of each form of energy [1]. The total width of the output arrows equals the width of the input arrow [1].

2

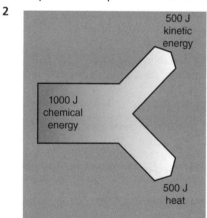

Give marks as follows: a double headed arrow [1]; equal width arrows [1]; correctly labelled energy forms [1]; correctly labelled amounts of energy [1].

Efficiency

1a efficiency = useful output energy/input forms of energy [1] = 60% [1]

b Allow sensible answers, for example: Yes [1], more efficient equipment saves money on running costs [1] so people should have this information [1].

OR No [1], Sankey diagrams are hard to interpret/contain too much information [1]; the efficiency could be stated as a percentage (%) [1].

2 No energy transfer can be more than 100% efficient [1]. There is more than one energy transfer if the motor is battery operated [1]. Energy is wasted at each transfer [1].

Electrical appliances

1a chemical (from food [1] -> kinetic (winding the recharger) [1] -> chemical (in battery) [1]

b i For example: The recharger does not use mains electricity [1], so it can be used where there is no mains supply [1].

ii For example: A battery is a portable supply of electrical energy [1], so the phone can be used on the move [1].

2a Credit valid answers, any two from: the expense of providing a mains supply of electricity to the garden [1]; safety issues if the supply is wrongly installed [1]; it is inconvenient to provide a mains supply [1]; there are portable alternatives [1].

b Any two from, for example: solar panels need no maintenance [1]; there is sufficient sunlight outside to power lights [1]; batteries are expensive to replace [1].

Energy and appliances

1 energy = power x time [1] = 850 W
850 x 15 x 60 seconds [1] = 765 000 J [1]

2 The microwave oven is less powerful than the electric oven [1] and uses a smaller current [1], so the cable doesn't heat up as much [1] so a thinner cable is sufficient [1].

Pages 190–191

The cost of electricity

1a 0.1 x 1500 [1] = 150 kWh [1]

b 150 kWh at 15p [1] = £22.50 or 2250p [1] x 5 years = £112.50 [1]

c For example: less electricity is used lighting homes [1]; generating less electricity reduces carbon emissions from power stations [1].

2 Credit valid responses, for example: This saves the householder money [1] as they are not paying for electricity used to keep equipment on standby [1]. Less electricity will be wasted so less electricity needs to be generated [1], which reduces carbon emissions [1]. However, there is a cost in buying new equipment [1]. Manufacturing new equipment uses electricity/causes carbon emissions, etc. [1]. (maximum 6 marks)

Power stations

1a chemical in gas -> thermal in water and kinetic in steam -> kinetic in turbine and generator -> electrical in generator (1 mark per correct transfer)

b Unwanted heat from combined heat and power stations is used to directly heat homes [1]. In power stations which only generate electricity, electricity is transferred to homes then used for heating [1] there are more stages in the energy transfer [1] energy is lost at each stage. [1]

2 One mark for stating an option and three marks for valid reasons. Credit valid responses, for example: Convert existing older power stations [1]: fewer resources will be needed [1]; quicker timescale [1]; lower overall cost [1].
or
Develop new technologies: [1] older power stations will never match the efficiencies of new ones [1]; existing fuels will run out in future [1]; so we need new/renewable/non-polluting sources of energy [1].

Renewable energy

1 Iceland uses geothermal energy [1]. This uses heat in volcanic rocks to produce steam to spin turbines [1]. Norway uses hydroelectricity [1]. This uses falling

water trapped behind dams to spin turbines [1].

2 Credit valid responses, for example: Advantages include being able to use free, renewable energy sources; using batteries increases our use of wind power which reduces our greenhouse gas emissions [1]. Disadvantages include the expense of installing batteries; difficulties of transferring energy from batteries to homes [1].

Electricity and the environment

1

Type of station	Advantage of the scheme	Disadvantage of the scheme
Coal-fired	a single power station in a relatively small land area generates large amounts of energy [1]	greenhouse gas emissions; mining damages the environment; large quantities of waste to dispose of [1]
Wind turbines	no pollution is emitted to the atmosphere [1]	wind farms take up large land areas; noise pollution [1]

2 Two reasons from: Harm caused to wildlife [1] as local habitats are destroyed to grow crops [1]; reduced biodiversity [1] caused by crop monoculture [1]; burning biofuels [1] releases greenhouse gases [1]. (maximum 4 marks)

Pages 192–193

Making comparisons

1a Credit sensible reason and explanation, for example: more demand for lighting at night because it is dark [1]; more demand for electricity for heating in winter because it is cold [1]; more demand for electricity at meal times for cooking [1]. (3 marks maximum)

b Waves, wind, tidal, gas. [3]

2 Credit sensible answers that are linked to the information. These are examples of possible responses: Wind farms are unreliable [1] and fossil fuel power stations are reliable [1]. It's more important to have a reliable supply of electricity [1] so we should only have a few wind farms [1] in carefully chosen places which are really windy [1]. OR It's more important to reduce reliance on fossil fuels [1] so we should have a large numbers of wind farms [1] but improve their reliability [1]. (maximum 5 marks)

The National Grid

1a Smaller current [1] so less energy is wasted as heat. [1]

b current = power/voltage = 200 000 000 / 400 000 [1] = 500 [1] A [1].

2 For example: Bury underground [1]: the extra cost is worth paying to reduce environmental impact [1]. It reduces the risk of damage during severe weather [1] and harm due to accidents [1].

OR Keep as overhead power lines [1]: it is too expensive/disruptive to change [1], the landscape will still be damaged with underground lines [1], easier to find and mend faults with overhead power lines [1]. (maximum 4 marks)

What are waves?

1 Credit valid answers

Difference 1: vibrations are perpendicular to direction of energy travel (transverse) or parallel to direction of energy travel (longitudinal). [1]

Difference 2: longitudinal waves cannot transfer energy through a vacuum but transverse waves can. [1]

Similarity 1: both waves transfer energy. [1]

Similarity 2: both types of wave travel from a source. [1]

2a Mechanical waves cannot transmit energy through a vacuum. [1]

b Waves transfer energy from the earthquake to other places [1]. The time taken depends on the distance from the earthquake to other places [1]. Waves travel at different speeds in different materials [1]. Longitudinal waves travel at different speeds to transverse waves [1].

Changing direction

1 Diffraction is the spreading of waves through a gap [1]. Diffraction is most pronounced when the wavelength of the wave is similar to the gap [1]. Doorways are a similar size to the wavelength of many sound waves [1].

2 When light is refracted, it changes direction at a boundary [1]. Light changes direction in water compared with air [1]. To the observer, the light ray appears to have travelled in a straight line rather than changing direction [1].

Pages 194–195

Sound

1a Any number between 441 Hz and 480 Hz. [1]

b It gets shorter. [1]

c The amplitude [1] gets bigger [1].

2a The trace shows more waves [1] that have a smaller amplitude [1]. The waves should be the same height and have the same wavelength throughout.

b Credit valid responses, for example: oscilloscopes provide information about different sounds [1]. Louder sounds will have a larger amplitude and taller trace on the screen [1]. The engineer can use the information to adjust the position/settings on the microphone near that player [1].

Light and mirrors

1 The image is upright, virtual and the same distance behind the mirror. [3]

2

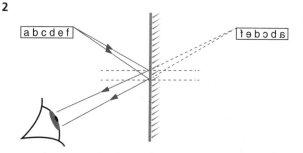

One mark for each ray linking the eye and the object [2]. (Credit rays correctly drawn from 2 different places on the object that link correctly to corresponding place on the image.) 1 mark for extending rays from mirror to image [1]. 1 mark for adding arrows to the rays [1].

Using waves

1

Type of electromagnetic radiation	Use
Radio wave	BBC TV broadcasts
Microwave	Satellite TV
Infrared radiation	Remote control
Visible light	Photography

(1 mark for each correct pair)

2 Mobile phones use microwaves [1]. Microwaves have shorter wavelengths than radio waves [1]. The wavelength of microwaves is much smaller than the size of hills and buildings [1]. Microwaves do not diffract round hills and buildings [1]. Phones must be in line-of-sight of transmitters for good reception [1].

The electromagnetic spectrum

1 wavelength = wavespeed/ frequency [1] = 300 000 000/ 2 700 000 000 [1] = 0.111 [1] m [1] (maximum of 3 marks if units are not correctly converted)

2 The electromagnetic spectrum is a continuous spectrum of waves whose wavelength varies from more than 10 000 m to 10^{-15} m. [2]

Pages 196–197
Dangers of radiation

1a Ultraviolet. [1]

b Skin cancer [1]; sunburn [1], premature aging on the skin [1].

c Wear a hat/cover up [1] so radiation cannot reach the skin [1]; stay inside between the hours of 11 am and 3 pm [1] to avoid very intense sunlight [1].

2a The DNA molecules are ionised. [1]

b Cancer cells are killed [1], healthy cells outside the treatment area receive a lower dose/no radiation [1]; healthy cells are less likely to be damaged [1].

Telecommunications

1 Microwaves transmit images to satellites [1] satellites transmit the images to receiving stations in Australia [1]. Microwaves travel so fast (300 million m/s) that the signal arrives in less than a second [1].

2 The satellite takes 24 hours for one complete orbit [1]. It orbits in the same direction as Earth rotates, so stays above the same point all the time [1]. The satellite dish can be fixed facing the direction of the satellite [1].

Cable and digital

1a Analogue. [1]

b The signal needs to be amplified/gets weaker when it is transmitted long distances [1]; when the signal is amplified, any interference is amplified too [1].

Searching space

1a Different objects emit different types of electromagnetic radiation [1]; the same detectors cannot detect all forms of electromagnetic radiation [1].

b It is larger [1] because radio waves are longer than visible waves [1].

Page 198
Waves and movement

1a Doppler effect. [1]

b The wave has a longer wavelength [1] but the same amplitude. [1]

2a The red shift is when light from an object moving away from an observer [1] appears to have a longer wavelength/appears redder [1].

b The larger the red shift, the faster the object is moving away [1]; more distant galaxies have larger red shifts than closer galaxies [1].

Origins of the Universe

1a All matter and energy in the Universe were in one place at one point in time [1]. About 14 billion years ago, a rapid expansion started to take place [1], which is still continuing [1].

b Cosmic microwave background radiation/the echo of the Big Bang [1]. The red shift which provides evidence that galaxies are all moving apart [1]. The larger red shift from more distant galaxies, which is evidence more distant galaxies are moving apart faster [1].

2 Cosmic microwave background radiation cannot be explained using other theories [1]. It is the same frequency we would expect radiation from the initial expansion to have if the theory is correct [1]. It is detected coming from all directions with equal intensity [1].

Page 199
Extended response question

5 or 6 marks:
A detailed description of relevant properties of radio waves, microwave, infrared and visible light (e.g. they all travel at the speed of light; some are absorbed by the atmosphere; diffraction effects are noticeable with radio waves). The communication use of these electromagnetic waves is linked to its properties, e.g. microwaves are not absorbed by the atmosphere so can be used for satellite communication. *All information in answer is relevant, clear, organised and presented in a structured and coherent format. Specialist terms are used appropriately. Few, if any, errors in grammar, punctuation and spelling.*

3 or 4 marks:
A limited description of relevant properties of radio waves, microwave, infrared and visible light. Some communication uses of members of the electromagnetic spectrum are stated, but may not always be linked to its properties. *For the most part the information is relevant and presented in a structured and coherent format. Specialist terms are used for the most part appropriately. There are occasional errors in grammar, punctuation and spelling.*

1 or 2 marks:
The answer includes an incomplete description of relevant properties, which may be linked to specific members of the electromagnetic spectrum. Some relevant uses of more than one type of electromagnetic wave should be stated. *Answer may be simplistic. There may be limited use of specialist terms. Errors of grammar, punctuation and spelling prevent communication of the science.*

B2 Answers

Pages 200–201

Animal and plant cells

1a plant [1]

b A: cell wall [1] B: vacuole [1]

c It's where protein synthesis takes place. [1]

d Chloroplasts absorb light energy to make food [photosynthesis] [1]. Roots are underground where there is no light so have no need for chloroplasts [1]. Leaf cells do receive light so require many chloroplasts [for photosynthesis] [1].

2 Viruses are very small [1] so need to be magnified many times in order to be studied [1]. Electron microscopes can magnify objects more than a light microscope [1].

Microbial cells

1a Similarity: Both have cell wall / nucleus / cytoplasm / cell membrane / vacuole / ribosomes. [1]
Difference: Plant cell wall is made of cellulose, yeast cell wall isn't / Yeast does not contain chloroplasts. [1]

b Similarity: Both have cytoplasm/cell membrane. [1]
Difference: Animal cell has a nucleus. Bacteria do not / Bacteria has a cell wall. Animal cells do not. [1]

2 20/0.02 = 1000 [1 mark]

Diffusion

1 The scent of the cake is caused by cake molecules [1]. The concentration of cake molecules will be high in the kitchen [1] but low in other parts of the house [1]. There will be a net movement of molecules from where there is a high concentration to a lower one [1] so the cake molecules will spread away from the kitchen into the rest of the house.

2a They use it up [for respiration]. [1]

b So the concentration remains higher in the blood [1] and therefore glucose keeps on diffusing into the cells [1].

c There will be glucose inside the Visking tubing [1] because glucose has diffused across the membrane of the tubing [1] from where it is in a high concentration to a lower one [1].

Specialised cells

1 It has haemoglobin [1] to transport oxygen [1] /a disc shape [1] for taking in and letting out oxygen [1].

2 It has a 'hair' protruding from the cell [1], which increases the surface area [1] to increase water absorption [1].

Pages 202–203

Tissues

1a muscle [1]

b to release energy [1] which the cells need to contract [1]

2a diffusion [1]

b They increase the surface area of the lungs. [1]

c Bacteria are just one cell [1] whereas humans are many cells, so the oxygen has to be transported around the body so all cells are supplied with oxygen [1].

Animal tissues and organs

1 One mark for each: 1A, 2D, 3E, 4B, 5C.

2 They protect the intestinal cells [1] from being digested by the juices that get mixed with the food [1]. By producing mucus which adds a barrier [1] this also makes it easier for the food to slide through the digestive system [1].

Plant tissues and organs

1a i It stops too much water leaving the leaf [1] and protects the underlying cells [1].

ii to carry out photosynthesis [1]

b xylem [1]

2a They allow gases to diffuse into and out of the leaf. [1]

b To stop too much water escaping from the leaf. [1]

Photosynthesis

1a The leaves will not go black. [1]

b No starch was present [1] because the plants were in the dark so could not carry out photosynthesis [1] so no glucose was made which is stored as starch [1].

c The part that went black was green [1] and had carried out photosynthesis to produce starch [1] because the cells contain chlorophyll [1]. The bits that stayed orange were white so did not carry out photosynthesis and produce starch [1].

2 It absorbs the red and blue part of white light [1] which are needed to drive photosynthesis [1] and reflects the green part [1].

Pages 204–205

Limiting factors

1a oxygen [1]

b As the light intensity increases, the rate of photosynthesis increases. [1]

c The rate of photosynthesis stops increasing. [1]

d Another factor is limiting the rate of photosynthesis, e.g. concentration of carbon dioxide [1] so even if the light intensity increases the rate of photosynthesis cannot [1].

2a Any sensible suggestion, for example: the plants would be taller / the tomatoes would be bigger / the tomatoes would taste sweeter. [1]

b The heater increases the temperature [1], and the automatic watering system ensures that plants always have water [1].

c a source [increased concentration] of carbon dioxide [1]

The products of photosynthesis

1 It adds nitrogen to glucose [1]. Nitrogen comes from nitrate ions [1] which are absorbed from the soil through the roots [1].

2a for the plant embryo to grow [1]

b Oil contains a high amount of energy per gram [1] so the seed is light and can be easily dispersed by the wind [1].

Distribution of organisms

1a There's not enough light [1] so plants cannot photosynthesise [1].

b It needs to be where it will be covered in water for some periods of the day [1] or it will get too hot and dry out [1].

2a The distribution of the grey squirrel has increased and the distribution of the red squirrel has decreased. [1]

b There is competition between the grey and red squirrels for food [1]. The grey squirrels are better adapted so will win [1], they will survive while the red squirrels will die [1].

Using quadrats to sample organisms

1a 4 × 16 [1 mark] = 64 [1 mark]

b Make sure that the quadrat is placed randomly. [1]

2 This size contained a reasonably high number of different species inside (fifteen) [1]. A larger quadrat would have contained more [1] but it would have made it more difficult for her to count the number of species inside it [1].

Pages 206–207
Proteins

1a amino acids [1]

b The order determines the shape of the protein [1]. The protein will only function correctly if the shape is right [1].

2 Protein in food gets broken down by enzymes [1] into amino acids [1], which are transported to leg muscle cells in the blood [1]. Muscle protein is made on ribosomes [1] using instructions from genes [1].

Enzymes

1a i starch [1]

ii sugars [1]

b temperature [1]

c Any one from: concentration of starch/amylase, volume of starch/amylase, room temperature. [1]

2 Between 0 °C and 40 °C, the rate of reaction increased [1] because the molecules of starch and amylase had increasing amounts of energy [1] and were colliding more frequently and with more energy so they reacted to break down the starch more often [1]. After 40 °C, the rate started to slow down [1] as the amylase started to change shape [denature] [1] so the starch no longer fitted into the active site so the reaction could not occur [1].

Enzymes and digestion

1a starch [1]

b salivary [1]

c protease [1]

d stomach [1]

e lipase [1]

f glycerol [1]

2a 7.5 [1]

b i so as to provide the optimum pH for the enzyme pepsin [1]

ii Bile [1] mixes with the food which is alkaline [1] and neutralises the acid [1].

Enzymes at home

1a It contains lipase [1] which will break down the fat [1] so it becomes soluble and washes away [1].

b The stain will not come out [1] because the lipase will not work at this temperature [1]. It will denature [1].

2a To a test tube of egg white, add some biological detergent [1]. To another test tube of egg white, add the same volume of distilled water [1]. Leave both in a water bath at 30 °C [1]. If the egg white has dissolved (gone clear), the detergent has broken it down [1].

b She is not using keratin. [1]

Pages 208–209
Enzymes in industry

1a by using carbohydrase enzymes [1]

b Starch solution is cheaper. [1]

c Fructose is sweeter than glucose [1] so less is needed [1], which lowers the energy [calorie] content of the food [1].

d isomerase [1]

2a fructose [1] and glucose [1]

b making soft-centre chocolates [1]

Aerobic respiration

1a For respiration [1], to release energy from glucose [1].

b Air enters the lungs [1] and oxygen travels into the blood [1].

2 Carbon dioxide is low at 6 a.m. and 6 p.m. because it is dark and the plant is not carrying out photosynthesis [1], so the leaves do not absorb carbon dioxide [1]. It is highest at 12 p.m. because this is the brightest part of the day and the rate of photosynthesis is at a maximum [1]. Oxygen uptake is lowest at 12 p.m. because the plant is using oxygen produced from photosynthesis [1] for respiration [1] [there is no need for it to absorb any from the air]. Oxygen is high at 6 a.m. and 6 p.m. because the plant is only carrying out respiration and so needs to absorb oxygen from the air [1].

Using energy

1a His muscles are contracting [1] and using up energy [1].

b His body will use up more energy [1] to maintain his body temperature [1].

2 The muscle has to contract at all times [1] so needs a large supply of energy [1]. Mitochondria are where energy is made via respiration [1].

Anaerobic respiration

1a Aerobic respiration releases more energy per gram of glucose [1], anaerobic produces lactic acid which is toxic [1].

b She is taking in extra oxygen [1] to break down the lactic acid that has formed in her muscles [1].

2a So he is fitter [can swim faster and for longer without getting tired]. [1]

b cross-country skiing [1]

c His heart and lungs will get stronger [1] and more efficient at getting oxygen to his muscles [1], so he can respire more efficiently [1] and swim faster and for longer.

Pages 210–211
Cell division – mitosis

1a The parent cell divides into two. [1]

b They have the same chromosomes. [1]

c To replace old skin cells that die/to repair cuts in the skin. [1]

2a A copy of each has been made. [1]

b The copies of chromosomes are being pulled apart [1] to form two cells with the same chromosomes [1].

Cell division – meiosis

1a ovaries [1]

b Cells: 1 = 64 [1], 2 and 3 = 64 [1], 4, 5, 6 and 7 = 32 [1]

c It forms gametes with half the normal number of chromosomes [1], so during fertilisation a zygote is formed with a full set of chromosomes [1].

2 Any four from: Mitosis forms new body cells and meiosis forms gametes [1]. Mitosis happens in all parts of the body and meiosis only happens in the testes and ovaries [1]. In mitosis the cell divides once but in meiosis there are two cell divisions [1]. In mitosis two cells are made and in meiosis four are made [1]. In mitosis the new cells have the same number of chromosomes as the original cell. In meiosis the new cells have half the number of chromosomes as the original cell. [1]

Stem cells

1 Stem cells could be taken from an embryo or adult bone marrow [1] and placed into the brain of the patient [1] where they differentiate to form new brain cells [1].

2 Any 3 points from: Their practice is probably unregulated and so potentially dangerous or will not work [1]/they may use embryonic stem cells which some people find morally wrong [1]/the stem cells may be rejected [1]/the stem cells they use may carry genetic mutations for disease [1]/the treatment may trigger side-effects [1].

Genes, alleles and DNA

1a possible father B [1]

b The child shares DNA fragments with him [1] which were transferred from the father in his sperm [1].

2a People who carry the BRCA mutation are more likely to develop breast cancer than the general population [1]. The older you get, the higher the chance of developing breast cancer [1].

b She may have inherited the BRCA mutation [1] which increases her risk of developing breast cancer [1].

Pages 212–213

Mendel

1 His ideas were only accepted when scientists had discovered how characteristics were inherited [genes/chromosomes]. [1]

2a The offspring would all be medium height. [1]

b The tallness allele was dominant over the shortness allele [1]. All the offspring have one tallness allele and one shortness allele [1] so they are all tall [1].

How genes affect characteristics

1a One from each parent. [1]

b bb [1]

c In either order [but always a capital letter first]: Bb [1] BB [1].

2 The Y chromosome [1] because only males have it [1].

Inheriting chromosomes and genes

1a The one for short hair. [1]

b Hh [accept any letter as long as there is an upper case and corresponding lower case] [1]

c

Parent	short-haired	long-haired	
	Hh	hh	[1]
Gametes	H and h	h and h	[1]
Offspring		H	h

		H	h
h		Hh	hh
h		Hh	hh

[1 mark]

d 1:1 [1 mark]

2 He could breed the green-seeded plant with a yellow-seeded one [1 mark]. If all of the offspring are green then the genotype is most likely to be homozygous dominant [GG] [1 mark]. If some of the offspring are yellow then it is most likely to be heterozygous [Gg] [1 mark].

How genes work

1 The allele is a section of DNA [1]. The order of the bases in the DNA is a code [1] to make a brown protein [pigment] [1] that the cells in her eye produce [1].

2 Joshua could have a mutation in the insulin gene [1], which means that the DNA no longer codes for, and so does not produce, insulin [1].

Pages 214–215

Genetic disorders

1a cc [or any answer with two lower-case letters] [1]

b Their child [E] has cystic fibrosis [1] so they must have each passed on the cystic fibrosis allele [1].

c 50 per cent/1 in 2 [1]

2 Use IVF to produce several embryos [1]. Check the DNA of the embryos [1]. Only implant the embryo[s] free of the alleles that cause the disease into Louise's uterus. [1]

Fossils

1 There are not many fossils of jellyfish [1] as they just decay [1], so scientists do not have a lot of evidence of how they used to look millions of years ago [1].

2a All life on Earth uses the same genetic code. [1]

b Valid and reliable evidence [1]. Agreement between most scientists [1].

Extinction

1 Any two from: The trout are eating them/The trout are better adapted at catching food/The trout are carrying a disease which they pass onto the fish. [2]

2a High numbers of species become extinct [1] over a short period of time [1].

b A higher number of fossils have been found dating from that time. [1]

c The small mammals were better adapted [1] to living at the colder temperatures [1] than the dinosaurs.

New species

1 On each island there is a different environment [1]. The finches show variation [1] and the ones better adapted to living on that island survive [1] to pass on these genes to their offspring [1]. The finches on each island change so much from the mainland species that they are a new species [1].

2 It can no longer breed with the other grass [1] because to breed together the flowers have to form at the same time [1]. As they can no longer breed, they are separate species [1].

Page 216

Extended response question

5-6 marks

A detailed description of how prolonged exposure to temperatures above 37 °C results in the denaturation of enzymes and how this effects their ability to function. The answer should contain details about the importance of the shape of the enzyme and why the functioning of enzymes is vital for life. All information in answer is relevant, clear, organised and presented in a structured and coherent format. Specialist terms are used appropriately. Few, if any, errors in grammar, punctuation and spelling.

3-4 marks

Limited description of how high temperatures affect enzymes. The answer will state that temperatures over 37 °C leads to a change in shape of the enzyme which inhibits its function. For the most part the information is relevant and presented in a structured and coherent format. Specialist terms are used for the most part appropriately. There are occasional errors in grammar, punctuation and spelling.

1-2 marks

An incomplete description, stating that enzymes do not work over 37 °C. Answer may be simplistic. There may be limited use of specialist terms. Errors of grammar, punctuation and spelling prevent communication of the science.

C2 Answers

Page 217

Investigating atoms

1a

particle	Relative mass	Relative charge
proton	(i)1	+1
(ii) neutron	1	(iii) none
electron	(iv) very small	-1

(for 1 mark each)

b   i 14 [1]

ii 14 [1]

iii 2,8,4 [1]

2a alpha particles [1]

b   thin layer of gold foil [1]

c   to show where the alpha particles are deflected [1]

d   D: undeflected, have missed the nucleus showing the atom is mainly empty [1]. E: those deflected forward have had their direction of travel influenced by a central nucleus/passing close to the nucleus [1]. C: those deflected backwards have been bounced back from a central nucleus [1].

Mass number and isotopes

1a An isotope is an atom of an element that has a different mass to other atoms. [1]

b   Potassium-40 has one [1] more neutron [1].

c $\frac{(90 \times 39) + (10 \times 40)}{100} = \frac{3910}{100} = 39.1$ [2]

Another answer, but showing correct method, [1].

2   Measure the carbon-14 present in the sample [1], compare it with how much should be present today/ use half-life of carbon-14 to find out how old the specimen is [1].

Pages 218–219

Compounds and mixtures

1a an aqueous or water solution [1]

b   137 + (35.5 × 2) [1 mark] = 208 [1 mark]

c   208 grams [1]

d   (27 × 2) + ((32+(16 × 4)) × 3) [1 mark] = 342 [1 mark]

Electronic structure

1a 2,8,8,2 [1] for diagram
[1] for written structure

2,8,8,2

b   2 [1] for diagram
[1] for written structure

2

c   2,8,3 [1] for diagram
[1] for written structure

2,8,3

d   2,7 [1] for diagram
[1] for written structure

2,7

2a i 7 [1]

ii 2 [1]

b   It tells you its reactivity/chemistry/ how it will react. [1]

c   They have full outer electron shells [1] so there are no electrons to share / gain or lose / to react. [1]

Ionic bonding

1a high melting or boiling point [1] and conducts in solution but not when solid [1]

b   i Either a diagram showing 2,8,8 configuration or a statement that it is 2,8,8. [1]

ii Either a diagram showing 2,8 configuration or a statement that it is 2,8. [1]

c   The potassium atom loses its outer electron to the fluorine atom, making two ions [1]. The oppositely charged ions then attract each other [1] [with an electrostatic attraction].

2a When solid, the sodium and chloride ions are not free to move [1]; when dissolved, they are free to move and so can conduct the electric current [1].

b   When molten, the sodium and chloride ions are free to move. [1]

Alkali metals

1a sodium + water → sodium hydroxide + hydrogen [1]

b   They all have a single outer electron [1], which they lose to become 1+ ions. [1]

c   It would react very violently [1] as it is lower in the group and so more reactive than potassium [1].

2a They are less reactive so easier to extract. [1]

b   by using electrolysis [1]

c   Electricity was only discovered then. [1]

Pages 220–221

Halogens

1a

chloride ion

[1]

b 1– [1]

c argon [1]

2a The chlorine in sodium chloride is reacted with the sodium [1]; this makes a new substance with different properties [1].

b Polyvinyl chloride has different properties from sodium chloride and doesn't dissolve in water. [1]

c Enough is used to kill bacteria, but not enough to kill people / people are larger than bacteria and need far more chlorine to be killed. [1]

Ionic lattices

1a 6 [1]

b Sodium chloride has a crystal lattice but hydrogen chloride does not [1]. It is easy to separate hydrogen chloride molecules from each other, but much harder to separate the sodium and chloride ions of sodium chloride [1].

c Sodium goes to the negative electrode and chlorine goes to the positive electrode [1]. At the negative electrode, sodium forms [1]; at the positive electrode, chlorine gas molecules are made [1].

2a Na^+ [1] SO_4^{2-} [1]

b $(23 \times 2) + 32 + (16 \times 4) = 142$ [2]
or $(23 \times 2) + 32 + (16 \times 4)$ [1]

Covalent bonding

1a

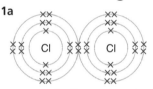

outer shells overlapping [1]
one pair of shared electrons drawn within the overlapping portion [1]

b a shared pair of electrons [1]

c Any two from: they only need to gain a few electrons [1] to achieve a noble gas structure [1], and sharing is the easiest way [1].

2a 2 [1]

b 4 [1]

c 0.232 nm [1]

Covalent molecules

1a fluorine and chlorine [1]

b 7 [1]

c As the molecule gets larger, the boiling point gets higher. [1]

d The boiling points get higher down the group [1], the higher the boiling point the greater the intermolecular forces [1], so the forces increase down the group [1].

2a Whilst the molecules have the same mass, and so should boil at same temperature [1], the intermolecular forces between water molecules are much stronger than between fluorine molecules [1].

b Iodine [1] because it has a boiling point closer to 100 °C than any other halogen has. [1]

Pages 222–223

Covalent lattices

1a i 4 [1]
ii 3 [1]

b Each atom is joined to four others making a very strong structure [1] that is very hard to distort [1].

c The layers of graphite are held together by weak forces [1], so they slide easily past each other [1].

2a They are electrons that are only loosely held in a lattice and are free to move around. [1]

b The delocalised electrons are free to move [1], carrying the electric current with them [1].

Polymer chains

1a Thermosetting polymers harden on heating [1]. Thermosoftening polymers soften or melt on heating [1].

b It can be easily recycled after use. [1]

c Thermosetting polymers have bonds between different polymer chains [1]; thermosoftening polymers have no bonds between chains [1].

2a Individual polymer chains are of different lengths [1], each chain has a different melting point making the polymer melt over a range of temperatures [1].

b The chain is shorter so will be less tangled with other chains, so will have a lower melting point or range. [1]

Metallic properties

1a Diagram with most of the atoms the same size as in the printed diagram; the extra atoms are sized larger or smaller; and they distort the regular arrangement of layers. (All **three** for 2 marks, **two** for 1 mark.)

b An alloy that returns to its original shape when heated. [1]

c It is stretched to fit tightly at room temperature [1], then when warmed in the mouth it shrinks [1] pulling the teeth into position.

2a The positive nuclei [1] repel each other and so space out evenly [1].

b The sea of electrons are free to move [1], carrying the electric current with them [1].

c The moving electrons allow heat energy [1] to quickly move throughout the metal [1].

Modern materials

1a up to 300 atoms [1]

b A nanoparticle is from 1–100 nm in size. [1]

c The smaller sized particles will be colourless [1], so thick applications will not be visible [1].

2a The toxic drug will be unable to attack healthy cells. [1]

b The buckminsterfullerene will break open [1], releasing the toxic drug at the cancer cell [1].

Pages 224–225

Identifying food additives

1a three [1]

b five [1]

c Manufacturer C [1]

2a pigment will have different retention factors in different solvents [1]

b 18/27= 0.67 for 2 marks, 18/27 for 1 mark.

Instrumental methods

1a using an instrument or machine to carry out the analysis [1]

b Any two from: they are quick to use [1], work with

small samples of material [1], are easy to standardise [1].

c   It allows the relative atomic or molecular mass to be found [1], which allows for identification of the substance [1].

2a positive [1]

b   different molecules will produce different fragments [1] that will be specific to that particular molecule [1]

Making chemicals

1a when an acid reacts with a base or alkali [1]

b   a reaction where a solid is made out of two solutions [1]

c   a reaction where a substance gains oxygen/loses electrons [1]

d   When a substance gains oxygen/loses electrons, another substance will have lost oxygen/gained electrons [1], so one reaction is the opposite of the other. [1]

2a 2 moles [1]

b   2 moles [1]

c   1 mole [1]

Chemical composition

1a 3.98 g [1]

b   3.18 g [1]

c   0.80 g [1]

d   Cu = 3.18/63.5 = 0.05 [1 mark]

O = 0.80/16 = 0.05 [1 mark]

so 0.05 : 0.05 = 1 : 1 ratio, formula = CuO [1 mark]

2a 12 + (16 × 2) = 44 [1 mark]

b   (1 × 2) + 16 = 18 [1 mark]

c   carbon is 10.56/44 = 0.24 [1 mark]

hydrogen is 4.32/18 = 0.24 [1 mark]

as water has two hydrogen atoms, reacting ratio is 1 : 2, or CH_2 [1 mark]

d   The compound is an alkene because alkenes have twice as many hydrogens as carbons. [1]

Pages 226–227

Quantities

1a 40 + 12 + (16 × 3) = 100 [1 mark]

b   40 + 16 = 56 [1 mark]

c   125/100 = 1.25 [1], so 1.25 × 56 = 70 tonnes [1 mark]

2a $Pb(NO_3)_2$ = 207 + ((14 + (16 × 3)) × 2) = 331 [1 mark], $NaNO_3$ = 23 +14 + (16 × 3) = 85 [1 mark],

so 16.5 g/331 = 0.05 [1 mark]. From the equation 1 unit produces 2 units of $NaNO_3$, so 0.05 × 2 × 85 = 8.5 g [1]

b   The yield is unlikely because not all of the products may be recovered, or not all of the lead nitrate may react. [1]

How much product?

1a More than enough of a reactant to ensure that the other reactant completely reacts. [1]

b CuO = 63.5 + 16 = 79.5, 1.59 g/79.5 = 0.02 [1 mark], $CuSO_4$ = 63.5 + 32 +(16 × 4) = 159.5 [1 mark], so 159.5 × 0.02 = 3.19g [1 mark]

c   (2.2/3.19) × 100 = 69% [1]

2a 2S, 100/(32 × 2) = 1.56 units [1]; $2SO_3$, (32+(16 × 3)) × 2 = 160 [1], so 1.56 units of 160 = 249.6 tonnes [1]

b   either 1.56 × (3 × (16 × 2)) = 149.7, or 249.6 – 100 = 149.6 tonnes [1]

c   (187.6/249.6) × 100 = 75.2% [1]

Reactions that go both ways

1a the reaction is reversible / it is an equilibrium reaction [1]

b   from blue to pink [1]

c   The wet pink cobalt chloride will lose water as it dries and go blue [1]. Adding more water will make it go back to pink. [1]

Rates of reaction

1a changes in mass [1] and time [1]

b   At the start there are lots of reactants so it is quick [1], as the reaction continues the reactants are used up so the reaction slows down [1].

c   12.6 cm³ [1]

2a rate = change in gas volume/time taken for the change [1], using tangent method, accept answers in range = 2–3 cm³ per minute (tangent method is the most reliable). [1]

b   More than enough calcium carbonate [1] was used to react with all the acid [1], so some remains; OR not enough acid was used [1] to react with all the calcium carbonate [1].

Pages 228–229

Collision theory

1a It would increases the rate [1], as the higher the temperature the more reacting collisions will occur [1].

b   It would increase the rate [1] as there are more particles on the surface able to react [1].

c   It would decrease [1] as there are less acid particles to react with [1], so frequency of reacting collisions is reduced.

2a a chemical that speeds up the rate of reaction but is unchanged at the end of the reaction [1]

b   2 cm³ has surface area of 2 × 2 × 6 = 24 cm², 1 cm³ has area of 1 × 1 × 6 = 6 cm², and as there are 8 of them, so 6 × 8 = 48 cm² [1]

so increasing the number of cubes has doubled the surface area and so will double the rate [1]

c   Changing the catalyst will affect the rate of reaction [1], so it is not possible to use a different catalyst to check repeatability. [1]

Adding energy

1a 9.6 cm³ [1]

b   i curve to left of the two curves [1], that plateaus at 9.6 cm³[1]

ii The rate will be faster as the particles are moving faster so will produce more reacting collisions [1]; height will be the same as there are no more particles to react than before [1].

2a Heat energy provides activation energy for the first atoms to burn [1], subsequent heat energy produced from burning magnesium provides activation energy for more atoms to burn [1].

b   Increasing temperature provides more energy to the particles [1], this means that more particles will possess enough energy to react and this is the activation energy [1].

Concentration

1a A [1]

b C [1]. The curve is lowest/takes longest time to reach the end point/plateau [1].

c This is the maximum amount of hydrogen gas that can be produced from the reactants [1]. Changing the rate only changes the time it takes to make the products, not the quantity of products [1].

d 3.6 min [1]

2a draw tangent at 2, and take readings, for example:

$\dfrac{7.4 - 4.8}{3 - 1} = 2.6/2 = 1.3$

accept answer between 1.1 and 1.5. (1 mark for tangent figures, and 1 mark for answer between 1.1 and 1.5)

b 1.8 minutes [1]. Doubling concentration will halve the time for the reaction to complete. [1]

Size matters

1a iron filings [1]

b The iron filings have the smallest particles [1], so will react fastest as there are more particles available to react at the same time [1].

c The nail did not react [1]. There were no brown bits on the mat at the end/it was still the same size at the end [1].

2a Sprayed fuel forms drops with a larger surface area [1], so it is easier for the oil molecules to burn [1].

b The smaller drops of oil in the spray burn quicker [1] this releases heat energy faster, raising the temperature quickly. [1]

Pages 230–231
Clever catalysis

1a Reaction is five time faster [1 mark] as 250/50 = 5 [1 mark]

b The reaction rate was quicker with manganese(IV) oxide present [1] and there was no change in the mass of manganese(IV) oxide during the experiment [1].

c It could have lowered the activation energy; or it could have formed an intermediate that reacted easily, [1] so making it easier to release the oxygen [1].

2a Metals like lead can coat the catalyst's surface [1]; this would stop the catalyst working [1].

b Although there is no lead present there may be other substances that could coat the catalyst's surface [1], these may be present in air or the fuel [1].

Controlling important reactions

1a nitrogen + hydrogen $\rightleftharpoons$ ammonia [1]

b It is very slow. [1]

c Increasing the temperature increase the rate of a reaction [1], so to make the ammonia quickly 450 °C is used [1]

d Increasing the pressure forces the molecules closer together [1], this means there will be more collisions so a quicker reaction [1].

2a The ammonia produced can be easily separated from the unreacted gases [1], which can then be recycled so eventually all the reactants will become products [1].

b Any three from: high temperature and pressure make the reaction quick [1], but high temperature gives a poor yield but quick reaction time [1]; high pressure increases yield, but is very costly[1]; the combination chosen gives best yield for least time and cheapest energy cost [1].

The ins and outs of energy

1a endothermic [1]

b The products need more energy than the reactants [1], so energy is taken from the solution so the temperature falls [1].

c quicklime and water [1]

d The quicklime and water have more stored energy than the products they make [1], the surplus energy is released as heat energy [1].

2a a reaction that can go in either direction [1]

b the reaction from left to right in the written equation [1]

c The backward reaction is the exact opposite of the forward reaction [1], whatever energy change happens with the forward reaction, the exact opposite will happen with the backward reaction [1].

Acid–base chemistry

1a An alkali can dissolve in water, a base cannot. [1]

b sodium nitrate and water [1]

c copper chloride [1] and carbon dioxide and water [1]

2a The universal indicator turned blue showing it is an alkaline solution. [1]

b NH_4^+(aq) [1] and OH^- (aq) [1]

c The hydroxide ions react with the hydrogen ions from the acid to form water. [1]

Pages 232–233
Making soluble salts

1a sulfuric acid [1]

b i warm the mixture [1]

ii When no more cobalt oxide dissolves [1], check the pH of the solution [1].

c Filter the solution. [1]

d Evaporate the solution [1] until crystals appear[1].

2a fizzing/a gas being produced [1]

b The calcium sulfate produced did not dissolve in the solution [1], it coated the calcium carbonate preventing the sulfuric acid from reaching the calcium carbonate so the reaction stopped. [1]

Insoluble salts

1a calcium carbonate and sodium chloride [1]

b Filter the solution. [1]

c Rinse the contents of the filter paper with water. [1]

2a i A yellow precipitates is formed. [1]

ii The precipitate is silver iodide. [1]

b white [1]

c $2AgNO_3$(aq) + $MgCl_2$(aq) $\rightarrow$ $Mg(NO_3)_2$(aq) + 2AgCl (1 mark for the unbalanced chemical equation, 1 mark for balancing the equation)

Ionic liquids

1a electrolysis [1]

b Dichromate ions are negatively charged [1] and will be attracted by the opposite charge of the positive electrode [1].

c The ion will gain an electron [1] becoming an atom of potassium or react with the water becoming potassium hydroxide [1].

2a Electricity is a flow of electrons, the potassium ion gains an electron at the negative electrode [1], the dichromate ions lose (two) electrons at the positive electrode; this allows for a flow of electrons in the circuit [1].

b In AC, the current rapidly changes polarity [1] which means that the ions will be attracted first one way, then the other, with the result that they will not move from the original position [1].

Electrolysis

1a the positive electrode [1]

b It loses two electrons [1] and dissolves in the solution [1].

c to allow copper to be deposited on the negative electrode [1]

d the impurities from the copper [1]

2a It only shows the reaction at one electrode, which is only half the reaction. [1]

b $2Cl^-(\ell) \rightarrow Cl_2(g) + 2e^-$ [1 mark]

c Oxidation is the loss of electrons: and the chloride ions lose electrons. [1]

d $2NaCl(\ell) \rightarrow 2Na(\ell) + Cl_2(g)$ [1 mark]

Page 234

Extended response question

5 or 6 marks:

There is a clear, balanced and detailed description of the positive and negative reasons for the use of brine rather than molten sodium chloride with 5–6 points from the **examples** given. The answer shows almost faultless spelling, punctuation and grammar. It is coherent and in an organised, logical sequence. It contains a range of appropriate or relevant specialist terms used accurately.

3 or 4 marks:

There is some description of the positive and negative reasons for the use of brine rather than molten sodium chloride with 3–4 points from the **examples** given. There are some errors in spelling, punctuation and grammar. The answer has some structure and organisation. The use of specialist terms has been attempted, but not always accurately.

1 or 2 marks:

There is a brief description of the reasons for the use of brine rather than molten sodium chloride with 1–2 points from the **examples** given. The spelling, punctuation and grammar are very weak. The answer is poorly organised with almost no specialist terms and/or their use demonstrating a general lack of understanding of their meaning.

possible points include:
positive

• it can be carried out at room temperature so saves energy

• hydrogen is produced at a low temperature

• so it will be harder to ignite

• sodium hydroxide is produced directly, with molten reagents a further reaction is needed.

• with brine the sodium immediately reacts with the water solution to make NaOH.

negative

• with molten sodium chloride the sodium made has to react with water
$2Na + 2H_2O \rightarrow 2NaOH + O_2$

• the more processes used the costlier it will be.

P2 Answers

Page 235

See how it moves

1a i travelling forward at a steady speed [1] ii stopped [1]

b section C [1]

c a curved line getting steeper [2]

2a Instantaneous speed is speed at a certain point in time (e.g. seen on a speedometer) [1]. Average speed is distance/time for the whole journey [1].

b The photographs are used to compare the position of the car relative to markings at known intervals [1]. The distance travelled by the car is calculated [1]. Use speed = distance/0.5 s to calculate speed [1].

Speed is not everything

1a Velocity is speed in a certain direction. [2]

b Since velocity depends on speed and direction [1], although speed remains constant, the car's direction changes [1].

c Acceleration is change in speed/time [1]

$\dfrac{-40}{5} = -8$ [1] m/s^2 [1]

2a Section A – steady speed [1]; section B – accelerates [1]; section C – decelerates [1] section D – slower steady speed [1]

b Work out the area under each section of the graph [1] and add the distances together [1].

Pages 236–237

Forcing it

1a The car and van pull in opposite directions [1] with equal forces [1].

b The tractor pulls harder / force is larger than the van [1].

2 Any five from: drag forces are larger if there is a roof rack/open window [1]; drag forces increase with speed [1]; larger forces from the engine are needed to overcome drag [1]; the engine provides a larger force when a car accelerates [1]; fuel provides energy for the engine to provide a larger force [1].

Force and acceleration

1a As the force increases, the acceleration increases. [2]

b Yes, they have [1], you can see a clear pattern [1] OR No, they haven't [1], you can only say that the pattern applies up to forces of 5 N [1].

c To improve the reliability of the data by identifying anomalous results [1], allowing random errors to be identified [1].

2a F = m x a [1] = 4 x 5 = 20 [1] N [1]

b Measure the size of the force applied to the rock using a forcemeter [1]. Measure its acceleration using the accelerometer [1]. Mass = force/acceleration [1]

Balanced forces

1a i For example: gravity/reaction forces [2]
ii tension in rope/force exerted by the person on the rope [2]

b   When the team moves, the weight/gravity stays fixed [1], and the forces exerted by the right-hand team become larger than the forces exerted by the left-hand team [1].

2   Any 5 from: the weight of the person stays constant [1] the force from the lift is greater than weight as the lift accelerates upwards [1] it equals weight as the lift travels at a steady speed [1] it is less than the weight as it slows down [1] forces are balanced when the lift travels at a steady speed /when it is stopped [1] direction of the resultant force is upwards at the start of the trip [1] direction of the resultant force is downwards at the end of the trip [1]

1a thinking distance [1], braking distance [1]

b   less experienced/may not recognise hazards [1]

c   i thinking distance increases with speed; or thinking distance doubles as speed doubles, etc [2]
    ii 9 m [1]
    iii The thinking distance is 6 m with no drink [1], so thinking distance has increased [1].

2   Credit valid points: 1 mark per example (maximum 2); plus 1 mark per explanation (maximum 2); plus 2 marks for choosing a factor and explaining why it has most effect.
    For example: worn tyres [1] have less grip/less friction on road/more likely to skid [1];
    worn brake pads [1] reduce stopping force applied to wheels [1]; worn windscreen wipers or dirty windscreen [1] reduces visibility increasing thinking distance [1].
    Evaluation factor, for example: the condition of tyres is most important, as these are the only point of contact a car has with the road [2] OR the condition of the brake pads as these control the force applied to the wheels [2].

Pages 238–239

Terminal velocity

1a Terminal velocity is the top speed reached by a moving object. [1]

b i Weight is constant and greater than drag forces, so the resultant force is down [2].
  ii Drag forces increase to match weight, which does not change. At terminal velocity, there is no resultant force. [2]

2a Weight remains constant [1]; drag forces increase with speed and surface area [1]; opening the parachute increases the surface area [1]; the drag forces at that speed are greater than weight [1]; the resultant force is upwards which causes deceleration [1].

b   The skydiver can only hover if upward forces match weight, i.e. there is no resultant force [1]; when they are not moving up or down [1] there is a drag force upwards only if skydiver is falling [1]. A very large surface area reduces the terminal velocity greatly but does not eliminate it [1]. (Note: credit correct and relevant discussion, for example upward thermal forces may match weight in some conditions.)

Forces and elasticity

1a independent variable – force applied [1]

b   dependent variable – extension of spring [1]

c   He may make a mistake in his calculations/it is harder to compare his data with the other groups [1]

d   He must control other variables, e.g. the spring used/ how measurements are taken [1] so only one factor is changed at a time [1].

2   (Credit correct responses – the test should relate to the quality being tested.)
    Suitable qualities include: how strong or elastic the rubber is; how consistent the batches of rubber are; whether it snaps easily. [3]
    Tests include: making catapults in an identical way using samples from different batches and using them to fire projectiles to see if the projectile travels the same distance each time; testing catapults to destruction. [2]

Energy to move

1a chemical [1] heat [1] sound [1]

b   The rate of energy transferred depends on how quickly work is done against friction [1], friction is less on a smooth road [1].

2   Credit correct points – relevant factors include: the proportion of energy transferred to the flywheel [1]; the time the energy can be stored [1]; how the energy is transferred from the flywheel [1]; physical size [1]. (maximum 2)
    An explanation of the impact of each factor should be included, for example: more energy transferred to the wheel means less transferred to electricity production [1]; the flywheel may store energy for minutes/hours, but not for days [1]. (maximum 2)

    A statement of which factor is considered most important with some explanation. [1]

Working hard

1a 12 J [1]

b   gravity [1] or weight [1]

c   work = force x distance [1] = 2 x 25 = 50 [1] J [1]

2a 5 x 30 = 150 m [1]

b   work = force x distance = 600 x 150 [1] = 90 000 J or 90 kJ [1]

c   The engine is still working against friction [1], and also working against gravity; the car is gaining gravitational potential energy but has the same kinetic energy [1].

Pages 240–241

Energy in quantity

1a mass x gravity x height = 6 x 10 x 0.8 [1] = 48 [1] J [1]

b   6 x 10 x 1.4 [1] = 84 [1] J [1]

2   KE =½ x mass x velocity^2 [1] = ½ x 80 x 36 [1] = 1440 J [1] for the runner
    KE = ½ x 50 x 64 [1] = 1600 J [1] for the skateboarder

    The child on a skateboard has more energy [1].

Energy, work and power

1a power = energy transferred/time [1] = 400 x 10 x 3/60 [1] = 200 [1] W [1]

b   More power means more energy is transferred in the same time [1]; so the motor could lift a greater weight in the same time, or lift the same weight a higher distance in that time [1]. More power means the same work is done in less time [1]; so the weight is moved in a shorter time [1].

2   Approximately/the same work is done in both cases [1]. When the piano is pulled up the ramp, the force

needed equals the component of its weight acting parallel to the ramp [1], so a small force acts over a large distance [1]. If the piano is lifted, the force needed equals its full weight [1], so a large force is applied over a shorter distance [1]. A force matching the weight of a piano is very hard for one person to produce [1].

Momentum

1a mass x velocity [1] = 6 [1] kg m/s [1]

b 6 kgm/s [1]

c momentum/mass [1] = 6/4 = 1.5 [1] m/s [1]

2 Conservation of momentum means momentum before and after a collision or explosion is the same provided no external factors act [1], gases are ejected from the jet packs [1], the momentum change of the moving gas in one direction equals the momentum change of the satellite in the opposite direction [1], the gas has small mass but large velocity; the satellite has a large mass and smaller velocity [1] in space, there is no air resistance so no external forces [1].

Static electricity

1a Electrons [1] are rubbed off the cloth [1] and move onto the balloon [1].

b positive [1]

c Hold the balloon near the object [1], it attracts a positively charged object [1] and repels a negatively charged object [1].

2a charge [1]

b The electroscope is given a negative charge and the gold leaf is repelled from the metal stem [1]. Then when a negatively charged object comes close, the golf leaf is repelled more [1]; when a positively charged object comes close, the gold leaf is repelled less [1].

Pages 242–243

Moving charges

1a electrostatic induction [1]

b Electrons in the wall are repelled by the negatively charged balloon [1], leaving an overall positive charge on the wall's surface [1].

2a Electric charge moves easily through electric conductors [1]; electric charge cannot move through insulators [1].

b Any three from: static electricity discharges as a spark [1]; the spark causes an electric shock [1]; at very high voltages the spark can travel through air [1]; the spark is more likely to travel to or from a point (such as a person's finger) [1].

Circuit diagrams

1a 1.6 A [1]

b 3 V [1]

c 1.5 V [1] in the series circuit, voltage is shared between components [1]

2 Any four points from: the cell transfers energy to electrons/electrical charge [1]; electrons transfer energy round the circuit/to the bulbs [1]; the bulbs transfer energy to the surroundings [1] as light/heat [1]; the total energy supplied equals the total energy used/transferred [1].

Ohm's law

1a resistance measures how easily electrons move through a material [1] OR resistance is voltage ÷ current [1]

b a number larger than 10 ohms [1]

c thickness of wire [1]; material the wire is made from [1]

2a As voltage increases, current increases [1] OR as voltage doubles, current doubles [1].

b 1.5 A [1]

c resistance = voltage/current [1] = 6/1.5 = 4 [1] ohms [1]

Non-ohmic devices

1a Change the light intensity it is exposed to. [1]

b For example: security lighting. [1]

c symbol 2 [1]

2 One mark for each example (maximum 2 marks), e.g. a thermistor's resistance falls with increasing temperature [1] an LDR's resistance falls with increases in light intensity [1] a diode has very high resistance in one direction.

One mark for each description of how the circuit responds to changes (maximum 2 marks): when the resistance falls below a certain value [1], a component (e.g. a light, a buzzer) turns on or off [1].

Pages 244–245

Components in series

1a They are all the same. [1]

b 3 V [1]

c 2 V [2]

d It increases [1]; OR it increases by a factor of 4/3 [2].

2a The potential difference of cells in series subtracts if they are connected in opposite directions [1], the total potential difference supplied is zero [1].

(Credit correct answers, for example the potential difference of two cells may be slightly different so there could be a small overall voltage.)

Components in parallel

1a 1.5 A [1]

b 6 V [1]

c e.g. brighter bulbs; if one bulb goes out the others stay on[1]; you can control individual bulbs[1]

2a i upper 2 ohms[1] ; ii lower 4 ohms [1]

b 4/3 ohms [2]

c upper branch [1] it has the lowest resistance [1]

Household electricity

1a A [1]

b The current repeatedly changes direction [1], there are 50 cycles per second [1].

2 Time for one cycle is 0.08s [1] frequency = 1/time per cycle [1] = 12.5 Hz [1]

Plugs and cables

1 Allow correct statements, any three from, for example: the wires are connected to the correct pins [1] (maximum two marks if colours and pins are correctly identified); the correct fuse is used; the outer part of the cable is gripped under the cable grip; the screws hold each wire tightly in place in each pin; check no bare wires are visible.

2 Any four from: wet skin has lower resistance than dry

skin [1]; electrocution is more likely if a person has wet skin than dry skin [1]; people are more likely to have wet skin, or be sitting/standing in water in a bathroom [1]; not allowing mains sockets means that electrical equipment is less likely to be used inside the bathroom [1]; battery-operated equipment is less dangerous as it uses a lower voltage [1].

Pages 246–247

Electrical safety

1a The fuse does not cut off the electric current [1]; the cable is still live [1]; there is a risk of electrocution [1].

b Credit correct answers: the RCCB cuts off the current very quickly [1] if there is a different current in different wires in the cable [1] so there is no risk of electrocution [1].

2a The fuse can carry a current of 13 A without melting [1]. At higher currents, the fuse should melt [1].

b 5 A [1]. Current flowing in the iron is 3.47 A [1]; the fuse with the rating that is closest to the current but larger than it should be used [1].

c To protect the flex [1], from overheating/prevent fires [1].

Current charge power

1a energy = power x time [1] = 5 x 60 x 1000 [1] = 300 000 J [1] or 300 kJ

b charge = current x time [1] = 4.3 x 5 x 60 [1] = 1290 [1] C [1]

2 The potential difference [1]; doubling the potential difference doubles the energy transferred by the charge [1]. If the current is large, more energy is transferred [1], because more charge flows [1].

Structure of atoms

1a i negligible or 1/2000 [1]; **ii** +1 [1]; **iii** neutron [1].

b i number of neutrons;

ii number of protons and electrons.

2 Any four from: electrostatic force [1]; like charged protons repel each other [1]; positively charged nucleus attracts negatively charged electrons [1]; neutrons do not feel electrostatic force [1]; strong nuclear force holds nucleons together [1].

Radioactivity

1a Radioactive – the nucleus emits ionising radiation, changing to a nucleus of a different element. [1]

b alpha [1]

c A neutron changes into a proton. [1]

2a X = 95 [1], Y = 237 [1]

b Alpha radiation cannot travel far in air/pass through skin [1]; it is only a danger if it is swallowed/inhaled/injected into the body [1].

Pages 248–249

More about nuclear radiation

1a positively charged [1]; central massive nucleus [1]; surrounded by negatively charged electrons [1]

b i deflection of some positively charged alpha particles [1]

ii most alpha particles not deflected [1]

iii electrons are relatively easy to remove from the atom [1]

2 Gamma radiation is not deflected as it has no charge [1]. Beta and alpha radiation are deflected in opposite directions as they have opposite charges [1]. Beta

radiation is deflected more [1] than alpha radiation as it has a smaller mass [1].

Background radiation

1a 37% [1]

b Credit correct examples: some sources of background radiation increase with lifestyle choices, e.g. the number of flights taken [1]; medical history, e.g. X-rays [1]; where you live [1].

2 Credit five correct statements (for either side of the argument), for example: the Earth has always been slightly radioactive/background radioactivity is normal [1]; radioactivity can increase the lifetime cancer risk [1]; lifestyle choices/many factors affect lifetime cancer risk [1]; natural sources of background radioactivity are more significant than man-made sources [1]; there are many other significant health risks, e.g. cardiovascular disease/accidents etc [1]; lifetime cancer risk is increased in smokers by exposure to radon and thoron [1].

Half-life

1a Half-life is the time [1] taken for the original count rate/activity/mass of radioactive atoms [1] to halve [1].

b The proportions will fall [1], the carbon atoms change into nuclei of other elements [1].

2a 600 counts per second [1]

b Three [1] half-lives [1], which is 15 years [1].

Using nuclear radiation

1a For example: reduce wastage/controls the quality. [1]

b It is absorbed by cardboard. [1]

c Less radiation detected means the cardboard is too thick [1]; so the rollers should move closer together [1].

2 Pupils should cover both sides of the argument. Credit five correct statements, for example: radioactive tracers help diagnose medical conditions while avoiding surgery [1]; they can be more accurate than other methods [1]; using tracers helps people choose the most effective treatment [1]; medical tracers can increase the lifetime cancer risk [1]; medical risks are reduced by choosing suitable tracers (e.g. short half-life; not alpha-emitting) [1]; tracers should be used if more serious conditions are suspected [1].

Pages 250–251

Nuclear fission

1a When a nucleus splits into two or more products. [2]

b Nuclear power/generating electricity. [1]

c 3 [1]

d More nuclei are involved at each stage of the chain reaction [1]; because each reaction produces more neutrons than are needed to cause fission [1]; too much heat/energy will be produced if too many nuclei are involved [1].

2 Pupils should suggest at least one advantage and one disadvantage with an explanation.

Advantages – they last for a long time so reducing the need for operations to replace the battery; they are compact (so not a nuisance to the patient); they are reliable (so people will not suffer side-effects). [2]

Disadvantages – disposal of radioactive material can be difficult/expensive; patient may be exposed to unacceptable levels of radiation; potential danger if medical staff are not aware that there is a nuclear battery inside the patient. [2]

Nuclear fusion

1a In nuclear fission, nuclei split; but in nuclear fusion, nuclei join [1]; nuclear fission involves large nuclei, but nuclear fusion involves light nuclei [1].

b the Sun [1]

2 Nuclear fusion creates different elements [1]; light elements form heavier elements [1]; in stars, elements up to iron are formed [1]; supernovas generate such immense energy and heat [1] that heavier elements can form [1].

Life cycle of stars

1 Credit correct references to changes in forces at specific stages. Any five from: gravity forces the star to collapse/increases the pressure inside it [1]; forces from heat generated in the star force it to expand [1]; during each stage the forces balance [1]; when the fuel runs out, less heat is generated and the star collapses [1]; because gravity is greater than forces from heat [1]; in very massive stars, the heat generated can be enough to cause a rapid expansion [1].

2 Any three from: more massive stars generate more heat in their core [1]; fusion reactions creating heavier elements take place in the core of heavier stars [1]; more stages in the life cycle can take place for more massive stars [1]; smaller stars may just cool and die as their fuel runs out [1]; more massive stars may explode as a supernova [1].

Page 252
Extended response question

5 or 6 marks:

A detailed description of two safety features, e.g. airbags, crumple zones, seat belts, and side impact bars. The answer should include a discussion of the momentum and energy changes taking place, link-ing increased impact duration with a reduced risk of serious injury (change in speed takes longer, so the deceleration and forces felt are smaller). The answer should explain how these safety measures increase the duration of the collision and/or absorb energy (e.g. seat belts stretch slightly, which extends the time a person takes to stop; crumple zones deform and absorb energy – this extends the time it takes to come to a complete halt). *All information in answer is relevant, clear, organised and presented in a struc-tured and coherent format. Specialist terms are used appropriately. Few, if any, errors in grammar, punc-tuation and spelling.*

3 or 4 marks:

A limited description of two safety features e.g. airbags, crumple zones, seat belts, and side impact bars. There should be a discussion, which describes what these safety measures do, and some attempt to link this to increased impact duration or reduced energy transfer. *For the most part, the information is relevant and presented in a structured and coherent format. Specialist terms are used for the most part appropriately. There are occasional errors in grammar, punctuation and spelling.*

1 or 2 marks:

An incomplete description of one or two safety features. There should be some description of energy changes or momentum changes during a collision. *Answer may be simplistic. There may be limited use of specialist terms. Errors of grammar, punctuation and spelling prevent communication of the science.*